Studies in Applied Philosophy, Epistemology and Rational Ethics

Volume 28

About this Series

Studies in Applied Philosophy, Epistemology and Rational Ethics (SAPERE) publishes new developments and advances in all the fields of philosophy, epistemology, and ethics, bringing them together with a cluster of scientific disciplines and technological outcomes: from computer science to life sciences, from economics, law, and education to engineering, logic, and mathematics, from medicine to physics, human sciences, and politics. It aims at covering all the challenging philosophical and ethical themes of contemporary society, making them appropriately applicable to contemporary theoretical, methodological, and practical problems, impasses, controversies, and conflicts. The series includes monographs, lecture notes, selected contributions from specialized conferences and workshops as well as selected Ph.D. theses.

More information about this series at http://www.springer.com/series/10087

Gordana Dodig-Crnkovic
Raffaela Giovagnoli

Editors

Representation and Reality in Humans, Other Living Organisms and Intelligent Machines

 Springer

Editors
Gordana Dodig-Crnkovic
Gothenburg
Sweden

Raffaela Giovagnoli
Faculty of Philosophy
Pontifical Lateran University
Vatican City
Italy

ISSN 2192-6255   ISSN 2192-6263   (electronic)
Studies in Applied Philosophy, Epistemology and Rational Ethics
ISBN 978-3-319-82909-8   ISBN 978-3-319-43784-2   (eBook)
DOI 10.1007/978-3-319-43784-2

Preface

This book project is based on several years of collaboration between two editors, starting with the organization of the symposium *Computing Nature* at the AISB/IACAP World Congress in 2012 in Birmingham. It continued with the AISB50 Convention at Goldsmiths, London in 2014 and the symposium *Representation: Humans, Animals and Machines* organized in cooperation with Veronica Arriola-Rios, University of Birmingham. During preparation, the book project was presented at the symposium *Representation and Reality* at UNILOG 2015, World Congress on Universal Logic in Istanbul. All those events, as well as our networks with research communities of cognitive scientists, computer scientists, philosophers, logicians, AI researchers, roboticists and natural scientists, connect us with the authors of the present volume, offering views on the topic of representation and its connections to reality in humans, other living organisms and machines. The choice of the title was made so as to refer to historical attempts at making connections, with Wiener's *"Cybernetics: Or Control and Communication in the Animal and the Machine"* (Wiener 1948), and Putnam's human-centric *"Representation and Reality"* (Putnam 1988).

How to address this vast topic and connect representation and reality in humans, other living beings and machines, based on the best contemporary knowledge? We invited prominent researchers with different perspectives and deep insights into the various facets of the relationship between reality and representation in those three classes of agents. How can we find a common link between reality-constructing agents like us humans with language ability and social structures that define our agency, other living organisms, from bacteria to plants and animals communicating and processing information in a variety of ways, with machines, physical and virtual? We have taken a cognitive, computational, natural sciences, philosophical, logical and machine perspective. Of course, no perspective is simple and pure, but rather a fractal structure in which recurrent mirroring of other perspectives at different scales and in different senses occurs. So in a contribution characterized predominantly as "cognitive perspective" there are elements of natural sciences, logic, philosophy and so on. Our aim is to provide a multifaceted view of the topic of representation and reality in the range of approaches from disciplinary (where a "discipline" is represented with its forefront outlooks) to multidisciplinary, interdisciplinary and cross disciplinary approaches. As a whole, the book presents a complex picture of a network of connections between different research fields addressing the state of the art of the topic reality versus representation.

We can see this book both as a collection of contributions of our authors to their fields of specialization and as an invitation to the reader to reflect on the kaleidoscope of domain-specific insights and their mutual connections in the context of the whole book. It is our common contribution to the continuous learning and shared knowledge about the nature of representation as found in the process of cognition (with "mind" as its philosophical reflection), cognition as it exists in different degrees in all living beings, including humans, and as it is recently being developed and constructed in machines in the field of cognitive computing.

Aspects of the relationship between representation and reality today connect networks of communities of philosophers, computer scientists, logicians, anthropologists, psychologists, sociologists, neuroscientists, linguists, information and communication scientists, system theoreticians and engineers, theoreticians and practitioners in computability and computing, information theory researchers, cybernetic systems researchers, synthetic biologists, biolinguists, bioinformaticians and biosemioticians, and many more. Current knowledge is distributed and no one's insight is total and exhaustive, even though some can see more and broader while others see sharper and in more detail within specific perspectives. Here is a short presentation of different perspectives as represented in the book's chapters, with an attempt to connect them into a common network.

Cognitive Perspectives

Connecting representation with reality, Terrence Deacon investigates the relation between information and reference, i.e. the relation of "aboutness". He starts by making the distinction between the Shannon model of information and semantic information. The Shannon model describes communication of information and its goal is to engineer the best ways of communicating information through a noisy channel, based on the quantitative measure of the amount of information. However, in everyday use, the most important characteristic of information is its *meaning*, that is information about something relates to something in the world. Deacon's analysis focuses on the capacity of a medium to provide reference and argues that this capacity can be seen in the difference between informational and thermodynamic entropy. This qualitative analysis shows that reference is a causally relevant physical phenomenon.

If representation is causally connected to reality, then how about hallucinations? In his contribution, Marcin Miłkowski studies models of visual hallucination in people with Charles Bonnet Syndrome and proposes to see them as illustrative cases where representations are not about anything in the real world. Miłkowski presents the computational model implementing neural network architecture with deep learning that can illuminate representational mechanisms for hallucination. It is interesting to observe that neural networks are often taken to be nonrepresentational models, in spite of the human brain being based on (biological) neural networks.

The old debate is still alive between representationalism and antirepresentationalism with the question of whether cognition relies on *representations mirroring* reality, or it is an *adaptive form of dynamics* based on the *interaction of an agent with the environment*. Typically, it is taken to be important for the critique of computational models of cognition as it is assumed that computations essentially depend on internal representations and that there is no computation without abstract symbol manipulation. This debate continues to this day because we still have no method to directly detect representations in a cognitive system.

In the opposition between *old ("orthodox") computationalism* (based on the idea of the abstract logical universal Turing Machine) and *enactivism* (that emphasizes the central importance of the body and the environment for cognition), Tom Froese connects *meaning* for a living being with its embodiment with the (ever present) possibility of death. The claim this article makes is that *orthodox* computational models cannot account for meaning. Meaning as based in the necessity for active autopoiesis and the struggle for survival is outside the scope of the abstract Turing Machine model of computation.

Computationalism as a theory of mind is often criticized in its classical/orthodox approach, such as the computational representational theory of mind presented by Fodor. Jesus Ezquerro and Mauricio Iza propose an effective alternative *computational model of embodied cognition*, understanding language as predicting sensorimotor and affective effects of the action described by a verb.

Computational Perspectives

Dynamical systems are abstract mathematical objects that are used to describe physical processes. They are defined as "A means of describing how one state develops into another state over the course of time. Technically, a dynamical system is a smooth action of the reals or the integers on another object (usually a manifold)" by Wolfram MathWorld. Dynamical systems are frequently believed to be opposite to and irreconcilable with computational models. However, in their chapter Jan van Leeuwen and Jiří Wiedermann use exactly the dynamic systems formalism to develop a new *dynamic knowledge-based theory of computation* capable of explaining computational phenomena in both living and artificial systems.

Even though many people, even among scientists and philosophers, would say that the process of life and computation have nothing to do with each other, Dominic Horsman, Viv Kendon, Susan Stepney and Peter Young have dedicated their study to computation in living systems. Within the framework of Abstraction/Representation theory (AR theory), computing is assumed to be *representational* activity. In this chapter, the AR approach is used to elicit conditions under which a biological system computes. The framework is developed in the context of a nonstandard human-designed computing and has already been applied.

Nicolas Gauvrit, Hector Zenil and Jesper Tegner take a computational stance and cover in their chapter the whole range of the information-theoretic and algorithmic approaches to human, animal and artificial cognition. They review existing models of computation and propose algorithmic information-theoretic measures of cognition.

Unlike the above computational approaches to cognition, the article by Dean Petters, John Hummel, Martin Jüttner, Ellie Wakui, and Jules Davidoff is of more empirical character. Computational models of object recognition are used to investigate representational change in the course of development in humans. By means of developmental studies in children, and computational modeling of their results by artificial neural networks, in comparison with the existing research on adults, it was possible to follow how visual representations mediate object recognition.

Natural Sciences Perspectives

Reductionism is one of the most severe sins one can commit, according to many philosophers and cognitive scientists. One should not even try to investigate a physical substrate on which cognition definitely relies. Gianfranco Basti is obviously untroubled by the danger of being accused of reductionism, and he examines the deepest roots of information processing in the physical world—the quantum field theory as a dual paradigm in fundamental physics—connecting it with semantic information in cognitive sciences. Basti makes us aware of the paradigm shift with respect to the Standard Model of physics, as well as quantum mechanics conceived as the many-body-dynamics generalization from classical mechanics. The basic assumption of the old paradigm about the closed physical system does not hold, as quantum field theory systems are inherently open to the background fluctuations of the quantum vacuum, and are capable of system phase transitions. It is interesting to observe that Prigogine had that insight in thermodynamics, making the step from closed thermodynamical systems or systems in thermodynamical equilibrium to the study of open thermodynamical systems with the inflow of energy that exhibited stable self-organized patterns. Maturana and Varela's work further developed our understanding of autopoiesis in living cells as processes essentially dependent on the openness of the system to the exchanges and interactions with the environment.

Irrespective of the representationalism versus antirepresentationalism discussion, today we have a broader concept of computation (computing nature) which posits that *every natural system computes*, with computation as its physical dynamics as presented in the chapter by Gordana Dodig-Crnkovic and Rickard von Haugwitz. That would mean that one does not need to search for representations in the brain just for the sake of settling the debate on whether the brain computes. Being a natural system, the human brain computes as well as the rest of nature in the framework of computing nature.

Dodig-Crnkovic and von Haugwitz search for answers to the questions: What is reality for an agent? What is minimal cognition? How does the morphology of a cognitive agent affect cognition? How do infocomputational structures evolve from the simplest living beings to the most complex ones? As a framework for answering these questions an infocomputational nature is assumed, constructed as a synthesis of (pan)informational ansatz (structures in nature are informational structures for us as cognitive agents) and (pan)computational view (dynamics of nature is computation), where information is defined as a structure (for an agent), and computation as the dynamics of information (information processing). Both information and computation in this context have broader meaning than in everyday use, and both are necessarily grounded in physical implementation, and the aim of this approach is to integrate computational approaches with embodiment and enactivism.

Philosophical Perspectives

Raffaela Giovagnoli presents the theme of language, classically central for the problem of representation. She focuses on inferential linguistic practices, which characterize human cognition. A pragmatic account helps us to understand what representational capacities are peculiar to humans, animals and machines. A study of the history of concept formation and use enables a clear distinction between humans and other animals. Starting from Frege, Giovagnoli presents the notion of representation theorized by Searle, and introduces the ideas proposed by the Brandom pragmatist order of explanation. The author points to the clear distinction between human and animal linguistic practices that can also be applied to the study of linguistic practices in machines.

Angela Ales Bello deals with the problem of consciousness in humans, animals and machines in a phenomenological account based on the hyletic dimension. The concept of hylé (a transliteration of Aristotle's concept of *matter*) was proposed by Husserl to denote the *nonintentional, direct sensory* aspects of living experience (*qualia*), shared by humans and animals. Ales Bello wishes to launch a challenge to those positions that ground themselves in the presumed objectivity of modern science. Starting from Husserl, Ales Bello examines the essential elements of phenomenology in order to understand how they developed and how they stand in relation to various scientific views. In this context the question "Can we conclude that the human being is like a machine?", looking at machines of today, is answered in the negative.

One of the arguments against computational models of mind is the so-called "hard problem of consciousness", which argues for irreducibility of the qualitative phenomenal features of mental experience to its functional cognitive features. Roberta Lanfredini presents the argument that "not only the classical cognitive pattern but also the classical phenomenological pattern give rise to a problem concerning the qualitative dimension." There is not only "the hard problem of consciousness" but "the hard problem of matter" as well. Lanfredini connects *representation* with *matter* and emphasizes the *central importance of time*, and *dynamics* in the phenomenology of the mind.

Logical Perspectives

The question of capturing homogeneity and heterogeneity is central for representation. Bateson's definition of information as a difference that makes a difference (for an agent) can be applied not only to perception but also to reasoning. How do we define differences and their role in reasoning? Henri Prade and Gilles Richard address comparative thinking as a basis of our comprehension of reality and present a description of logical proportions in comparing two objects or situations in terms of Boolean features within a cube of opposition. Homogeneous and heterogeneous logical proportions are important for classification, completion of missing information and anomaly detection, and they provide insights into human reasoning.

As already mentioned, among the strongest and recurrent objections against computational models of cognition is their alleged incapability to account for meaning (semantics). Ferdinando Cavaliere offers a remedy for this problem in the case of online search engines by means of adding the semantics with the help of a hypothetical program "Semantic Prompter Engine", based on the "Distinctive Predicate Calculus" logic, designed by the author. The aim of this approach is to bridge the gap between natural and artificial representations of concepts and reality by providing the semantics for the latter.

In his chapter Jean-Yves Béziau continues this tradition of critical analysis of human rationality, offering an original comparison between human and animal intelligence. Rationality in language is the logical way we have to represent knowledge but the richness of our emotional life is worth further exploration.

Machine Perspectives

Even Matej Hoffmann and Vincent Müller, like Jan van Leeuwen and Jiří Wiedermann in their chapter, find the dynamical systems framework as the most suitable, in their case for connecting computation controllers and robotic bodies. They introduce the concept of *morphological computation* in the sense it is used in robotics, which refers to "offloading" computational processing from a central controller to the body of a robot, exploiting the use of self-organization of the physical system in its interaction with the environment.

David Zarebski distinguishes between three types of ontology: the mind-independent structure of reality studied in philosophy and formal ontology, the structure of the human representation of the world studied in the cognitive sciences, or the structure of knowledge representation investigated in data engineering. Zarebski describes interactions between those three research fields – philosophy with formal ontology, cognitive science and data and knowledge engineering – as consisting in cognitive science explaining how human cognitive capacities can affect metaphysics, or in what way information systems ontologies can learn from formal ontologies. While cognitive science typically makes no distinction between ontological and epistemological realism, Zarebski defends the position that it is possible, in the Kantian tradition, to be realistic about the epistemology without the need for naturalization of the ontology.

A basic question regarding intelligent machinery concerns the nature of machine intelligence. Philip Larrey's chapter compares human intelligence and machine intelligence, as it looks today and as anticipated in the future. Based on Bostrom's classification of future-envisaged machine superintelligence into speed-, collective- and quality-superintelligence, Larrey presents critical views of all three types, proposing a fourth type that would combine human and machine intelligence, under the assumption that humans will do the really intelligent part while machines will continue as before to provide different services without being conscious of what they are doing. It is hard not to agree with the author's conclusion: "Ours is truly an 'unknown future'".

It remains to see what future developments of machine intelligence, robotics and cognitive computing will bring, and if perhaps in the future machines get one more capacity, "machine consciousness", which will not be like human consciousness, in the same way that "machine learning" is not like human learning, and machine walking is not like human walking but fulfills that function for a machine. Larrey's chapter considers the possibility of building (super)intelligent conscious machines. The next question, whether it is a good idea, or under what constraints is it justified to build possibly (super)intelligent conscious artifacts, is a different one, addressed both by Bostrom as well as Hawkins, Tegmark, Boden and many other prominent scientists. It might be considered a topic for ethicists or for political decision-makers, but the existing knowledge of those communities must be constantly updated through insights from researchers dealing in depth with the phenomena of natural and artifactual cognitive systems.

To sum up, the aim of this book is to enrich our views on representation and deepen our understanding of its different aspects. It is seldom the case that one discipline can exhaust all the complexity of a real-world phenomenon. The historical divisions formed deep trails that even the coming generations of researchers tend to follow, and existing divisions built into academic institutions, publications and research funding often present obstacles that prevent us from researching the big complex picture and our possible role in the overall network of knowledge. We are trying to provide a glimpse behind the mirror of our own specialist views on the phenomenon of representation and its connections to reality, written by researchers with different commitments and preferences regarding divisions into the computationalist versus enactivist approach, cognitive science versus phenomenology, logic versus. emotions, the "hard problem of mind" versus the "hard problem of matter", and so on. It should be emphasized that in spite of seemingly impermeable barriers between well-known opposing choices there is a natural building of networks that produces hybrids—enactivist studies based on computational simulations and physical computation models that take embodiment as fundamental, computational models of dynamical systems as well as dynamic systems formulations of computational frameworks, reductionism with the character of deconstruction, opening new views on emergence, the list goes on. We hope with this to contribute to the dialogue and further mutual connections between the research communities involved.

In conclusion we want to thank our authors for their excellent contributions, as well as for their involvement in the open and completely transparent review process done in a collegial and constructive spirit, where each chapter got at minimum three and at most eight well-informed and helpful reviews. We are thankful to Robert Lowe for his contribution to the review process.

Last but not least we want to thank our publisher, Ronan Nugent at Springer, for his continuous support and friendly advice in this project.

Gothenburg, Sweden   Gordana Dodig-Crnkovic
Vatican City, Italy   Raffaela Giovagnoli

Contents

Contributors

Gianfranco Basti Faculty of Philosophy, Pontifical Lateran University, Rome, Italy

Angela Ales Bello Pontifical Lateran University Rome, Rome, Italy

Jean-Yves Beziau University of Brazil, Rio de Janeiro, Brazil; University of San Diego, San Diego, CA, USA

Ferdinando Cavaliere Circolo Matematico Cesenate, Cesena, (FC), Italy

Jules Davidoff Goldsmiths College, London, UK

Terrence W. Deacon University of California, Berkeley, USA

Gordana Dodig-Crnkovic Chalmers University of Technology and University of Gothenburg, Gothenburg, Sweden

J. Ezquerro Institute for Logic Cognition Language and Information (LCLI), University of the Basque Country, UPV/EHU, Donostia-San Sebastián, Spain

Tom Froese Institute for Applied Mathematics and Systems Research, National Autonomous University of Mexico, Mexico City, Mexico; Center for the Sciences of Complexity, National Autonomous University of Mexico, Mexico City, Mexico

Nicolas Gauvrit Human and Artificial Cognition Lab, Université Paris 8 & EPHE, Paris, France

Raffaela Giovagnoli Faculty of Philosophy, Lateran University, Rome, Italy

Matej Hoffmann Faculty of Electrical Engineering, Department of Cybernetics, Czech Technical University in Prague, Prague, Czech Republic; iCub Facility, Istituto Italiano di Tecnologia, Via Morego 30, 16163 Genoa, Italy

Rickard von Haugwitz Chalmers University of Technology and University of Gothenburg, Gothenburg, Sweden

Dominic Horsman Department of Computer Science, University of Oxford, Oxford, UK ; Department of Physics, Durham University, Durham, UK

John Hummel University of Illinios at Urbana-Champaign, Champaign, USA

M. Iza Department of Psychology, University of Malaga, Málaga, Spain

Martin Jüttner University of Aston, Birmingham, UK

Viv Kendon Department of Physics, Durham University, Durham, UK

Roberta Lanfredini Dipartimento di Lettere e Filosofia, Florence, Italy

Philip Larrey Pontifical Lateran University, Vatican City, Italy

Marcin Miłkowski Institute of Philosophy and Sociology, Polish Academy of Sciences, Warszawa, Poland

Vincent C. Müller Anatolia College/ACT, Pylaia, Greece; Department of Philosophy, IDEA Centre, University of Leeds, Leeds, UK

Dean Petters Birmingham City University, Birmingham, UK

Henri Prade IRIT, Université Paul Sabatier, Toulouse Cedex 09, France

Gilles Richard IRIT, Université Paul Sabatier, Toulouse Cedex 09, France

Susan Stepney Department of Computer Science and York Centre for Complex Systems Analysis, University of York, York, UK

Jesper Tegnér Unit of Computational Medicine, Department of Medicine Solna, Centre for Molecular Medicine & SciLifeLab, Karolinska Institutet, Stockholm, Sweden

Jan van Leeuwen Department of Information and Computing Sciences, Utrecht University, CC Utrecht, The Netherlands

Elley Wakui University of East London, London, UK

Jiří Wiedermann Institute of Computer Science, Academy of Sciences of the Czech Republic, Prague 8, Czech Republic; Czech Institute of Informatics, Robotics, and Cybernetics, Czech Technical University in Prague, Prague 6, Czech Republic

J.P.W. Young Department of Biology, University of York, York, UK

David Zarebski IHPST, University Paris 1 Pantheon Sorbonne, Paris, France

Hector Zenil Unit of Computational Medicine, Department of Medicine Solna, Centre for Molecular Medicine & SciLifeLab, Karolinska Institutet, Stockholm, Sweden; Department of Computer Science, University of Oxford, Oxford, UK

Part I
Cognitive Perspectives

Information and Reference

Terrence W. Deacon

Abstract The technical concept of information developed after Shannon [22] has fueled advances in many fields, but its quantitative precision and its breadth of application have come at a cost. Its formal abstraction from issues of reference and significance has reduced its usefulness in fields such as biology, cognitive neuroscience and the social sciences where such issues are most relevant. I argue that explaining these nonintrinsic properties requires focusing on the physical properties of the information medium with respect to those of its physical context—and specifically the relationship between the thermodynamic and information entropies of each. Reference is shown to be a function of the thermodynamic openness of the information medium. Interactions between an informing medium and its physical context that drive the medium to a less probable state create intrinsic constraints that indirectly reflect the form of this extrinsic influence. This susceptibility of an informing medium to the effects of physical work is also relevant for assessing the significance or usefulness of information. Significance can be measured in terms of work "saved" due to access to information about certain contextual factors relevant to achieving a preferred target condition.

1 Introduction

I didn't like the term Information Theory. Claude didn't like it either. You see, the term 'information theory' suggests that it is a theory about information – but it's not. It's the transmission of information, not information. Lots of people just didn't understand this... information is always about something. It is information provided by something, about something. (Interview with R. Fano, 2001)

What I have tried to do is to turn information theory upside down to make what the engineers call "redundancy" [coding syntax] but I call "pattern" into the primary phenomenon... (Gregory Bateson, letter to John Lilly on his dolphin research, 10/05/1968)

T.W. Deacon (✉)
University of California, Berkeley, USA
e-mail: deacon@berkeley.edu

© Springer International Publishing AG 2017
G. Dodig-Crnkovic and R. Giovagnoli (eds.), *Representation and Reality in Humans, Other Living Organisms and Intelligent Machines*, Studies in Applied Philosophy, Epistemology and Rational Ethics 28, DOI 10.1007/978-3-319-43784-2_1

3

In common use and in its etymology the term "information" has always been associated with concepts of reference and significance—that is to say it is *about* something for some *use*. But following the landmark paper by Claude Shannon [22] (and later developments by Wiener, Kolmogorov and others), the technical use of the term became almost entirely restricted to refer to signal properties of a communication medium irrespective of reference or use. In the introduction to this seminal report, Shannon points out that, although communications often have meaning, "These semantic aspects of communication are irrelevant to the engineering problem," which is to provide a precise engineering tool to assess the computational and physical demands of the transmission, storage and encryption of communications in all forms.

The theory provided a way to precisely measure these properties as well as to determine limits on compression, encryption and error correction. By a sort of metonymic shorthand this quantity (measured in bits) came to be considered synonymous with the meaning of "information" (both in the technical literature and in colloquial use in the IT world) but at the cost of inconsistency with its most distinctive defining attributes.

This definition was, however, consistent with a tacit metaphysical principle assumed in the contemporary natural sciences: the assertion that only material and energetic properties can be assigned causal power and that appeals to teleological explanations are illegitimate. This methodological framework recognizes that teleological explanations merely assign a locus of cause but fail to provide any mechanism, and so they effectively mark a point where explanation ceases. But this stance does not also entail a denial of the reality of teleological forms of causality nor does it require that they can be entirely reduced to intrinsic material and energetic properties.

Reference and significance are both implicitly teleological concepts in the sense that they require an interpretive context (i.e. a point of view) and are not intrinsic to any specific physical substrate (e.g. in the way that mass and charge are). By abstracting the technical definition of information away from these extrinsic properties, Shannon provided a concept of information that could be used to measure a formal property that is inherent in all physical phenomena: their organization. Because of its minimalism, this conception of information became a precise and widely applicable analytic tool that has fueled advances in many fields, from fundamental physics to genetics to computation. But this strength has also undermined its usefulness in fields distinguished by the need to explain the non-intrinsic properties associated with information. This has limited its value for organismal biology, where function is fundamental, for the cognitive sciences, where representation is a central issue, and for the social sciences, where normative assessment seems unavoidable. So this technical redefinition of information has been both a virtue and a limitation.

The central goal of this essay is to demonstrate that the previously set aside (and presumed nonphysical) properties of reference and significance (i.e. normativity) can be reincorporated into a rigorous formal analysis of information that is suitable for use in both the physical (e.g. quantum theory, cosmology, computation theory) and semiotic sciences (e.g. biology, cognitive science, economics). This analysis builds on Shannon's formalization of information but extends it to explicitly model its link to the statistical and thermodynamic properties of its physical context and to the physical work of interpreting it. It is argued that an accurate analysis of the nonintrinsic attributes that distinguish information from mere physical differences is not only feasible, but necessary to account for its distinctive form of causal efficacy.

Initial qualitative and conceptual steps toward this augmentation of information theory have been outlined in a number of my recent works [8–12]. In these studies we hypothesize that both a determination of reference and a measure of significance or functional value can be formulated in terms of how the extrinsic physical modification of an information-bearing medium affects the dynamics of an interpreting system that exhibits intrinsically end-directed and self-preserving properties.

2 Background

The problems posed by the concepts of reference and significance have been the subject of considerable philosophical debate over many centuries. It would probably not be hyperbole to suggest that they are among the most subtle and complex issues in the history of philosophy. So it might seem presumptuous to imagine that these questions can be settled with a modest extension of formal information theory. Though it is often assumed that the current technical treatment of information either resolves these issues or shows them to be irrelevant, the fact that concepts of meaning and interpretation are treated as irrelevant to physical science and in biology indicates that a major gap in understanding still separates these technical uses from the traditional meaning of the term. Though it is not possible to even come close to doing justice to this long and byzantine philosophical history of analysis in this essay, it is worth setting the stage with a fundamental insight gleaned from this perspective.

The challenge posed by these ideas was eloquently, if enigmatically, formulated by Franz Brentano's [4] use of the terms "intention" and "inexistence" (borrowed from Medieval philosophy) when describing the referential property of information.

> Every mental phenomenon is characterized by what the Scholastics of the Middle Ages called the intentional (or mental) inexistence of an object, and what we might call, though not wholly unambiguously, reference to a content, direction toward an object (which is not to be understood here as meaning a thing), or immanent objectivity

… Brentano's use of the curious term "inexistence" points to the fact that informational content is not exactly an intrinsic physical property of the medium that conveys it, even though it somehow inheres in it as well. The reference conveyed is typically displaced, abstract, ambiguous, or possibly about something nonexistent. And in any case it is dependent on interpretation. Excluding this attribute from the technical concept of information appears to have been necessary to ground information theory in physics and make it suitable for engineering applications, but *this was bought at a cost of ignoring the very attributes that distinguish information from other physical phenomena.*

I argue that the key to formulating a more adequate concept of information that includes these "inexistent" properties is to be found, ironically, in more carefully attending to the physicality of information media. A hint that this is important is captured in two distinctively different uses of the concept of entropy (informational entropy and thermodynamic entropy). The term "entropy" was, of course, originally coined by Rudolph Clausius in [7] in the context of the early development of a concept of the relationship between heat and physical work. In 1874 Ludwig Boltzmann [3] formalized it further in his H-theorem that represents entropy in terms of the sum of the probabilities of microstates of a dynamical system. In Shannon's [22] analysis (building on the work of his predecessors at Bell Labs, Nyquist and Hartley), this same formula is used to represent the information of a given communication medium, also describing its distinguishable states. Apparently, following a suggestion from the mathematician John von Neumann, he also called this value "entropy". Many writers have cogently argued that these two concepts should not be confused (e.g. [15, 20, 24, 26]), and that Shannon's choice of this term to describe a statistical property of information media was a colossal mistake (e.g. [25]). And there is, of course, much to distinguish these two uses of the same term beyond this abstract mathematical similarity. For example, there is no informational analogue to Clausius' theorem that the total entropy of an isolated physical system can only increase (the second law of thermodynamics). Nor is informational entropy a dynamical concept associated with energy and work. Nevertheless, many have attempted to discern a deeper linkage underlying these statistical analogies (e.g. [2, 14, 21]).

The idea that there is a link between information and thermodynamics has a long history. Ways of demonstrating such a linkage have been proposed in many forms from Maxwell's [19] famous demon [18] to Landauer's [17] argument about the thermodynamic cost of information erasure. However, these approaches focus primarily on the energetic and thermodynamic "costs" of manipulating physical markers or taking measurements and how this might alter system entropy. So they have largely ignored the problem of how reference and significance are physically instantiated, to instead defend the belief that information cannot be used to violate the second law of thermodynamics.

Both notions of entropy can be applied to physical features of an information medium. This is because any physical medium capable of conveying information must be able to exhibit different states. As Shannon demonstrated, the number and relative probabilities of these different states (and, in a continuous communication, their rates of change) are what determines the capacity for that medium to "store" or "convey" information. Following insights from statistical mechanics, he argued that this could be measured in terms of the value of $-\sum p_i \log p_i$, where p_i is the probability of the occurrence of a given state i of the information-bearing medium. Any process, structure, or system that can be analyzed onto component states, each able to be assigned a probability of being exhibited, can in this way be described in terms of this measure of information. This insight at the foundation of Shannon's approach also provided one of the first widely accepted model-independent measures of complexity [11]. Most subsequent means of measuring form or complexity have been developed with respect to this basic insight (as in the work of [6, 16], etc.). Information theory thus became more than just a tool for analyzing communication.

Besides providing a measure of the information capacity of a given medium, Shannon's analysis also demonstrated that the amount of information provided by a *received* signal (i.e. a message) can be measured as a function of the amount of uncertainty that the received signal removes. This can be measured as a difference between the prior (or intrinsic) uncertainty (the Shannon entropy) of the signal medium being in a given state and its current received state (e.g. in a received message). This necessarily relational nature of information is an important distinction that is often overlooked, and it effectively distinguishes two interdependent and inter-defined uses of the concept: the potential information capacity of a given medium and the information provided by a specific message conveyed by that medium. Both states can be assigned an entropy value. Using this relative measure, problems of noise, error correction and encryption can be likewise analyzed in terms of differences or changes in component signal state probabilities.

The relational nature of information, even in this technically minimalistic form, is an important clue to the fact that information is not a simple intrinsic property of things. Interestingly, this is also a feature that both concepts of entropy share. The third law of thermodynamics likewise asserts that entropy is a relational measure: a difference between states of a system. The relational nature of thermodynamic entropy was not fully appreciated until 1906 when Walter Nernst augmented thermodynamic theory establishing an absolute reference point at $0°$ K. Likewise the relational nature of informational entropy is also often overlooked.

The entropy of a signal is also not an intrinsic property but a relational property. This relational character of both concepts of entropy is a clue that the statistical signal properties of a medium are linked to its ability to convey reference. Both Shannon's analysis and thermodynamic theory depend on comparing the relative degree of constraint on current entropy with respect to what is minimally and/or maximally possible for a given system. A received signal that exhibits constraint in its information entropy is more predictable even if it is not reduced to a single fixed value. Thus even a noisy signal reduces uncertainty by this difference, so long as it is not fully random.

Noise, which is the corruption of a signal, increases uncertainty by reducing this constraint. But noise is often the result of physical degradation of the conveying medium. This is yet a further clue to the intrinsic interrelationship between thermodynamic and information entropy.

The distinction between signal and noise is an important clue to how an information-bearing medium can be linked to some nonintrinsic object, event, or property. Noise is a difference or change of the information entropy of a message. Often, as in the case of radio transmission, noise is due to an increase in thermodynamic entropy of the conveying medium as a result of interference or simple signal degradation, i.e. to factors extrinsic to the information-bearing medium that affect its physical attributes. Any such physical influence will entail physical work, and work entails a change in thermodynamic entropy. This suggests that what at first appears to be an unrelated and superficial parallelism between these two concepts of entropy is instead a critical clue to how a signal medium can be about something that it is not.

Noise and signal are *both* linked to something extrinsic that has affected the information medium. The difference is due to interpretive assessment, not anything intrinsic to the informing medium. The nonintrinsic distinction between signal and noise is an additional relational attribute indicating that, even though specific reference and significance can be set aside to analyze the statistical properties of an information-bearing signal, they are nevertheless assumed.

3   Physicality of Aboutness

To exemplify the way that the statistical properties of a medium can provide the potential for reference, consider the use of information-theoretic analyses in molecular biology. The statistical properties of nucleotide sequences can provide critical clues to potential biological functions even though the sequencing of genomes is largely accomplished in ignorance of any specific gene function. Indeed, exploring the statistical structure of these sequences has provided many important clues to unanticipated functional properties of DNA, RNA and the organic properties they specify. For example, the presence of constraints on the possible statistical entropy of a gene sequence, exhibited in sequence redundancy across species, tends to predict that the sequence in question plays a functional role in the organism, i.e. that it encodes information "about" that function. This is because constraint in the form of redundancy provides evidence that nonrandom—i.e. functional—influences are at work. Simply scanning gene sequences for constraint, then, provides a tool for discovering sequences that probably contain information "about" some function, even though no specific functional information is provided.

The concept of mutual information is also a clue to the way that signal constraint relates to reference. The mutual information between two signals or sequences of alphanumeric characters is a measure of their statistical nondifference. This can even be assessed despite nonidentity of any of the components in the two, as for example exists in the statistical parallels between nucleotide sequences in DNA and amino acid sequences in proteins. Shannon demonstrated that any degree of signal noise less than total noise can be compensated for by a comparable level of signal redundancy. In this respect, redundancy—a constraint on possible variety—is what preserves signal reference. In genetic evolution, sequence redundancy and mutual information are clues that a given genetic signal carries functional information and provides a kind of reference to cellular-molecular dynamics that were useful in the past and are likely useful in the present. As the epigraph to this essay from Gregory

Bateson suggests, reference is ultimately embodied in signal constraint (i.e. redundancy and the regularity or pattern that this produces).

So implicit in the technical sense of "information" is an understanding that statistical properties of the signal medium are in some way related to its referential function. The question that I want to address is how it might be possible to more precisely characterize this relationship.

To address this let us begin with a focus on the relationship between thermo-dynamic and information entropy. Worries about the relationship between infor-mation and thermodynamics emerged in parallel with the kinetic theory of gasses and long predate Shannon's use of the entropy concept. They are best exemplified by the many analyses that developed in response to Maxwell's [19] famous thought experiment, which subsequently came to be known as Maxwell's demon. Though he was one of the major contributors to the formalization of the second law of thermodynamics, Maxwell pondered the possibility that molecular-level informa-tion might be able to drive an otherwise isolated thermodynamic system from a higher to a lower entropy state in violation of the second law. He described this problem in terms of a fanciful microscopic "demon" able to judge the relative momentum and/or velocity of gas molecules in either side of a two-chambered container with a closable passage separating them. He wondered whether the demon could use these measurements to determine whether or not to let a molecule move from one container to the other. Such a demon (or a device that acted in the same way) could selectively allow only fast-moving molecules to pass one way and only slow-moving molecules to pass the other way, thus progressively reducing the entropy of the whole system, in violation of the second law. But if this were even logically conceivable (even if not achievable with any physical mechanism), then the ubiquity and ineluctable directionality of the second law would be questioned.

Subsequent analysts have variously attempted to answer this conceptual chal-lenge and preserve the second law by analyzing the thermodynamic aspects of the information assessment process. This has been explored in simplified abstract mechanical terms involving one or just a few moving particles (e.g. [23]), or by considering the relative entropic "cost" of acquiring this information (e.g. [5]), or in terms of the need to erase information from previous measurements in order to repeat the process (e.g. [17]). The intention has been to show that, even if one were able to create a device capable of such actions, the second law of thermodynamics would be preserved because there would be more entropy generated by obtaining and using this information than would be reduced by its actions.

In contrast, I propose to analyze the referential property of information directly and irrespective of the energetic cost of generating or erasing bits of data (though ultimately this cost is a relevant factor as well). For this purpose I instead focus on one of the most basic principles of Shannon's analysis: the role played by statistical constraint in the assessment of the amount of uncertainty removed by receipt of a given message or introduced by thermodynamic noise in a transmission. This allows us to make an abstract, but direct, comparison between the entropy of a given information-bearing medium in Shannon's terms and the entropy of that same medium in physical terms (though not necessarily just in thermodynamic terms).

The unifying factor is that both are expressions of the physical attributes of the information-bearing medium.

The essential point is this: every medium for storing or conveying information is constituted physically and its distinguishable states are physical states. So any change in that medium's statistical physical properties (e.g. its thermodynamic entropy) could potentially also change its informational properties. This is not a necessary relationship, since the distinguishable states used to convey information in any given case are inevitably a very small subset of the total range of different states that the physical medium can assume. However, because of its physicality, any change in the informational entropy of a given medium must necessarily also entail a change in its physical statistical properties. And, following the strictures of the second law of thermodynamics, any physical medium will only tend to be in an improbable constrained state if it has in some way been driven away from its more probable state by the imposition of physical work or prevented from achieving it by some extrinsic restriction. In other words, the relationship between the most probable state of the medium and the observed state at any particular moment is a reflection of its relationship with its physical context. Its intrinsic statistical properties are therefore clues to factors that are extrinsic to it.

In this analysis I build on this insight to argue that *referential* information is based on the constraints generated by physical work introduced due to the thermodynamic openness of an information-bearing medium and its susceptibility to contextual modification. The next section provides a point-by-point outline of the logic that formally defines the referential aspect of information in terms of the relationship between Shannon (information) entropy and Boltzmann (thermodynamic) entropy.

4   Steps to a Formalization of Reference

A. General case: passive information medium near equilibrium (e.g. geological formation, crime scene evidence, data from a scientific experiment, text, etc.)
1. Information (e.g. Shannon) entropy is not equivalent to thermodynamic (e.g. Boltzmann–Gibbs) entropy (or to the absolute statistical variety of physical states). [For convenience these entropies will be provisionally distinguished as Shannon vs. Boltzmann entropy, though recognizing that each includes multiple variant forms.]
2. However, for any physical signal medium, a change in Shannon entropy must also correspond to a change in Boltzmann entropy, though not vice versa because the distinctions selected/discerned to constitute the Shannon entropy of a given signal medium are typically a small subset of the possible physical variety of states—e.g. statistical entropy—of that medium.
3a. The Shannon information of a received message is measured as a reduction of signal uncertainty (= a reduction of Shannon entropy).

3b. For a simple physical medium, reduction of Shannon entropy must also correspond to a reduction of the Boltzmann entropy of that medium.

3c. This can be generalized as "any deviation away from a more probable state" (which can violate 3b in the case of media that are actively maintained in an improbable state, such as maintained far from equilibrium). **(See B below.)**

4a. A reduction of Boltzmann entropy of any physical medium is exhibited as constraint on its possible states or dynamical "trajectories".

4b. The production of physical constraint requires physical work in order to produce a decrease of Boltzmann entropy, according to the second law of thermodynamics.

5a. For a *passive medium* the physical work required to reduce its Boltzmann entropy must originate from some physical source *extrinsic* to that medium.

5b. Generalization: Constraint of the Shannon entropy of a passive medium = constraint of its Boltzmann entropy = the imposition of prior work from an external source.

6. An increase in constraint (i.e. deviation away from a more probable state) in the information medium literally "re-presents" the physical relationship between the medium and the extrinsic contextual factors (work) that caused this change in entropy (= what the information embodied in the constraint can be "about").

7. Since a given constraint has statistical structure, its *form* is a consequence of the specific structure of the work that produced it, the physical susceptibilities of the information-bearing medium and the possible/probable physical interactions between that medium and this extrinsic contextual factor.

8. The form of this medium constraint therefore corresponds to and can indirectly "re-present" the *form* of this work (i.e. in-*form*-ation).

9. Conclusion 1. The possibility of reference in a passive medium is a direct reflection of the possibility of a change in the Boltzmann and Shannon entropies of that medium due to a physical interaction between the information-bearing medium and a condition extrinsic to it.

10. Conclusion 2. The possible range of contents thereby referred to is conveyed by the form of the constraint produced in the medium by virtue of the form of work imposed from an extrinsic physical interaction.

11. Conclusion 3. The informing power of a given medium is a direct correlate of its capacity to exhibit the effects of physical work with respect to some extrinsic factor.

12. Corollary 1. What might be described as the referential entropy of a given medium is a function of the possible independent dimensions of kinds of extrinsically induced physical modifications it can undergo (e.g. physical deformation, electromagnetic modification, etc.) multiplied by the possible "distinguishable" states within each of these dimensions.

13. Corollary 2. Having the potential to exhibit the effects of work with respect to some extrinsic physical factor means that even no change in medium entropy or being in a most probable state still can provide reference (e.g. the burglar alarm that has not been tripped, or the failure of an experimental intervention to make a difference). It is thus reference to the fact that *no work to change the signal medium has occurred.*

In addition, since not all information-bearing media are inert physical structures or otherwise passive systems at or near thermodynamic equilibrium, we need to modify certain of these claims to extend this analysis to media that are themselves dynamical systems maintained far from equilibrium. This yields the following additional claims:

B. Special case: nonpassive information medium maintained far from equilibrium (e.g. metal detector or organism sense organ)
1. A persistently far-from-equilibrium process is one that is maintained in a lowered probability state. So certain of the above principles will be reversed in these conditions, specifically, those that depend on extrinsic work moving a medium to a lower probability, lower entropy state.
2. Maintenance of a low Boltzmann entropy dynamical process necessarily requires persistent physical work or persistent constraints preventing an increase of Boltzmann and Shannon entropies.
3. Any corresponding increase in Shannon entropy therefore corresponds to a disruption of the work that is maintaining the medium in its lower entropy state. This can occur by impeding the intrinsic work or disrupting some dissipation-inhibiting constraint being maintained in that system.
4a. An increase in the Shannon entropy of a persistently far-from-equilibrium information medium can thereby "indicate" extrinsic interference with that work or constraint maintenance.
4b. A persistently far-from-equilibrium dynamical medium can be perturbed in a way that increases its entropy due to contact with a passive extrinsic factor. Any passive or dynamic influence that produces a loss of constraint in such a system can provide reference to that extrinsic factor.
5a. Since work requires specific constraints and specific energetic and material resources, these become dimensions with respect to which the change in entropy can refer to some external factor.
5b. The dynamical and physical properties of a far-from-equilibrium information-bearing medium determine its "referential entropy".
6. Corollary 3. This can be generalized to also describe the referential capacity of any medium normally subject to regular end-directed influences that tend to cause it to be in an improbable or highly constrained state. This therefore is applicable to living systems with respect to their adaptations to avoid degradation and also to far more complex social and cultural contexts where there is active "work" to maintain certain "preferred" orders.

5 Active Acquisition of Information

The far-from-equilibrium case is of major importance also because it provides the foundation for an analysis of the nature of an interpretive process, such as might be applied to simple organism adaptations and genetics (see next section).

A simple exemplar of a far-from-equilibrium information medium is a metal detector. Metal detectors typically operate by constantly maintaining a stable electromagnetic field. This signal medium requires the work provided by the constant flow of an electric current through a coil. This magnetic field is easily distorted by the presence of a conducting object, thus interfering with the electronic work otherwise maintaining a redundant signal also linked to this work (often the tone produced by an oscillator sensitive to a change in current).

This is roughly analogous to the way that living processes gain information about their world. Whether by virtue of a ligand binding to a specific receptor molecule on a cell surface and changing its conformation or a photon modifying the dynamics of signal processing in retinal neurons, it is ultimately interference with an ongoing living process requiring metabolic work that provides referential information to an organism.

Recall that, according to Shannon's analysis, the measure of information in a message is proportional to the reduction of entropy in the received signal compared with the potential entropy of the channel (i.e. the medium). Now we can see that reduction of the Boltzmann entropy $(-\Delta S_b)$ of the passive information medium is proportional to a reduction of its Shannon entropy $(-\Delta S_s)$ by some proportionality constant m (usually far below 1.0) that determines what portion of its Boltzmann entropy is used as Shannon entropy. In the far-from-equilibrium case the situation is reversed: some fraction of the increase in a medium's Boltzmann entropy $(+\Delta S_b)$ is proportional to an increase of its Shannon entropy $(+\Delta S_s)$. In both cases, information about the source of an external disturbance is negatively embodied in the Shannon entropy $(\pm\Delta S_s)$ of the medium as a change in intrinsic constraint.

In general this means that a medium in its most probable state, exhibiting no change in entropy, can also provide information about the *absence* of a given specific referent. This is nevertheless a form of reference. A smoke alarm that remains silent in the absence of smoke still provides information. So both a high probability state and a low probability state of an information medium are potentially referential, demonstrating that every referential relationship corresponds to a reduced probability state of Shannon entropy. Thus we can conclude that reference is made possible by the susceptibility of a given information medium to reflect the effect of work with respect to an extrinsic context, and that the sign of this effect— i.e. whether there is an increase or decrease of medium constraint—will depend on whether this work originates in the interpretive process or in its extrinsic physical context.

6   Conclusion

The above analysis demonstrates that the capacity of an informing medium to provide reference (i.e. "aboutness") derives from a linkage between its information entropy and thermodynamic entropy. Despite the fact that these are nonequivalent statistical measures of different properties (e.g. free energy vs. formal complexity, respectively) they can both reflect related changes introduced by physical work. So although reference is not an intrinsic physical property of any information-bearing medium, this linkage to work renders the referential property of information susceptible to exact formal and empirical analysis. Thus reference is not a subjective, heuristic, or epiphenomenal product of prescientific theorizing, like phlogiston, able to be dispensed with as more precise physical science comes to explain it away. Rather, it is a causally relevant physical property affecting systems that depend on selected contextual features for their operation, but lack direct access to them.

This qualitative analysis does not provide a way to quantify something like referential capacity. Indeed, it is not clear what such a measure might consist in. But it does suggest that to utilize the state of a mediating substrate to access the reference afforded by that medium's intrinsic constraints, an interpreting system must do so with respect to the consequences of work. For such a system, functional significance can be assigned to this referential information with respect to work "saved" due to access to information about contextual factors relevant to achieving a preferred target condition. This implies that a system capable of assessing reference must be organized to achieve or maintain a far-from-equilibrium state, such as in a living system.

Although the qualitative form of this analysis may still limit technical applications, such as in molecular biology, cognitive neuroscience, or artificial intelligence, it should nevertheless be sufficient to serve as a framework for the future development of a precise formal theory of reference and functional significance.

References

1. Bateson, G.: Upside-Down Gods and Gregory Bateson's World of Difference. Fordham University Press, New York (1968)
2. Ben-Naim, A.: A Farewell To Entropy: Statistical Thermodynamics Based on Information. World Scientific Publishing Co, Singapore (2008)
3. Boltzmann, L.: The second law of thermodynamics. Populare Schriften, Essay 3, address to a formal meeting of the Imperial Academy of Science, 29 May 1886, reprinted in *Ludwig Boltzmann, Theoretical Physics and Philosophical Problems*, S. G. Brush (Trans.). Boston: Reidel (1874)
4. Brentano, F.: Psychology From an Empirical Standpoint. Routledge & Kegan Paul, London, pp. 88–89 (1874)
5. Brillouin, L.: Science and Information Theory. Academic Press, New York (1962)
6. Chaitin, G.: Algorithmic Information Theory. IBM J. Res. Develop. **21**(350–359), 496 (1977)

7. Clausius, R.: The Mechanical Theory of Heat: With its Applications to the Steam Engine and to Physical Properties of Bodies. John van Voorst, London (1865)
8. Deacon, T.: Shannon-Boltzmann-Darwin: Redefining Information. Part 1. Cogn. Semiot. **1**, 123–148 (2007)
9. Deacon, T.: Shannon-Boltzmann-Darwin: Redefining Information. Part 2. Cogn. Semiot. **2**, 167–194 (2008)
10. Deacon, T.: Incomplete Nature: How Mind Emerged from Matter. W. W. Norton & Co., New York (2012)
11. Deacon, T., Koutroufinis, S.: Complexity and dynamical depth. Information **5**, 404–423 (2014)
12. Deacon, T., Srivastava, A., Bacigalupi, J.A.: The transition from constraint to regulation at the origin of life. Front. Biosci. **19**, 945–957 (2014)
13. Fano, R.: Quoted from *Information Theory And The Digital Age*. Compiled by Aftab, Cheung, Kim, Thakkar, Yeddanapudi, in 6.933: Project History, Massachusetts Institute of Technology (2001). SNAPES@MIT.EDU. http://web.mit.edu/6.933/www/Fall2001/Shannon2.pdf
14. Jaynes, E.T.: Information theory and statistical mechanics. Phys. Rev. **106**, 620 (1957)
15. Kline, S.J.: The Low-Down on Entropy and Interpretive Thermodynamics. DCW Industries, Lake Arrowhead (1999)
16. Kolmogorov, A.N.: Three approaches to the quantitative definition of information. Probl. Inf. Transm. **1**, 1–7 (1965)
17. Landauer, R.: Irreversibility and heat generation in the computing process. IBM J. Res. Dev. **5**, 183–191 (1961)
18. Leff, H.S., Rex, A.F. (eds.): Maxwell's Demon, Entropy, Information, Computing. Princeton University Press, Princeton, NJ (1990)
19. Maxwell, J.C.: Theory of Heat. Longmans, Green and Co, London (1871)
20. Mirowski, P.: Machine Dreams: Economics Becomes a Cyborg Science. Cambridge University Press, New York (2002)
21. Seife, C.: Decoding the Universe: How the New Science of Information is Explaining Everything in the Cosmos, from Our Brains to Black Holes. Penguin Books, London (2007)
22. Shannon, C.: The mathematical theory of communication. *Bell Syst. Tech. J.* **27**, 379–423, 623–656 (1948)
23. Szillard, L.: On the decrease of entropy in a thermodynamic system by the intervention of intelligent beings. In: The Collected Works of Leo Szilard: Scientific Papers (MIT Press, 1972), pp. 120–129 (1929)
24. Ter Haar, D.: Elements of Statistical Mechanics. Rinehart Press, Boulder (1954)
25. Thims, Libb: Thermodynamics ≠ Information Theory: Science's Greatest Sokal Affair. J. Hum. Thermodyn. **8**(1), 1–120 (2012)
26. Wicken, J.: Entropy and information: suggestions for a common language. Philos. Sci. **54**, 176–193 (1987)

Modelling Empty Representations: The Case of Computational Models of Hallucination

Marcin Miłkowski

Abstract I argue that there are no plausible non-representational explanations of episodes of hallucination. To make the discussion more specific, I focus on visual hallucinations in Charles Bonnet syndrome. I claim that the character of such hallucinatory experiences cannot be explained away non-representationally, for they cannot be taken as simple failures of cognizing or as failures of contact with external reality—such failures being the only genuinely non-representational explanations of hallucinations and cognitive errors in general. I briefly introduce a recent computational model of hallucination, which relies on generative models in the brain, and argue that the model is a prime example of a representational explanation referring to representational mechanisms. The notion of the representational mechanism is elucidated, and it is argued that hallucinations—and other kinds of representations—cannot be exorcised from the cognitive sciences.

1 Introduction

Contemporary discussion on the use of the notion of representation for explanatory purposes in cognitive science focusses mainly on the controversy between representationalism and anti-representationalism. Representationalism claims that cognitive representations are at least sometimes relevant for cognition, explanatorily and causally. There are numerous criticisms of positing representations in a hasty manner. Critics point out that some representations are not supposed to fulfil any particular representational role [1], can be treated merely instrumentally [2, 3], or lack naturalistic credentials, especially when it comes to showing how they can have satisfaction conditions [4]. Although I argue for representationalism, the case for representationalism requires that we consider relevant alternatives. Hence, discussing anti-representational alternatives is important. Anti-representationalism claims that, at least in some domains, positing representations that have satisfaction

M. Miłkowski (✉)
Institute of Philosophy and Sociology, Polish Academy of Sciences, Warszawa, Poland
e-mail: mmilkows@ifispan.waw.pl

© Springer International Publishing AG 2017
G. Dodig-Crnkovic and R. Giovagnoli (eds.), *Representation and Reality in Humans, Other Living Organisms and Intelligent Machines*, Studies in Applied Philosophy, Epistemology and Rational Ethics 28, DOI 10.1007/978-3-319-43784-2_2

conditions is unnecessary (however, they rarely deny all kinds of representations; see [5] for more detail).

In this paper, I argue that there are no relevant non-representational explanations of episodes of hallucination. These are usually explained by positing empty representations, or representations that have no actual reference. To make the discussion more specific, I focus on visual hallucinations occurring in people with Charles Bonnet syndrome. I claim that the character of such hallucinatory experiences cannot be explained away in a non-representational manner, for they cannot be taken as simple failures of cognizing or as failures of contact with external reality —such failures being the only genuinely non-representational explanations of hallucinations and cognitive errors in general. Then I briefly introduce a recent computational model of hallucination, which relies on generative models in the brain, and argue that the model is a prime example of a representational explanation referring to representational mechanisms for which there is simply no non-representational alternative. The notion of the representational mechanism is elucidated, and it is argued that hallucinations—and other kinds of empty representations—cannot be exorcised from the cognitive sciences.

2   Charles Bonnet Syndrome and Representations

There are numerous kinds of hallucinatory episodes, and the number of studies on these phenomena is virtually countless (for a comprehensive review, see [6]; for an accessible introduction, see [7]). To make my discussion more specific, in this paper I discuss Charles Bonnet syndrome (henceforth: CBS). CBS is usually a complex visual hallucination in people with some impairment of vision. CBS hallucinations are frequently bizarre in nature: they include figures in elaborate costumes, human beings of non-natural size, fantastic creatures, or extreme colours, which may partly overlap with real visual perception. Yet there is nothing to which these hallucinations correspond; in other words, they are not visual illusions, even if, owing to the impairment of vision, these hallucinations occur at the same time as various other abnormal phenomena.

Importantly, CBS subjects usually (but not always) realize that the visual episodes they are experiencing are not real, and sometime may even think that their unusual perception is a result of their hallucinations. One reason it is easy for them to understand that they are experiencing a hallucination is that CBS is merely visual, and there is usually discrepancy with auditory or tactile perception.

There are two main competing neurophysiological explanations of CBS. The first classifies CBS as release hallucinations, i.e. 'hallucinations mediated by spontaneous electrophysiological activity originating from subcortical brain areas such as the thalamus, the pedunculus cerebri and the limbic system' [6, p. 93]. Another attributes CBS to increased excitability of the visual pathways or the visual cortex, owing to a lack of inhibitory afferent impulses. Brain regions considered capable of mediating spontaneous visual percepts include the retina, the lateral geniculate nucleus, the primary visual cortex and the visual association cortex [6, p. 94].

These features of CBS syndrome make it difficult to explain away in a non-representational manner, because the content of hallucinations is apparently decoupled from perception. At the same time, these hallucinatory representations thereby satisfy the requirement proposed by several theorists as particularly important for representations, namely decouplability or detachment of the representation from its target [8, 9]. While Andy Clark does not see decouplability as a necessary feature of representation, it is reliably present in such hallucinations. In this case, detachment may actually justify the use of representational talk. Note that a generic move recommended by proponents of anti-representationalism in response to Clark's decouplability argument, namely to introduce time-extended perceptual processes instead of decoupling [10], will not work for CBS. Simply put, there was no point of contact between a hallucinated entity—for example, a fantastic creature—and the CBS subject in the past, so extending the perceptual process still cannot reach the hallucinated creature.

This, however, is not enough to show that representations cannot be avoided. First, anti-representationalism can appeal to an empiricist argument that all ideas stem from sensual impressions; in a Humean manner, they can appeal to elementary building blocks of perception that are recombined to build a complex hallucination. For example, you could combine an elementary perception of a white horse, and a perception of a horn, and get an image of the unicorn. The problem of course is that the 'solution' sounds fairly incompatible with the general approach of most anti-representationalists, which is the dynamical account of cognition. Dynamicists seem to embrace the claim that there are no elementary primitives of concepts or perception at all. The assumption of elementary building blocks in the traditional symbolic approach (or cognitivism) was criticized by Dreyfus [11]. In other words, the price of the classical empiricist move may be too high for most anti-representationalists to pay. But the recombination approach, with some time-extended perception, might seem to work if one believes that there is a credible empiricist answer to Berkeley's puzzle of how one can imagine things one has never seen. While CBS subjects have not perceived miniature people, they have perceived people, and they have perceived small entities, so it is possible for them to combine the two.

For the sake of argument, let me suppose that some solution like this might be put to work. However, the perceptual *recombination* would still seem to produce a representation—something semantic, something about something else—even if particular elementary perceptions are held to be non-representational points of contact with reality. Their recombination is additionally non-veridical for the CBS subject. Simply put, it is easier to eliminate veridical perception and replace it with 'direct' or 'representationally unmediated' contact with reality than to replace content-rich hallucinations with time-extended contact with reality. It is the recombination, if it actually occurs, that drives a wedge between the representation and reality. Hence, the recombination 'solution' is merely verbal—representation has been merely rebranded as recombined perception but retains the essential features of representation: aboutness and satisfaction conditions (which are never satisfied in the case of hallucinations).

But there is another option still open for anti-representationalists: to apply a neo-Gibsonian analysis of errors in cognition [12]. James J. Gibson, the founder of ecological psychology, insisted that hallucinations differ from imaginations in that they are passively experienced, and from perceptions in that they are not 'made of the same stuff' [13], p. 425). Namely, *'a person can always tell the difference between a mental image and a percept when a perceptual system is active over time'* (ibid., italics in original).[1] His general rule for distinguishing perception is as follows:

> Whenever adjustment of the perceptual organs yields a corresponding change of stimulation there exists an external source of stimulation and one is *perceiving*. Whenever adjustment of the perceptual organs yields *no* corresponding change of stimulation there exists no external source of stimulation and one is imagining, dreaming or hallucinating. [13], p. 426)

Gibson may be roughly right about some kinds of hallucination,[2] including most cases of CBS: subjects usually discover CBS hallucinations just because they are not accompanied by proper adjustments of auditory or tactile stimulation. At the same time, in another passage, he seems to contradict himself by claiming:

> One perceptual system does not *validate* another. Seeing and touching are two ways of getting much the same information about the world. [14, pp. 257–8]

In CBS, it is exactly the case that discrepancy between different perceptual systems allows hallucinators to understand that their visual experiences are not entirely veridical; and touching does provide *different* information to the subject. All in all, Gibson seems to ignore the complex role of multimodal integration, for example, the role of the vestibular system in seeing [15].

However, for our purposes, what is important is the question of whether hallucinations are to be understood as representations. To this question, unfortunately, Gibson gives no clear answer, but his theory is usually interpreted as a form of direct realism (however, see [16] for a representational reading of Gibson). In direct realism, error is understood as a failure to cognize in some way, or a failure to cognize that one fails to cognize [12]. This latter, hierarchical solution is an account of false beliefs and similar errors, so it should work for bizarre CBS hallucinations as well. What this account claims, basically, is that people who have CBS, during hallucinatory episodes, fail to cognize that they do not have perceptual states. But the truth is exactly the opposite, and Gibson himself stresses that one discovers that

---

[1]Note that this is an extreme empirical claim, and a false one, and CBS subjects can not only take hallucination to be veridical but also sometimes mistake veridical perception for hallucination [7]. Such a mistake might go easily undetected forever.

[2]Only in some hallucinations can a person tell the difference between the hallucination and perception. There might be scenic, multimodal and persistent hallucinations [6], which can be confused with perceptions by a subject.

hallucinations are not perceptions. They usually *know* that they are hallucinating, so this is not a failure to cognize that the subject fails to cognize. Neither can the contents of their hallucination be naturally accounted for in terms of a simple failure to perceive, which may happen when you cannot find your keys in a drawer. It cannot be accounted for by stipulating a hierarchy of failures to cognize, as there is no explanation of the rich *content* of bizarre CBS hallucinations (subjects stress the richness as striking; cf. [7, pp. 4–5]). For this reason, a neo-Gibsonian alternative is doomed to fail.

A third explanation of hallucination in broadly non-representational terms appeals to a sense of real presence, which is supposedly present during perception or hallucination, and absent in imagining [17]. The experienced presence of objects, even when they are occluded but accessible on further exploration, is supposed to help explain hallucination's content non-representationally. The illusory presence of hallucinatory objects is supposed to stem from the skilful exercise of perceptual skills: "The hallucinator acts out the same sensorimotor repertoire as the perceiver" [17, p. 249].[3] But this does not explain why CBS subjects describe their hallucinations as involving bizarre figures. There are no real sensorimotor skills that involved the hallucinator's previous perception of bizarre figures (and if you suppose that bizarre figures are recombinations of previous skills, you presuppose experiential primitives that the enactive approach rejects explicitly). The sensorimotor account seems, rather, to presuppose intentionality of these visual episodes:

> [T]he approach actually makes it easier to envisage brain mechanisms that engender convincing sensory experiences without any sensory input, since the sensation of richness and presence and ongoingness can be produced in the absence of sensory input merely by the brain being in a state such that the dreamer implicitly "supposes" (in point of fact incorrectly) that if the eyes were to move, say, they would encounter more detail. [18, pp. 66]

The word 'supposes' is obviously intentional, and it is not easy to eliminate or explain it away from this passage. For this reason, the sensomotoric account is actually representational, and cannot be considered an alternative to the representational explanation. Any talk of sensory experiences being about anything implies representationalism.

Critics of representationalism sometimes complain that illusions, hallucinations and misperceptions are the focus of representational explanations in psychology, and that correct perception could be understood in terms of contact. A milder non-representational position might be that of disjunctivism,[4] or the claim that perceptual processes are essentially different from misperceptual processes. Of course, in hallucinations, perceptual processes are different in that hallucinations are

---

[3]There are further problems with the sensorimotor account of CBS; it may be accompanied with partial or total paralysis, which makes any exercise of motor skills simply impossible.

[4]Note that there are representationalist versions of disjunctivism. I discuss only a possibility of disjunctivist anti representationalism above.

not perceptions. But that's trivially true.[5] The problem for the disjunctivist is that the underlying brain and bodily machinery (including active exploration) recruited for perception are the same [20]; it is not just that the subjective experience may be the same. Additionally, the difference between perception and hallucination is likely a matter of degree rather than of quality [21]. For the agent and its subpersonal processes, until discrepancy is detected between various sources of information, hallucinatory episodes may seem perceptual. When discrepancy is detected, though, non-veridical representations are considered to be such, so that the cognitive process is actually different but shares a common core with the standard perceptual process, as can be witnessed in the model analyzed below. So while there is a grain of truth that these processes differ for the subject (at least when the subject is not delusional or not confabulating), it is not the case that the differences can substantiate a non-representational position. But they may seem the same for the subject, in which case the disjunctivist claim has no explanatory role to play at all. Non-veridical hallucinations can explain behaviour just as well as veridical perception can; when the deluded subject takes hallucinations to be real, disjunctivism has yet another fact to explain: why a completely different process leads to the same behaviour as perception would. In brief, the presupposition that there is a single process underlying hallucination and perception is more parsimonious than a disjunctivist proposal that adds unnecessary complexity.

All three anti-representational alternative explanations of CBS therefore fail, and disjunctivism is unnecessarily complex. I know of no successful non-representational attempt to explain the content of hallucinations away, but my argument is based only on negative induction. Representational explanations seem to be much more plausible. Let me turn to a computational model of CBS, which will be used to offer a representationalist explanation.

3   Computational Modelling of Hallucination

Existing computational models of hallucination are all representational. In general, recent work on predictive coding in the brain [22–24] includes also some suggestions regarding hallucinations and psychosis [25]. Here, I will focus on a similar but non-Bayesian model of CBS [26], according to which there are hierarchical, generative models in the brain that control perception. These models work homeostatically; i.e. the bias varies proportionally to input strength (as such, the model assumes a neurophysiological explanation close to the second type mentioned in Sect. 2, which appeals to the increased excitability of the visual system). There are two pieces of empirical evidence that support the bias hypothesis. One is synaptic

---

[5]Only representationalists who endorse methodological solipsism [19] might deny this, as they cannot include the relationship with the environment in their explanations. But methodological solipsism is long dead. Content does *not* locally supervene on the brain alone.

(a)   **(b)**

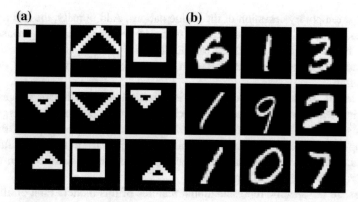

Fig. 1 a A custom data set of simple shapes at various positions. **b** The MNIST data set of handwritten digits, a standard benchmark in machine learning [26] doi:10.1371/journal.pcbi. 1003134.g003

scaling, which is a change in synaptic efficacy that is thought to affect all synapses in a neuron together, keeping their relative strengths intact. Synaptic scaling is known to occur in the neural system. Another is the fact that a neuron's intrinsic excitability can be regulated by changing the distribution of ion channels in its membrane.

The model has been implemented by a deep Boltzman machine (DBM), which is a connectionist architecture supporting deep learning. This means that there is no need for biologically implausible learning algorithms such as backpropagation [27]; the machine can find the model from the data. As the name implies, DBM is also probabilistic, but not classically Bayesian.

The network contains three hidden layers, and has been trained to recognize input figures. Notably, these input figures were not geometrically complex (simple figures and numbers, see Fig. 1), which might seem a huge idealization vis-à-vis the real phenomenon. However, in mild cases of CBS, hallucinations may consist in simpler shapes; at the same time, the model is unable to reconstruct complex, bizarre images typical of CBS. After the training, the input was clipped.

After the loss of input, owing to the homeostatic nature of the network, the output was restored. This happened as early as after 20 cycles of operation. But this result is not the only reason to believe that the model is correct. There is also a structural match of the model to the neural process (in other words, the model can be said to be structurally valid in terms of [28]).[6] The evidence for the structural match is that clamping the first hidden layer to zeros stopped hallucinations appearing homeostatically, and that this same behaviour has been observed in humans; namely applying transcranial magnetic stimulation (TMS) to early areas

---

[6]For this reason, the model is a prime example of the computational mechanistic explanation: the computational process is thought to correspond to the actual neural process (for more on computational explanation and mechanism, see [29]).

leads to a temporary cessation of the hallucinations. Additionally, the hallucinatory episodes are more likely to appear in states of drowsiness or low arousal. The authors believe that this is related to acetylcholine (Ach) dysfunction; and this is modelled as the balance factor between the feedforward and feedback flows of information.

All in all, while this model is an idealization it nevertheless suggests that interfering with cortical homeostatic mechanisms might prevent the emergence of hallucinations in CBS. This means that it provides novel predictions. DBM also uses top-down interactions during inference, not just learning, which fulfils an important requirement for modelling the role of hierarchical bottom-up and top-down processing in hallucination.

What are the specific representational features of this model? First of all, DBM learns the patterns in the input data instead of memorizing them: "rather than just memorizing patterns, BMs can learn internal representations of sensory data" [26]. These patterns are amplified by bias owing to homeostasis, which leads to hallu-cinations in drowsiness or low arousal. The model, however, does not explain why subjects realize that the experienced images are hallucinatory rather than perceptual. This is not included in the intended scope of the model—i.e. the model simply does not represent such perceptual evaluation processes at all. In other words, it is not a model of all the relevant processes responsible for CBS hallucinations but of one critical phenomenon, namely the mechanism of varied sensory excitability that may lead to visual hallucinations in case of no visual input.

4 Representational Mechanisms and Hallucinations

The discussion so far has not referred to any substantial theory of representation; I only mentioned in passing that there are two features that have been considered essential: decouplability and satisfaction conditions. Although I do not think that potential decoupling is necessary for representation as such, it is certainly one of the essential properties of hallucinatory representations. But more needs to be said here.

I do not assume that all computational models are representational; on the contrary, I think that proving that a computational model is a model of represen-tation is difficult (see [29], Chap. 4). There are obvious necessary conditions that the DBM satisfies, such as having some tokens (in the generative model in the DBM) that play the role of representation vehicles.[7] This is satisfied just because DBM is a computational mechanism, and such mechanisms process information, which means that they manipulate certain information vehicles. Another satisfied condition is that we talk of representation targets in this case: the simple figures are not just vehicles of information; they are supposed to refer to perceptually given

---

[7]Note that I do not claim that the input layer contains any representational vehicles at all. They are merely input information, not representation. For more on this distinction, see [30, 29].

entities, while there are no relevant perceptual inputs in reality, just as in CBS episodes. Representational targets would be found in reality were this a perceptual process—if only in a person's visual field, there would be entities with visual characteristics furnished by the visual system. In traditional terminology, one could say that there is some content of visual representations, or some intension of these representations. In the DBM, these characteristics are represented as the configuration of the network output layer.

These conditions—referring to targets (which may fail), having characteristics, having satisfaction conditions, containing vehicles of information—do not suffice for something to count as a representation, however. In general, without any reference to agency and evaluation of the representation, it is still unclear whether these vehicles of information are *mental* representations at all [31]. I suggest that one approach to this problem is to apply the neo-mechanistic account of explanation to see what is lacking in the model. The advantage of this approach is that the mechanism always posits mechanisms in the context of other mechanisms, so that representation won't float freely without being part of an organized cognitive system.

Before I systematically introduce the notion of a representational mechanism, the notion of a mechanism needs to be elucidated. While definitions of mechanisms offered by various authors accentuate different aspects, the main idea can be summarized as follows: mechanisms are complex structures, involving organized components and interacting processes (or activities) that contribute jointly to a capacity of the structure. Mechanistic explanation is a species of causal explanation, and interactions of components are framed in causal terms (for the main proponents, see [32–35]).

What is important is that there are no mechanisms per se; there are only *mechanisms of something*. In other words, it is critically important to specify the phenomenon to be explained, or the capacity of the mechanism. The mechanism is defined by its capacity, or individuated with respect to its capacities. For example, the capacity of the mechanism of a mousetrap is to catch mice. The mousetrap may be physically connected to a table, but as long as this connection makes no difference to its exercising the capacity to catch mice, the connection to the table does not make the table a component of the mechanism. Briefly, only those activities and components that contribute to the mechanism's exercising its capacity count as belonging to the mechanism. Hence, mere spatiotemporal co-occurrence does not make anything a component of the mechanism; the notion of the mechanism, which is a spatiotemporal entity, is defined via its capacity (or function). For this reason, it is also theory dependent [36, 37].

A representational mechanism will be one that has the above-mentioned necessary properties and that makes the information contained in the vehicles available to the cognitive system by modifying the system's readiness to act. The notion of information used here requires more elucidation. DBM states can be treated as states of the physical medium; and as long as the medium has different physical states, as distinguished by the machinery of the DBM, it contains information. In this case, one can call this information *structural* (following [38]). Minimally, the physical medium needs to have at least one degree of freedom distinguished by the whole system. Now, the structural information becomes *semantic* as soon as it modifies the system's readiness to act, more precisely as soon as the conditional probabilities of actions of the system are changed accordingly when the information vehicles change [38].

However, mere semantic information in the above sense does *not* make a representation. The representation needs to play a representational role in the system, and for that it needs to have satisfaction conditions—truth or veridicality conditions for descriptive representations, and success or failure conditions for directive representations (some representations may have both kinds of satisfaction conditions, in particular pushmi-pullyu representations sensu Millikan; cf. [39]). These satisfaction conditions, additionally, have to be evaluable by the cognitive system itself. Only then may the contents be said to be available for the system. Such evaluation requires more than negative feedback in the information-processing mechanism: negative feedback might merely modify the system's input value. What is required instead is that the error is detected by the system; note that this is what CBS subjects normally do, although error detection was not included in the DBM model. For this reason, the DBM is not a complete model of a representational mechanism. The idea that system-detectable error gives rise to genuine representationality is by no means new (for an extended argument, see [30]).

By framing the capacity of the representational mechanism as the modification of the readiness to act, based on the information available to the cognitive system, this framework is committed to a claim that representation is essentially action-oriented. However, this orientation does not mean that all representations directly activate effectors of the system, or that representation simply controls the motor activity of the system. There might be content that is not exploited in action; what is altered is just the readiness to act. The notion of action is to be understood liberally to include cognitive operations.

Summing up, we can define the complex capacity of the representational mechanism as follows:

1. Having information vehicles that modify the cognitive system's readiness to act
2. Referring to the target (if any) of the representation
3. Identifying the characteristics of the target
4. Having satisfaction conditions based on these characteristics
5. Evaluating the epistemic value of information, or checking the satisfaction conditions

In research practice in cognitive sciences, a variety of different representational mechanisms have been posited, and the current proposal is neutral with regard to empirical questions such as whether there might exist mechanisms that only deal with the "language of thought" (usually dubbed "symbolic"), or whether there are also imagistic formats of representation; it does not decide the nature of concepts, or non-conceptual thinking, either. In other words, the account is very liberal. However, just because it requires more than negative feedback, it will not license representations in a thermostat or, Watt governor or any other simple control system [40]. It also will not license content attributions to creatures capable only of taxes, such as phonotaxis, because taxes do not imply any satisfaction conditions available to the system itself [29, 41]. But it does ascribe content to theories that talk of sensomotoric contingencies as modifying readiness to act [42].[8] This, contra Hutto [43], is not a disadvantage of the sensomotoric account of vision. Hutto is right that the account is merely *verbally* non-representational, if we apply his minimal understanding of representation as having content with satisfaction conditions. Noë [17], however, presupposes that the representationalist claim is that it is perceptual experience *as a whole* that is representational and fully detailed. But in reality there are many other options available for a representationalist, and representations can be sketchy and highly action-oriented. Experience can be considered an ongoing interaction that creates multiple representational states; it may not be reducible to any particular representation among them.

Indeed, it need not be presupposed that the goal of perception is to create detailed, rich visual representations [44]. But saying that the goal of perception is not to build detailed percepts but to drive further exploration and modify the readiness to act is not the same as saying that vision does not require representation or that percepts do not exist (*contra* [45]). The latter does *not* follow from the former. However, adversaries of the representational account of perception seem to presuppose that this account implies a kind of detailed mental image, sense data and so forth, and that such entities are end products of perception. But it does not imply anything like that. Why should it? It implies much less; that there are entities that are perceived (targets), that they have perceived properties (characteristics), that they can be veridical or not, and that they can be evaluated, sometimes successfully, by the cognitive system, just by acting and exploring further.

Radical enactivism proposes that the notion of perceptual representation can be replaced with one of contact [4, 17]. In this respect, it shares the presupposition of crude causal theories of reference.[9] However, under the assumption that content is reducible to causal contact, hallucinations or misperceptions could not exist, as they are not *about* the entities that caused them. They are non-veridical or at least not entirely veridical (some of their contents may match reality). Hutto and Myin are right in saying that neither causation, correlation, or similarity constitute content; of course falsity would be impossible under such a model of representation. But hallucinations are about something, and they are false, so they do have satisfaction conditions, which can be known to the hallucinators. So how do such satisfaction conditions arise? For Hutto and Myin, this is the hard problem of content, which presumably cannot be solved. They think that there is no naturalistically kosher explanation of how content with satisfaction conditions may emerge from

---

[8]O'Regan and Noë even appeal to MacKay's [38] theory of information in their account of vision.

[9]Tom Froese suggested in his review of the previous version of this paper that direct realism does not share this trouble, as the world "directly shapes experience like a mold shapes clay". This is still causation if anything is, and direct realism faces the same troubles as crude causal theories of reference [46].

non-cultural and non-social processes. In particular, they think that there is no naturalistic explanation of the emergence of content from information (as constituted by causation, correlation, similarity, or learning).

Before I go on, it is important to note that the main argument for anti-representationalism given by Hutto and Myin [4] is a striking case of a straw man. Beside the crudest version of the causal account of reference—Fodor [47] half-jokingly attributes it to B.F. Skinner—no current account in naturalized semantics actually claims that content is constituted merely by a tracking (causation or covariation) or similarity relation. But content is *not* constituted by tracking or similarity. If a relation (in a strict logical sense) between the vehicle and the representation's target had constituted content, then false content would have been impossible. Relations obtain only when relata exist, and in the case of intentionality, the targets, or what the representation is about, might not exist.

However, Dretske, Millikan, Fodor and other proponents of naturalized semantics do not treat intentionality as a relation. For this reason, in their accounts, intentionality is not reduced to tracking or similarity relationships. First of all, the problem of the impossibility of falsehoods would reappear. In addition, we know that not all tracking or similarity relationships constitute mental representations. They are necessary but not sufficient for representation. For Dretske and Millikan, another crucial factor of content determination is the notion of teleological function; for Fodor, the important role is assigned to counterfactual considerations (in this paper, I barely touch upon the notion of function; see however [48] for a full account of a complex notion of observer-independent teleological function). Briefly, according to Dretske's account, a certain activation of neurons in the visual pathway has the function of indicating the properties of the perceived scene. In the case of biological dysfunction (such as in people with visual impairment), the visual system may still seem to indicate bizarre figures even though there is nothing in the visual field that corresponds to them. But then of course there is no real indication; the system uses the visual pathway *as though* it were indicating visual properties. The content is not determined by mere indication but by a *function* of indication.

One fact that is frequently missed in polemics against teleofunctional theories of content is that indication is for Dretske a basic form of predication. Let us see how Dretske defines functional meaning (meaning$_f$):

> (M$_f$) d's being G means$_f$ that w is F = d's function is to indicate the condition of w, and the way it performs this function is, in part, by indicating that w is F by its (d's) being G. [49, p. 22]

Indication is truth-functional; a property F is ascribed to w, and this can be spelled out in basic logical terms as ascribing a predicate to a subject. Hence, indication has satisfaction conditions. At the same time, indication cannot be false; it cannot fail to indicate that w is F. To make this possible, Dretske makes falsehood asymmetrically dependent on truth by introducing the notion of function. The entity d has the function of indicating that w is F, but as soon as it malfunctions, the indication is false. But the content is not lost; if it were an indicator, it would truly indicate that w is F.

There might be various problems with Dretske's account of content, but it solves the hard problem of content—at least in principle. The satisfaction conditions are determined by the indication relation *cum* teleological function, and there is nothing non-naturalistic about the account. While various accounts of naturalized semantics differ in many regards, they usually recruit a similar solution. What is particularly interesting is that the solution does not treat truth and falsehood symmetrically: falsehood is dependent on truth but not vice versa. For example, an account closer in spirit to the account of representational mechanisms is the interactivist model [50, 51]. In interactivism, information relationships—such as those constituted by causation, correlation, or similarity—are recruited for action, and they are used to build indications of possible actions. These indications say that such-and-such an action would be successful in such-and-such circumstances, while the circumstances are determined, inter alia, by information relationships with the environment.[10]

To sum up, there are several ways that one could analyze the hallucinatory episodes simulated by the DBM model, and there is no real difficulty with solving the hard problem (see also [5]). On the contrary, as soon as the DBM model is framed in terms of representational mechanisms, the representational role of some states of the network becomes clear; these states have satisfaction conditions just because they fail to refer. The representational explanation given by the DBM model of CBS is not complete with regard to the requirements specified by the current proposal: the scope of the model does not include evaluation processes, as I have stressed several times, so it does not fully license representational explanations. However, it clearly conforms to the general scheme proposed here: it relies on vehicles of information, and they are about (non-existing) targets, identified via visual characteristics; in the verbal gloss on the model, researchers add that subjects are usually aware that these representations are not veridical, so they are evaluated as such, and for that, they need to have satisfaction conditions. Note that, even though I rephrased the description of the model to show that it involves representational mechanisms, I do not see a plausible way to rephrase it in a non-representational way, by supposing that there is a failure to cognize along neo-Gibsonian lines (all other non-representational explanations turn out to be only verbally non-representational, so they are not even candidates for paraphrasing). Put simply, a non-representational explanation of CBS episodes does not seem to be forthcoming at all.

---

[10]Note that the content that emerges first in the interactivist model is—just like the notion of affordance in Gibson [14]—egocentric, and it involves an indication about the agent. However, via a hierarchy of differentiations—information relationships recruited for action—it is supposed to provide other kinds of information. It is not entirely clear, however, if this model can supply allocentric representation, i.e. representation that does not relate immediately to a cognitive system or to its actions. In contrast, the account of representational mechanism does not claim that the basic form of content needs to be egocentric.

5 Conclusion

Real content does involve satisfaction conditions, and hallucinations have them. They fail to be veridical, and having them does not require one to speak a language. Hence, they are empirical counterexamples to the claim of radical enactivism, namely that rich content with satisfaction conditions requires language [4]. Additionally, there is currently no viable non-representational explanation of hallucinations, and offering one remains an open problem for non-representationalism.

Visual hallucinations of the kind experienced under CBS have now only representational explanations. This means that they involve the work of representational mechanisms that are responsible for the hallucinatory episodes. As for other perceptual impairments, an ideal explanatory text would contain all relevant causal factors, including the detail of neural mechanisms, relevant computational processes and the environmental context. The DBM model included epistemic evaluation only as a verbal gloss. Thus, it is definitely not an ideal explanation but a partial one.

Representations cannot be easily exorcised from the cognitive sciences, and there is no need to exorcise them unless one is ready to defend extreme reductionism. Explanations that involve representations are parsimonious and general; changing one's mental representations is the easiest way to intervene in one's own behavior; representations also play essential heuristic roles [52]. Put simply, representations are here to stay, and the news of representationalism's death was greatly exaggerated.

Acknowledgements The work on this paper was financed by the National Science Centre under the program OPUS, grant no. 2011/03/B/HS1/04563. The author wishes to thank reviewers, in particular Tom Froese, and other careful readers of the draft, in particular Joe Dewhurst, Krzysztof Dołęga, Gualtiero Piccinini and Paweł Zięba for their very helpful comments.

References

1. Ramsey, W.M.: Representation Reconsidered. Cambridge University Press, Cambridge (2007)
2. Chemero, A.: Anti-representationalism and the dynamical stance. Philos. Sci. **67**(4), 625–647 (2000)
3. Chemero, A.: Radical Embodied Cognitive Science. The MIT Press, Cambridge, Mass. (2009)
4. Hutto, D.D., Myin, E.: Radicalizing Enactivism: Basic Minds Without Content. MIT Press, Cambridge Mass. (2013)
5. Miłkowski, M.: The hard problem of content: solved (Long ago). Stud. Grammar, Logic Rhetoric **41**(54), 73–88 (2015)
6. Blom, J.: A Dictionary of Hallucinations. Springer, New York (2010)
7. Sacks, O.: Hallucinations. Alfred A. Knopf, New York (2012)
8. Clark, A.: Being There: Putting Brain, Body, and World Together Again. MIT Press, Cambridge, Mass. (1997)

9. Smith, B.C.: On the Origin of Objects. English. MIT Press, Cambridge, Mass. (1996)
10. Garzon, F.C.: Towards a general theory of antirepresentationalism. Br. J. Philos. Sci. **59**(3), 259–292 (2008)
11. Dreyfus, H.: What Computers Can't Do: A Critique of Artificial Reason. Harper & Row, New York (1972)
12. Rantzen, A.J.: Constructivism, Direct Realism and the Nature of Error. Theory & Psychology **3**(2), 147–171 (1993)
13. Gibson, J.J.: On the relation between hallucination and perception. Leonardo **3**(4), 425–427 (1970)
14. Gibson, J.J.: The Ecological Approach to Visual Perception. Psychology Press, Hove (1986)
15. Berthoz, A.: The Brain's Sense of Movement (G. Weiss, Trans.), Harvard University Press, Cambridge, Mass (2000)
16. Bickhard, M.H., Richie, D.M.: On the nature of representation: a case study of James Gibson's theory of perception. Praeger, New York (1983)
17. Noë, A.: Real Presence. Philosophical Topics **33**(1), 235–264 (2005)
18. O'Regan, J.K.: Why Red Doesn't Sound like a bell: Understanding the feel of Consciousness. Oxford University Press, New York (2011)
19. Fodor, J.A.: Methodological solipsism considered as a research strategy in cognitive psychology. Behav. Brain Sci. **3**(01), 63 (1980)
20. Ffytche, D.H.: The hodology of hallucinations. Cortex **44**(8), 1067–1083 (2008)
21. Collerton, D., Mosimann, U.P.: Visual hallucinations. Wiley Interdisc. Rev. Cognit. Sci. **1**(6), 781–786 (2010)
22. Clark, A.: Whatever next? Predictive brains, situated agents, and the future of cognitive science. Behav. Brain Sci. **36**(3), 181–204 (2013)
23. Friston, K., Kilner, J.M., Harrison, L.: A free energy principle for the brain. J. Physiol. Paris **100**(1–3), 70–87 (2006)
24. Hohwy, J.: The Predictive Mind. Oxford University Press, New York (2013)
25. Adams, R.A., Stephan, K.E., Brown, H.R., Frith, C.D., Friston, K.J.: The computational anatomy of psychosis. Front. Psychiatry **4**, 47 (2013). doi:10.3389/fpsyt.2013.00047
26. Reichert, D.P., Seriès, P., Storkey, A.J.: Charles Bonnet syndrome: evidence for a generative model in the cortex? PLoS Comput. Biol. **9**(7), e1003134 (2013)
27. Crick, F.: The recent excitement about neural networks. Nature **337**(6203), 129–132 (1989)
28. Zeigler, B.P.: Theory of Modelling and Simulation. Wiley, New York (1976)
29. Miłkowski, M.: Explaining the Computational Mind. MIT Press, Cambridge, Mass. (2013)
30. Bickhard, M.H.: Representational content in humans and machines. J. Exp. Theor. Artif. Intell. **5**(4), 285–333 (1993). doi:10.1080/09528139308953775
31. Morgan, A.: Representations gone mental. Synthese **191**(2), 213–244 (2013)
32. Bechtel, W.: Mental Mechanisms. Routledge, New York (2008)
33. Craver, C.F.: Explaining the Brain. Mechanisms and the mosaic unity of neuroscience. Oxford University Press, Oxford (2007)
34. Glennan, S.S.: Modeling mechanisms. Stud Hist. Philos. Sci. Part C Stud. Hist. Philos. Biol. Biomed. Sci. **36**(2), 443–464 (2005)
35. Machamer, P., Darden, L., Craver, C.F.: Thinking About Mechanisms. Philos. Sci. **67**(1), 1–25 (2000)
36. Craver, C.F.: Functions and mechanisms: a perspectivalist view. In: Hunemann, P. (ed.) Functions: Selection And Mechanisms, pp. 133–158. Springer, Dordrecht (2013)
37. Pöyhönen, S.: Carving the mind by its joints: culture-bound psychiatric disorders as natural kinds. In: Miłkowski, M., Talmont-Kaminski, K. (eds.) Regarding the Mind, Naturally: Naturalist Approaches to the Sciences of the Mental, pp. 30–48. Cambridge Scholars Publishing, Newcastle upon Tyne (2013)
38. MacKay, D.M.: Information, Mechanism and Meaning. MIT Press, Cambridge (1969)
39. Millikan, R.G.: Pushmi-pullyu representations. Philos. Perspect. **9**, 185–200 (1995)
40. Miłkowski, M.: Satisfaction conditions in anticipatory mechanisms. Biol. Philos. **30**(5), 709–728 (2015). doi:10.1007/s10539-015-9481-3

41. Burge, T.: Origins of Objectivity. Oxford University Press, Oxford (2010)
42. O'Regan, J.K., Noë, A.: A sensorimotor account of vision and visual consciousness. Behav. Brain Sci. **24**(5), 939–73; discussion 973–1031 (2001)
43. Hutto, D.D.: Knowing what? Radical versus conservative enactivism. Phenomenol. Cognit. Sci. **4**(4), 389–405 (2006)
44. Churchland, P.S., Ramachandran, V.S., Sejnowski, T.J.: A critique of pure vision. In: Large-Scale Neuronal Theories of the Brain, pp. 23–60. MIT Press, Cambridge, Mass (1994)
45. Neisser, U.: Cognition and Reality: Principles and Implications of Cognitive Psychology. W. H. Freeman, San Francisco (1976)
46. Bielecka, K.: Spread mind and causal theories of Content. Avant **V**(2), 87–97 (2014). doi:10.12849/50202014.0109.0004
47. Fodor, J.A.: A Theory of Content and Other Essays. MIT Press, Cambridge, Mass. (1992)
48. Miłkowski, M.: Function and causal relevance of content. New Ideas Psychol. **40**, 94–102 (2016)
49. Dretske, F.I.: Misrepresentation. In: Bogdan, R. (ed.) Belief: Form, Content, and Function, pp. 17–37. Clarendon Press, Oxford (1986)
50. Bickhard, M.H.: The interactivist model. Synthese **166**(3), 547–591 (2008)
51. Campbell, R.J.: The Concept of Truth. Palgrave Macmillan, Houndmills, Basingstoke, New York (2011)
52. Bechtel, W.: Investigating neural representations: the tale of place cells. Synthese (2014). doi:10.1007/s11229-014-0480-8

Life is Precious Because it is Precarious: Individuality, Mortality and the Problem of Meaning

Tom Froese

Abstract Computationalism aspires to provide a comprehensive theory of life and mind. It fails in this task because it lacks the conceptual tools to address the problem of meaning. I argue that a meaningful perspective is enacted by an individual with a potential that is intrinsic to biological existence: death. Life matters to such an individual because it must constantly create the conditions of its own existence, which is unique and irreplaceable. For that individual to actively adapt, rather than to passively disintegrate, expresses a value inherent in its way of life, which is the ultimate source of more refined forms of normativity. This response to the problem of meaning will not satisfy those searching for a functionalist or logical solution, but on this view such a solution will not be forthcoming. As an intuition pump for this alternative perspective I introduce two ancient foreign worldviews that assign a constitutive role to death. Then I trace the emergence of a similar conception of mortality from the cybernetics era to the ongoing development of enactive cognitive science. Finally, I analyze why orthodox computationalism has failed to grasp the role of mortality in this constitutive way.

1 Introduction

Computationalism tries to explain natural phenomena in terms of the concept of computation, where "a computation is a set of objects and relations within the domain of abstract entities (as described in the logical formalisms of theoretical computer science)" [39]. In this chapter I will present criticisms of attempts to

T. Froese (✉)
Institute for Applied Mathematics and Systems Research,
National Autonomous University of Mexico, Mexico City, Mexico
e-mail: t.froese@gmail.com

T. Froese
Center for the Sciences of Complexity, National Autonomous
University of Mexico, Mexico City, Mexico

© Springer International Publishing AG 2017
G. Dodig-Crnkovic and R. Giovagnoli (eds.), *Representation and Reality in Humans,*
Other Living Organisms and Intelligent Machines, Studies in Applied Philosophy,
Epistemology and Rational Ethics 28, DOI 10.1007/978-3-319-43784-2_3

develop a general theory of life and mind based on the concepts of computation and information processing, for example, info-computationalism [22].[1] In essence, I argue that such attempts fail to account for the meaningful perspective that we normally experience in our lives, and contend that this is because they ignore life's deathly underpinnings, its irreducible *precariousness*.

This argument will not be a formal solution to the problem of explaining how there can be a subjective perspective in an objective world (which would amount to finding an algorithmic solution to the perennial problems of cognitive science, e.g. the frame problem, explanatory gap, the hard problem of consciousness, etc.). The aim of this chapter is to come up with a compelling alternative argument that does not rely on the concepts of computation and information processing. Instead the argument is based on assuming continuity between life and mind, such that having a mind is dependent on biological embodiment [55]. Specifically, I will be making two claims: First, the potential to die is constitutive of an individual life, where the individual's process of living is considered to be a process of sense-making in its most basic sense [64]. And second, the precariousness that is intrinsic to all organismic, and therefore also of all mental, existence is the original reason why things matter to that individual being [43].

Here we therefore assume at the outset that the phenomenon to be explained is that a living being is immediately presented with a meaningful world (and not a physicist's set of objective facts about the environment, or a computer's bits of formal information). In contrast to modern views that continue the Cartesian legacy, which does not distinguish between living beings and merely mechanical objects, we assume that to live is to always be concerned with something, most fundamentally with the continuation of one's individual manner of living. Closely related, therefore, is the classic problem of biological *individuality*. For it is only when we have the concept of an individual that we can start thinking about the role that the individual's potential to transform into an irrevocable absence, its mortality, has in shaping its lived presence [42].

These considerations are applicable to even the most basic forms of life, which like all organisms must continually struggle to stave off death for yet another moment.[2] In the most minimal cases, meaning will be wholly determined by basic metabolic needs. But in the specific case of human existence it also takes on a highly symbolic dimension, and well-defined needs are replaced by open-ended desire [4]. Awareness of our own finitude has been the inspiration for some of the

---

[1]Note that there are alternative accounts of computation that go beyond formal structure and also appeal to properties of the physical mechanisms that realize the computations [47, 48, 51]. It remains to be seen if such mechanistic accounts can sidestep my criticisms, so from now on I use computation to refer specifically to accounts that do not bottom out in non-computational mechanisms.

[2]Certain kinds of organisms, especially single-cell organisms, can be considered to be "immortal" in the sense that they are not susceptible to deleterious effects of aging and will live as long as favorable circumstances persist. But they are still mortal in the sense that matters here: they live because they actively avoid disintegration and will die if outside their viability range.

oldest expressions of human culture, particularly graves, and continues to provide a dramatic source of creativity, such as Shakespeare's famous soliloquy by Hamlet or Heidegger's existential phenomenology. Indeed, Jonas [41] argues that the presence of tombs and other expressions of concern for the deceased is the most incontrovertible archaeological evidence of a fully developed human mind, more so than tool or image production, because it points to an incipient metaphysics that reaches beyond life as such. Death marks the ultimate limits of living and therefore also of natural sense-making, requiring of those who desire to grasp what happens after our death the capacity to make sense of non-sense by clothing it in symbolism [12].

If individuality, mortality and meaning are interconnected concepts then they must lie at the heart of cognitive science, at least given an ambition to account for our meaningful first-person perspective. However, so far computationalism has been unable to offer a coherent approach to any of these three concepts, let alone their interdependence. While the problem of meaning has been widely discussed in cognitive science and philosophy of mind, where it has given rise to several famous thought experiments that continue to provoke debate, there is no agreed upon solution in sight. The problem of individuality also occasionally receives attention, but mainly in the philosophy of biology because evolutionary theory requires the concept of a unit of selection. As we will see, the concept of an agent in cognitive science is used frequently but has no clear definition. However, the problem of mortality has received almost no attention at all. In fact, for cognitive science it may not seem like a problem at all, but rather as a contingent fact of life on Earth without any philosophical relevance. Yet from the point of view I will present in this chapter the two concepts of individuality and mortality are both essential for coming up with new ways of addressing the problem of meaning.

2 The Problem of Meaning

Computational theories of mind already have a long history of struggling with the problem of meaning [24], famously expressed in a number of ways, such as the Chinese room argument [52], the frame problem [19] and the symbol grounding problem [37]. Essentially, the problem is to understand how things can be meaningful for a system from its own perspective, and not just from the perspective of the human observer of that system. Another way of phrasing this is that we still have not found a response to Hume's claim that we cannot derive a value from a fact.

Nevertheless, for many researchers the notion of information seems to offer an appealing solution to this conundrum, given that it has both a strictly technical definition (following Shannon's information theory derived from the concept of entropy) and a common folk psychological interpretation (i.e. that which means something or is informative for someone). Given this terminological ambiguity it is tempting to make use of a sleight of hand, whereby the same word is used but the former concept is somehow identified with or transformed into the latter concept. Let us take a look at an example taken from info-computationalism (IC):

Information is also a generalized concept in the context of IC, and it is always agent-dependent: *information is a difference (identified in the world) that makes a difference for an agent*, to paraphrase Gregory Bateson [8]. For different types of agents, the same data input (where data are atoms of information) will result in different information. [...] Hence the same world for different agents appears differently. [22]

I take it that when Dodig-Crnkovic refers to the world appearing *for* the agent she means that it appears in a meaningful way given that the information makes a difference *for* the agent. Presumably this happens via information processing, i.e. by somehow transforming the external atoms of "information" into internal meaningful "information". In other words, the mind is (metaphorically?) thought of as a container located inside the agent into which informational content (i.e. "atoms of information"), originating from the external environment, can be transferred and then manipulated. But how is it possible for environmental information, such as covariance, to be turned into mental content? There seems to be no compelling response to this problem from the perspective of traditional cognitive science [40]. But even if the problem of content could be solved, there is still an additional problem: at what point does this content become meaningful for an agent? How do we go from the fact of there objectively being a difference in the environment to the subjective event of that dissimilarity making a difference for an agent? Even Bateson [7] could not say much more than that the agent must be "responsive" to environmental difference.[3]

Dodig-Crnkovic struggles precisely with this crucial philosophical point while trying to find a definition of information that is sufficiently broad so as to include both its technical and folk psychological meanings. She adopts Hewitt's [38] definition and attempts to integrate it with Bateson's [8] definition:

"Information expresses the fact that *a system is in a certain configuration that is correlated to the configuration of another system.* Any physical system may contain information about another physical system." [38], my emphasis) Combining Bateson's and Hewitt's insights, on a basic level we can state: *Information is the difference in one physical system that makes a difference in another physical system.* [22]

First, we see that Hewitt assumes that one system's co-variation with another system is tantamount to the one system *containing* informational content *about* the other system. However, the latter does not straightforwardly follow from the former, as has been argued extensively by Hutto and Myin [40]. Second, there is an inherent ambiguity in the notion of "making a difference". Dodig-Crnkovic appeals to a causal interpretation, whereby one physical system causes changes to happen *in* another physical system. At least this definition is more consistent because it

---

[3]Hutto and Myin [40] also opt for such a non-autopoietic, behavior-based approach to basic minds. However, responsiveness to environmental difference (or to covariance) is not sufficient to account for the emergence of meaning (see also [30]. Interestingly, Bateson [6] avoided this problem because he assumed that the environment itself embodied a larger God-like Mind of which an individual mind is only a subsystem. Forms of panpsychism have also been attractive for contemporary information-theoretic approaches to consciousness [56, 13]), although it has also been used as a *reductio ad absurdum* [10].

describes both systems from the perspective of an external observer. However, leaving aside the problem of interpreting correlation as causation, the idea of cause and effect is still not enough to explain how making a difference *in* one of the systems could make a difference *for* that system.

One of the key problems is that information theory is incapable of providing a coherent definition of an individual agent. Dodig-Crnkovic adopts Hewitt's actor model of computation, and claims that "Hewitt's 'computational devices' are conceived as computational agents—informational structures capable of acting on their own behalf" [22]. This brings us to the deep problems of defining agency, action and even responsibility. Is a computer an agent in any relevant sense? Or a thermostat? Such loose definitions have been widely adopted in AI, but they are unsatisfactory for a number of reasons [34]. We might as well ask ourselves: what is *not* an agent on that view? If a computer can be said to be acting on its own behalf, can we not say the same thing about a planet moving around the sun or about any other physical system? Indeed, info-computationalism does not hesitate to adopt a definition of agency that applies to physical systems at all scales: "an agent can be as simple as a molecule" [22]. But to say that every physical system is an "agent" in some sense brings us no closer to explaining why there is a meaningful world for us and other living beings. For if every difference of any system is information, and any change in that information is computation, and any system undergoing that computation is an agent, then we have managed to unify everything in general, but at the steep cost of failing to explain anything in particular. Without a story about how the agency of living individuals, including ourselves, differs in essential aspects from the dynamics of mere objects we are forced to either side with some version of panpsychism by elevating objects to the status of genuine individuals or embrace a form of nihilism by reducing living individuals to the status of mere objects.

3 The Problem of Individuality

Computationalism struggles to come up with a coherent notion of individuality, which would require it to determine the other's boundaries in a manner that would allow that individual to transcend our determination from the outside, i.e. for the individual to at least partially escape complete reduction to an observer's perspective. Both computation and information are inherently observer-relative concepts that preclude them from being intrinsic properties of the phenomena [18, 25], and are therefore unsuitable for this task. We will return to this point at the end of this section. For now it is crucial to note that this criticism of relativity should not be misunderstood with reference to an absolute reality in itself. The point is not to remove the role of the observer altogether and adopt a view from nowhere, but to make space in the relationship that the observer has with the observed to allow for that other end of their relation to at least appear to have some intrinsic properties that are self-determining.

This is why enactive cognitive science is founded on the concept of autopoiesis [54], which can be loosely defined as a network of processes that form a whole because the processes are enabled by each other. In other words, this is a low-level concept of individuality in the form of a living system's self-organized identity, which can be realized in such a diverse and nested manner that even a single organism, including ourselves, can be thought of as a "meshwork of selfless selves" [57]. The key advantage of autopoietic theory is that it allows any living system, if we distinguish its boundaries appropriately, to appear to us as being autonomous, i.e. as spontaneously self-distinguishing. This is the first step toward a flexible and operational theory of individual agency, which additionally includes asymmetrical regulation of the autonomous system's environmental coupling in accordance with its own normativity [3]. When starting from such a definition of agency, based on the concept of a precarious self-producing network of processes, the system's emergent behavior is an expression of its ongoing metabolic self-realization and is therefore intrinsically related to satisfying the needs that allow the individual to maintain its way of living [33]. This inner relation between being and doing is one reason why the world as it appears from the perspective of the agent makes sense to it. And this is also why a living being is always situated in a meaningful world, whereas an artificially intelligent system, whose systemic identity is completely and arbitrarily defined from the outside, has to face the problem of meaning [21, 34].

A corollary of this account of individuality is that by definition it does not permit the formulation of a complete model of any specific example of an individual, at least not as long as it can be said to be alive. This applies as much to real as to virtual organisms. In the case of a real organism, all measurement depends on an interaction between an observer and the system, and a full determination of a living system from the outside perspective would only be possible by engaging in inter-actions that destroy the self-determination that is autonomously enacted by that system from the inside, i.e. by killing it. In the case of a virtual organism we do not have to kill it in order to know it completely, since we have full access to the code which implements it, but the final result is the same. Having a complete simulation model of an individual running on a computer would amount to that individual not transcending our determination from the outside, and thus failing to overcome the limitation of pure observer-relativity.

To be fair, more work needs to be done on how this concept of an autonomous systemic identity, which applies even to the most basic of living systems, scales up to the individual self that is characteristic of human existence [44]. And it is still not clear if even the basic concept can respond to all the challenges that have been associated with the notion of an individual in contemporary biology [14]. Similarly, the concept of agency requires further work. For example, the phenomenologically inspired concept of normativity, which lies at the root of the distinction between intentional action and passive movement, is not without its critics even among researchers who are otherwise sympathetic to an enactive approach [5, 63]. I mention these issues here to highlight that the enactive theory of individual agency is far from complete and is an ongoing project.

At the same time it must be acknowledged that research on computational and information-theoretic measures of aspects of biological and mental organization, for example, emergence, self-organization, homeostasis, autopoiesis and even consciousness, continues to advance (e.g. [26, 50]). But so far these measures are limited by the lack of a coherent concept of individual agency. And there are compelling reasons to think that they cannot account for this individuality even in principle due to their reliance on principles that are inherently observer-relative.

Information theory can only account for a system's information from the perspective of the external observer of that system, but this says nothing about any putative intrinsic perspective enjoyed by that system for itself [9]. Moreover, this dependence on the external observer entails that the reference (i.e. the "aboutness") of information is not an intrinsic property of the information-bearing medium, either [18]. Related worries about observer-relativity also apply to computational accounts, which suffer from a reliance on interpretation from the outside to determine the specific form of the computational process and its particular meaning [10, 25].

I agree that this reliance on observer-relative notions is problematic, but it is not fatal since computationalism can follow the enactive approach in adopting the presence of autopoietic self-distinction as the mark of an autonomous individual. Info-computationalism, for example, cites the work by Maturana and Varela on autopoiesis as one important influence [23]. However, a simulation model of autopoiesis is not sufficient for genuine individuality given that a model allows complete external determination.

4 The Problem of Mortality

The more serious problem that I am concerned with is thus computationalism's abstraction from the concreteness of biological existence, which prevents it from grasping the precariousness of an irreplaceable living being. Information theory is one way of formalizing this abstraction. I therefore agree with the assessment of Gershenson [36], who argues that "considering only information, one cannot distinguish the physical from the virtual" and that to have a complete scientific account "it is not enough to consider only the organization/information of systems; their substrate and their relation must also be considered".

To be fair, it is true that processes involving information and computation are necessarily also dependent on some physical implementation that realizes them [23]. But the concrete materiality of their implementation does not necessarily shape the form of these processes, which can after all be multiply realizable and substrate independent. Given some specification of a computation such as OR (True, False) it does not matter whether it happens to be realized as physical processes, or executed by a virtual machine on my laptop, or by my use of pen and paper, or in my imagination. At the level of computation the process of a logical OR is identical across all cases.

Given that computationalism does not distinguish the physical from the virtual it comes as no surprise that death, as the irrevocable disintegration of autonomous individuality, has hardly ever been problematized from an information-theoretic perspective (although there is an exception, [35], to which we will return later on). In brief, virtual agents are immortal because their existence is fully exhausted by informational structures that can be indefinitely recreated in an absolutely identical manner. Death is therefore relegated to an unfortunate fact of life on Earth that could conceivably be avoided under other circumstances, such as with more advanced medicine and technology.

To be sure, computationalism is certainly not alone in this neglect of death. Apart from mortality's role in population statistical considerations and the principles of evolutionary biology [53], it is a marginalized topic in the mainstream sciences of life and mind in general. Various reasons can be offered for this neglect, both cultural and theoretical. Bateson [6] relates it to modernity's rejection of all religious narrative which nevertheless lingers as a culture steeped in mind-body dualism: "It is understandable that, in a civilization which separates mind from body, we should either try to forget death or to make mythologies about the survival of transcendent mind". Indeed, death has been abstracted away by computationalism as irrelevant to understanding the basic principles of the mind. For example, even the most realistic simulations of the brain to date basically treat the neural network is if it were as immortal as the mathematical equations that model its activity. No relevance is seen in treating the brain as an organ of a precarious body in need of metabolic and dynamic self-renewal and thus forced to be an open system in interaction with the world.

It is not my intention to provide a more detailed analysis of the reasons for this scientific neglect of death here (but see [60], pp. 131–136). Instead I only indicate that the current scientific perspective is rather unusual when compared with many traditional worldviews, which assign a constitutive role to mortality in their representations of reality. I will consider two such worldviews in order to help us to bracket our modern attitudes toward the uselessness of death. Then I will return to the scientific perspective and highlight some defining moments in the history of systems biology and enactive cognitive science, which reveal that some of these aspects of traditional worldviews are currently being recovered, in particular that mortality plays a constitutive role in an individual's life. Finally, I conclude with an analysis of the limitations of computationalism with respect to coming to terms with this kind of perspective on mortality.

5 The Role of Death in Traditional Worldviews

Mortality may seem quite useless to us today in our youth-obsessed culture, but this stance is not universally shared across cultures. Death may also be considered to play an essential role in life. I will briefly illustrate this alternative perspective with two examples of foreign ancient cultures.

The family of cultures present throughout ancient Mesoamerica recognized that there was a circular interdependence between life and death. We can see this clearly in people's relationship with maize. Maize cultivation was a necessary condition for the rise of civilizations in this area, primarily because it allowed populations to expand to sufficiently large numbers [16]. Yet people were a necessary condition for the survival of maize, too. It required human help to free the seeds from the ears of corn, for otherwise they have problems germinating and fail to spread sufficiently. So while humans are the maize plant's principal cause of death (via harvest), humans are also an essential condition of its long-term survival and flourishing (via sowing). This unification of the duality of life and death into a circular, dynamically integrating whole was culturally manifested in a variety of ways.

This Mesoamerican relationship between life and death is often described as a form of duality, although it must be emphasized that no independence of the two terms is implied. It is a duality that recognizes the essential interdependence of opposites and thus implies complementarity. It can be traced as far back as the early formative period of central Mexico, from which ceramic masks of a half living, half skeletal face have been uncovered [49].

The use of such dualistic pairings is one of the basic principles of Mesoamerican thought. The interaction between the two halves of a duality was what we would call nonlinear, in that something new would emerge from their coupling. For example, in Nahuatl ritual speech the phrase "fire-water" signified war. The principle of duality was so important that it was deified as *Ometeotl* (the "two god") and assigned to the highest level of heaven, *Omeyocan* ("place of duality") in the form of a couple, *Ometecuhtli* and his consort *Omecihuatl*. The Aztecs venerated *Ometeotl* as the supreme creative principle, a self-generating being, in which male and female principles were joined. These principles in turn belonged to a larger group of oppositions where, for example, one side would include male, life and day, while the other side would include female, death and night [49]. This suggests that the creative principle of self-generation is itself also co-constituted as one complex unity by the interdependence of the specific principles of regeneration (life) and decay (death).

In Hinduism we find that time is circular (like it was in ancient Mesoamerica), and that each cycle of cosmic time, known as a *kalpa*, features a tripartite pattern of maintenance, creation and destruction that is enacted by the *trimurti* of gods [15]. Vishnu, Brahma and Shiva all had relatively distinct roles, namely to preserve, to create and to destroy the world, respectively. The unity of the *trimurti* can be seen in one of Hinduism's most important objects of worship, the male *linga*, a symbolic phallus: the top of the *linga* symbolizes Shiva, the middle Brahma and the base Vishnu. Interestingly, the most important god of the three is Shiva, who is also symbolized by the *linga* as a whole.

We see the principle of destruction at work as well in the idea of *samsara*, the circle of conditioned human existence. The organization of the cosmic cycles is matched by a belief in reincarnation, the cycle of personal life and death. In fact, according to Buddhist philosophy, we can even find it within the timescale of our lives: it is inherent in every moment of our existence. What we experience as the constant present is actually dynamically maintained as a "circle of arising and decay of experience [that] turns continuously" [62], p. 80. The concept of death plays a key role in this process:

> Whenever there is birth, there is death; in any process of arising, dissolution is inevitable. Moments die, situations die and lives end. Even more obvious than the uneasiness of birth is the suffering (and lamentation, as is said) experienced when situations or bodies grow old, decay and die. In this circular chain of causality, death is the causal link to the next cycle of the chain. The death of one moment of experience is, within the Buddhist analysis of causality, actually a causal precondition for the arising of the next moment. [62]

What this quotation makes clear is that a generalized concept of death can be taken as one of existence's essential principles. I also note that this quotation was taken from the foundational text of the enactive approach to cognitive science, namely *The Embodied Mind: Cognitive Science and Human Experience*. This was the first but not the only way in which the enactive approach began to recognize that death is an essential explanatory principle for its theory of life and mind. Let us take a closer look at its history.

6   The Role of Death in Enactive Cognitive Science

Enactivism is the latest installment in a long line of intellectual movements that built on each other and were equally shunned by mainstream thinkers. Many important ideas and methods of enactive theory can be traced back to the early cybernetics era and especially to that era's end in the work of Ashby, and to that work's later expansion and refinement in Maturana and Varela's biology of cognition, until we finally arrive at the first formulations of an enactive cognitive science [27, 28, 31]. Each of these phases contributed essential insights, some of which I want to highlight.

One of the key insights of cybernetics was that it is possible to devise a systems theory of *self-maintenance* based on the principle of negative feedback. Famous examples include the Watt's governor and the thermostat. Maturana and Varela's [46] theory of autopoiesis built on these insights, and added the crucial insight that living systems are not only self-maintaining, but also *self-producing*, which distinguishes them from AI and robotics [34].

In these two stages of conceptual development a positive role of death starts to be prefigured. Ashby [1] founded his ultrastability theory on the idea of a system *breaking* and thereby losing its original systemic identity (due to changes of its parameters) in the process. Maturana and Varela recognized that *decay* is an indispensable property of the components for the formation of the autopoietic system. However, ultimately neither the breaking nor the decay were conceived of as somehow affecting the identity of the whole system.

Ashby's [2] homeostat was built precisely so as to remain a homeostat, even while specific parts of it were occasionally "breaking" (undergoing random parameter changes). Note that the homeostat's systemic identity and the changes it could undergo were pre-defined externally by Ashby, and this lack of autonomy precludes a genuine role of precariousness [29]. And Maturana and Varela concluded, in line with the abstractness of cybernetics and general systems theory, that the "properties of the components of an autopoietic system *do not* determine its properties as a unity" [61], p. 192, thereby banishing the effects of decay from the domain of the system as a unity. I agree with Di Paolo [20] that Maturana's doctrine of non-intersecting domains, although a well-intentioned reaction against physicalist reductionism, has had the unfortunate side-effect of preserving mind-body dualism in another format. Following Bateson, it is therefore not surprising that death was once again neglected. Importantly, this doctrine prevents decay, as an inherent property of the chemical components, to be meaningfully related to the mortal existence of the living, as if the instability of the components had nothing to do with the precariousness of the whole.[4]

Varela later overcame this mere contingency of the principles of breaking and disintegration in two important ways. He came to see the breakdowns of animal behavior and human experience as the "birthplace of the concrete" in which the cognitive agent and their immediate world become spontaneously reconstituted and creatively rearticulated in an action-appropriate manner [58]. To illustrate this idea, Varela asks us to imagine what happens when we reach for our wallet and realize that it is no longer there—after a transitory moment of confusion we will find ourselves in a new task-specific being-in-the-world geared toward the rapid recovery of our wallet. We can see how this idea of breakdowns as the birthplace of the concrete resembles the Buddhist idea of the death of one moment causing a new one to emerge.

More generally, Varela also acknowledged the essential role of mortality in his later enactive theory of the organism, following a close reading of Jonas' [43] phenomenological philosophy of life [64]. And he was also forced to recognize its importance personally as he dealt with the rapidly deteriorating state of his own body toward the end of his life. He concluded the phenomenological analysis of his harrowing experiences of undergoing organ transplantation with the poignant statement: "Somewhere we need to give death back its rights" [59]. This intertwining between third-person scientific theory and first-person existential insight is a general characteristic of enactivism, for example, as practiced in neurophenomenology [11], but it is certainly at its most demanding and intimate when the phenomenon under consideration is death.

For Jonas [43] life and death are two sides of the same coin, and out of this complementary unity arises something novel: individual beings who are concerned

---

[4]In more recent formulations of autopoietic theory, Maturana [45] has started to emphasize that autopoietic systems are a kind of molecular system. But more needs to be done to unpack the implications of this restriction to the chemical domain in terms of our understanding of the phenomenon of life, implications that enactive theory is unfolding [32].

about maintaining their own form of being. By constructing their own boundaries under far-from-equilibrium conditions, living beings determine their own individuality and their relationship with the world, and they do so in a way that gives them intrinsic value. Starting from the phenomenological insight that we know ourselves to be more than pure mechanisms devoid of a meaningful perspective, and accepting this as a fundamental fact needing to be explained, he set out to argue for the essential role of mortality in accounting for that meaning:

> with metabolizing existence not-being made its appearance in the world as an alternative embodied in the existence itself [...]: intrinsically qualified by the threat of its negative it must affirm itself, and existence affirmed is existence as a *concern*. Being has become a task rather than a given state, a possibility ever to be realized anew in opposition to its ever-present contrary, not-being, which inevitably will engulf it in the end. [...]

> Are we then, perhaps, allowed to say that mortality is the narrow gate through which alone *value* [...] could enter the otherwise in different universe? [...] Only in confrontation with ever-possible not-being could being come to feel itself, affirm itself, make itself its own purpose. [42] pp. 35–36

Note that this is not a causal relationship between death and life. The upshot is that we cannot separate an organism's systemic identity from its precarious material realization without losing the capacity to account for the meaning and the intrinsic teleology of life we all know from our personal experience. Death is dependent on a certain material configuration: "Because form that desires itself in a purposeful manner is happening only in matter to which form is not its entropically 'natural' state, there is always the possibility, and final certainty, of death" [64]. On this view, a meaningful perspective and mortality are inextricably linked in their material embodiment.

Accordingly, we seem to find in recent enactivism something akin to the ancient Mesoamerican principle of complementarity, *ometeotl*, as applicable to death and life. And in its historical development there were three principles of particular significance that resemble the elements of the Hindu *trimurti*: self-maintenance (Ashby's cybernetics), self-production (Maturana's autopoietic theory), as well as death and precariousness (Varela's enactivism). This convergence to similar principles of human existence, under such hugely diverse circumstances, makes sense if we consider that all major worldviews are shaped by universally shared aspects of human existence. Moreover, enactivism was also inaugurated with the explicit aim of incorporating phenomenological invariants of human experience into cognitive science. In other words, perhaps this is a case of intersubjective validation at the intercultural level.

Yet this rediscovery of precariousness as an explanatory principle only occurred after around half a century of cognitive science and only on the sidelines of the mainstream. Is the problem of mortality in a blindspot of computationalism? If we can better understand this neglect we will get a better grasp of the limitations of the computational theory of mind as an account of human existence.

7 On the Impossibility of a Virtual Death

From the perspective of computationalism it is a purely contingent fact about a physical computational system that it can be destroyed, for example, by smashing it to bits and pieces. But this *potential* destructibility of the implementation at the physical level is completely irrelevant for the functions that these systems are abstractly realizing at the computational level.[5] In this respect I disagree with Gershenson's [35] analysis of death, in which no ontological distinction is drawn between the disintegration of a real individual and of a virtual "agent":

> If we can create again a living system with the same organization, did it die in the first place? I think the answer should be in the affirmative. The fact that an organism—artificial or natural—can easily be replaced or regenerated does not mean that the particular instantiation of its organization is not lost. [35], p. 3

In the case of a real organism, an organizational perspective on life, whereby the identity of a living system is defined only by its organization, makes it tempting to assume that death can be cheated by re-creating that same organization at a later point in time. Early conceptions of autopoietic theory can be criticized as promoting such an abstract stance in which the organization was considered independently from its material realization [31]. However, to identify the actual living being with its description as a living system simply confuses the description of its organization with the reality of its being. The description can never exhaust the actual reality because describing a physical phenomenon's organization depends on an act of abstraction that by definition distinguishes the abstract organization from the concrete materiality. A better way to think about the relationship between a real individual organism and its systemic organization is in terms of instantiation of a species. A particular instantiation is irreplaceable; on this point I agree with Gershenson, although it would have been more precise for him to say that it is the species, or at least a category, of organism that could be easily replaceable.

However, in the case of a virtual agent the reality does not exceed its abstract organization; the two are one and the same because in the computational realm there are nothing but abstract entities in the first place. In this case Gershenson's argument about the death of distinct material instantiations no longer applies because the computational level of the agent can be implemented so as to be formally independent from the underlying material substrate. In that case there is nothing more to a virtual agent but its organization at any point in time, and thus at the level of computation nothing could in principle distinguish it from a later,

---

[5]It might be argued that the biological phenomenon known as programmed cell death, whereby cells within a multicellular organism spontaneously disintegrate, is an exception: their lifespan is related to properties of their chromosomes and this relationship can be analyzed in information-theoretic terms [66]. Moreover, such disintegration of cells plays a variety of functions in a multicellular organism. Clearly, information theory can therefore help in analyzing some of the causes and functions of cellular death. But here death is approached as a contingent fact of life on Earth rather than as something essential to it.

independent instantiation of that very same identity. It is as if there is only one virtual agent defined by its purely logical, and therefore immortal, identity, which can be realized again and again in the form of indistinguishable clones.[6]

There is therefore a crucial difference between the existence of a real material organism and the persistence of a virtual agent in a computer simulation. Only the latter can return from disintegration to exactly its former being as if nothing had happened. Given that the virtual agent's identity is completely exhausted by its formal organization, that identity remains what it is even if it is not currently realized. Its state of death, or not-being, is only relative to the end of a particular instantiation from which it ultimately remains independent. Conversely, the real organism is a unique and irreplaceable individual, its future horizon necessarily limited in principle by the inevitable possibility of irrevocable death.

Following this contrast between the death of a real organism and the ending of an instantiation of a virtual agent we must refine Jonas' concept of mortality to deal with the technological advances of our times. Virtual agents can disintegrate but this is not sufficient for considering them to exist precariously. It is not just the potential for disintegration that is essential, but also the fact that this event is irrevocable once it occurs.[7] An eternal non-being rather than just a temporarily non-realized being must follow death. In other words, in order to give rise to a meaningful perspective, the precariousness of an individual cannot be separated from its uniqueness. Only a real organism in its continual struggle to continue its way of life can therefore be genuinely concerned with its own existence, with the world, and with the lives of others. It may be strange to consider the potential for irrevocable non-being as constitutive of meaning, but this kind of change in our explanatory framework may be exactly what we need in order to explain how mind emerged from matter [17].

8 Conclusions

Since the beginnings of the cybernetics era over half a century ago, system-based approaches to life and mind have been recovering the essential role of death in the modern scientific worldview. From the cybernetics of self-maintenance, to the biology of cognition of self-production, to contemporary enactive theory of pre-cariousness, we rediscovered the same interlinked principles that have been at the

---

[6]Incidentally, this is why no matter how hard copyright enforcement authorities try to convince people that the unpermitted replication of digital products is the same as stealing, there will always be an essential difference between stealing a car (or a handbag, a television, a movie—all distinct physical items) and downloading copies of movies (all indistinct virtual clones).

[7]Di Paolo [20] highlights that this constitutive role of precariousness marks a break with the tradition of functionalism: death is not a function, which could be reverted, but it is rather the cessation of all function.

core of important traditional worldviews for millennia. I believe this is a good thing, for it indicates that cognitive science is once again becoming aligned with human experience. This gives hope that the general crisis of the sciences, which Husserl diagnosed in the first half of the last century, is coming to an end. We are finally returning to the concrete domain from which all scientific activity must start in the first place, our pre-theoretical lifeworld.

Yet enactive theory is not going to be welcomed by the majority of researchers because it implies uncomfortable rethinking of basic assumptions, and because it cannot be separated from our personal ideas about life and death. For example, it implies that popular ideas about how to make people immortal by turning them into purely virtual selves are misguided. While those ideas may appear to be life affirming, they are actually stripping life of its essential nature—its precarious and therefore meaningful existence. Conversely, taking seriously the biologically embodied mind cannot avoid bringing us face to face with the inevitability of our own finitude, which conflicts with the transhumanist goal of defeating death by engineering our bodies to stay forever young (e.g. [65]. This conflict does not have to be situated at the ontological level, since even living bodies that stop aging and never get sick would still be mortal. And human existence can be precarious in more respects than just at the basic bio-logical level; it also has a lot to do with the continuation of a way of life rather than just with life itself. But there is a mismatch at the conceptual level: tran-shumanism views mortality as a burden to be removed or at least as something to be postponed indefinitely by scientific progress, rather than as constitutive of a meaningful way of life.

We may speculate that a human being who is leading a way of life without any real sense of finitude will face serious existential issues in the long run. And the alternative is actually not as bad as it may seem. For as Jonas, following his mentor Heidegger, emphasized: facing up to our own inevitable death is only a burden as long as we ignore mortality's role in making our life meaningful in the first place. Moreover, as conscious beings we enjoy the additional privilege of being able to take advantage of this insight into our finitude in order to realize the full potential of our lives with the awareness that each moment is as precious as it is precarious.

> As to our mortal condition as such, our understanding can have no quarrel about it with creation unless life itself is denied. As to each of us, the knowledge that we are here but briefly and a nonnegotiable limit is set to our expected time may even be necessary as the incentive to number our days and make them count. [42]

Acknowledgements Ezequiel Di Paolo provided constructive feedback on an earlier version of this manuscript. I thank the many reviewers whose detailed comments and criticisms helped to substantially sharpen the final version. This research was realized with support from UNAM-DGAPA-PAPIIT project number IA102415.

References

1. Ashby, W.R.: The nervous system as physical machine: With special reference to the origin of adaptive behavior. Mind **56**(221), 44–59 (1947)
2. Ashby, W.R.: Design for a Brain: The Origin of Adaptive Behaviour, 2nd edn. Chapman & Hall, London, UK (1960)
3. Barandiaran, X., Di Paolo, E.A., Rohde, M.: Defining agency: individuality, normativity, asymmetry, and spatio-temporality in action. Adapt. Behav. **17**(5), 367–386 (2009)
4. Barbaras, R.: Life and exteriority: the problem of metabolism. In: Stewart, J., Gapenne, O., Di Paolo, E.A. (eds.) Enaction: Toward a New Paradigm for Cognitive Science, pp. 89–122. The MIT Press, Cambridge, MA (2010)
5. Barrett, N.F.: The normative turn in enactive theory: An examination of its roots and implications. Topoi (2015)
6. Bateson, G.: Form, substance, and difference. Gen. Semant. Bull. **37**, 221–245 (1970)
7. Bateson, G.: The cybernetics of self: a theory of alcoholism. Psychiatry **34**(1), 1–18 (1971)
8. Bateson, G.: Steps to an Ecology of Mind: Collected Essays in Anthropology, Psychiatry, Evolution, and Epistemology. Ballantine Books, New York (1972)
9. Beaton, M., Aleksander, I.: World-related integrated information: enactivist and phenomenal perspectives. Int. J. Mach. Conscious. **4**(2), 439–455 (2012)
10. Bishop, J.M.: A cognitive computation fallacy? Cognition, computations and panpsychism. Cogn. Comput. **1**, 221–233 (2009)
11. Bitbol, M.: Science as if situation mattered. Phenom. Cogn. Sci. **1**, 181–224 (2002)
12. Cappuccio, M., Froese, T.: Introduction. In: Cappuccio, M., Froese, T. (eds.) Enactive Cognition at the Edge of Sense-Making: Making Sense of Non-Sense, pp. 1–33. Palgrave Macmillan, Basingstoke (2014)
13. Chalmers, D.J.: Panpsychism and panprotopsychism. In: Alter, T., Nagasawa, Y. (eds.) Consciousness in the Physical World: Perspectives on Russellian Monism. Oxford University Press, New York (2015)
14. Clarke, E.: The problem of biological individuality. Biolog. Theory **5**(4), 312–325 (2010)
15. Coe, M.D.: Angkor and the Khmer Civilization. Thames & Hudson, London, UK (2003)
16. Coe, M.D., Koontz, R.: Mexico: From the Olmecs to the Aztecs. Thames & Hudson, London, UK (2013)
17. Deacon, T.W.: Incomplete Nature: How Mind Emerged from Matter. W. W. Norton & Company, New York, NY (2012)
18. Deacon, T.W.: Information and reference. In: Dodig-Crnkovic, G., Giovagnoli, R. (eds.) Representation and Reality in Humans, Other Living Organisms and Intelligent Machines. Springer (2017)
19. Dennett, D.C.: Cognitive wheels: The frame problem of AI. In: Hookway, C. (ed.), Minds, Machines and Evolution: Philosophical Studies, pp. 129–152. Cambridge University Press, Cambridge (1984)
20. Di Paolo, E.A.: Extended life. Topoi **28**(1), 9–21 (2009)
21. Di Paolo, E.A.: Robotics inspired in the organism. Intellectica **1–2**(53–54), 129–162 (2010)
22. Dodig-Crnkovic, G.: Info-computational constructivism and cognition. Constr. Found. **9**(2), 223–231 (2014)
23. Dodig-Crnkovic, G., von Haugewitz, R.:. Reality construction in cognitive agents through processes of info-computation. In: Dodig-Crnkovic, G., Giovagnoli, R. (eds.) Representation and Reality in Humans, Other Living Organisms and Intelligent Machines. Springer (2017)
24. Dreyfus, H.L.: What Computers Can't Do: A Critique of Artificial Reason. Harper and Row, New York, NY (1972)
25. Eden, Y.: Being 'simple-minded': models, maps and metaphors and why the brain is not a computer. In: Dodig-Crnkovic, G., Giovagnoli, R. (eds.) Representation and Reality in Humans, Other Living Organisms and Intelligent Machines. Springer (2017)

26. Fernández, N., Maldonado, C., Gershenson, C.: Information measures of complexity, emergence, self-organization, homeostasis, and autopoiesis. In: Prokopenko, M. (ed.) Guided Self-Organization: Inception, pp. 19–51. Springer, Berlin (2014)
27. Froese, T.: From cybernetics to second-order cybernetics: a comparative analysis of their central ideas. Constr. Found. **5**(2), 75–85 (2010)
28. Froese, T.: From second-order cybernetics to enactive cognitive science: Varela's turn from epistemology to phenomenology. Syst. Res. Behav. Sci. **28**, 631–645 (2011)
29. Froese, T.: Ashby's passive contingent machines are not alive: living beings are actively goal-directed. Constr. Found. **9**(1), 108–109 (2013)
30. Froese, T.: Radicalizing Enactivism: Basic Minds without Content. Daniel D. Hutto and Erik Myin. Cambridge, Massachusetts: MIT Press. J. Mind Behav. **35**(1–2), 71–82 (2014)
31. Froese, T., Stewart, J.: Life after Ashby: Ultrastability and the autopoietic foundations of biological individuality. Cybern. Hum. Knowing **17**(4), 83–106 (2010)
32. Froese, T., Stewart, J.: Enactive cognitive science and biology of cognition: A response to Humberto Maturana. Cybern. Hum. Knowing **19**(4), 61–74 (2012)
33. Froese, T., Virgo, N., Ikegami, T.: Motility at the origin of life: its characterization and a model. Artif. Life **20**(1), 55–76 (2014)
34. Froese, T., Ziemke, T.: Enactive artificial intelligence: investigating the systemic organization of life and mind. Artif. Intell. **173**(3–4), 366–500 (2009)
35. Gershenson, C.: What does artificial life tell us about death? Int. J. Artif. Life Res. **2**(3), 1–5 (2011)
36. Gershenson, C.: Info-computationalism or materialism? Neither and both. Constr. Found. **9** (2), 241–242 (2014)
37. Harnad, S.: The symbol grounding problem. Physica D **42**, 335–346 (1990)
38. Hewitt, C.: What is commitment? Physical, organizational, and social. In: Noriega, P., Vazquez-Salceda, J., Boella, G., Boissier, O., Dignum, V. (eds.) Coordination, Organizations, Institutions, and Norms in Agent Systems II, pp. 293–307. Springer, Berlin (2007)
39. Horsman, D.C., Kendon, V., Stepney, S., Young, J.P.W.: Abstraction and representation in living organisms: when does a biological system compute? In: Dodig-Crnkovic, G., Giovagnoli, R. (eds.) Representation and Reality in Humans, Other Living Organisms and Intelligent Machines. Springer (2017)
40. Hutto, D.D., Myin, E.: Radicalizing Enactivism: Basic Minds without Content. The MIT Press, Cambridge, MA (2013)
41. Jonas, H.: Werkzeug, Bild und Grab: Vom Transanimalischen im Menschen. Scheidewege **15**, 47–58 (1985/86)
42. Jonas, H.: The burden and blessing of mortality. Hastings Cent. Rep. **22**(1), 34–40 (1992)
43. Jonas, H.: The Phenomenon of Life: Toward a Philosophical Biology. Northwestern University Press, Evanston, IL (2001)
44. Kyselo, M.: The body social: An enactive approach to the self. Front. Psychol. **5**(986) (2014).
45. Maturana, H.R.: Ultrastability autopoiesis? Reflexive response to Tom Froese and John Stewart. Cybern. Hum. Knowing **18**(1–2), 143–152 (2011)
46. Maturana, H.R., Varela, F.J.: Autopoiesis: the Organization of the Living Autopoiesis and Cognition: The Realization of the Living, pp. 59–140. Kluwer Academic, Dordrecht (1980)
47. Miłkowski, M.: Beyond formal structure: a mechanistic perspective on computation and implementation. J. Cognit. Sci. **12**, 359–379 (2011)
48. Miłkowski, M.: Explaining the Computational Mind. MIT Press, Cambridge, MA (2013)
49. Miller, M., Taube, K.: An Illustrated Dictionary of the Gods and Symbols of Ancient Mexico and the Maya. Thames & Hudson, London, UK (1993)
50. Oizumi, M., Albantakis, L., Tononi, G.: From the phenomenology to the mechanisms of consciousness: Integrated information theory 3.0. PLoS Comput. Biol. **10**(5), e1003588 (2014)
51. Piccinini, G.: Computation in physical systems. In: Zalta, E.N. (ed.), The Stanford Encyclopedia of Philosophy (Summer 2015 Edition) (2015)
52. Searle, J.R.: Minds, brains, and programs. Behav. Brain Sci. **3**(3), 417–424 (1980)

53. Sterelny, K., Griffiths, P.E.: Sex and Death: An Introduction to Philosophy of Biology. The University of Chicago Press, Chicago (1999)
54. Thompson, E.: Mind in Life: Biology, Phenomenology, and the Sciences of Mind. Harvard University Press, Cambridge, MA (2007)
55. Thompson, E., Varela, F.J.: Radical embodiment: neural dynamics and consciousness. Trends Cognit. Sci. 5(10), 418–425 (2001)
56. Tononi, G.: Consciousness as integrated information: a provisional manifesto. Biol. Bull. **215**, 216–242 (2008)
57. Varela, F.J.: Organism: a meshwork of selfless selves. In: Tauber, A.I. (ed.) Organism and the Origins of Self, pp. 79–107. Kluwer Academic Publishers, Dordrecht, Netherlands (1991)
58. Varela, F.J.: The re-enchantment of the concrete: Some biological ingredients for a nouvelle cognitive science. In: Steels, L., Brooks, R. (eds.) The Artificial Life Route to Artificial Intelligence, pp. 11–22. Lawrence Erlbaum Associates, Hove, UK (1995)
59. Varela, F.J.: Intimate distances: Fragments for a phenomenology of organ transplantation. J. Conscious. Stud. **8**(5–7), 259–271 (2001)
60. Varela, F.J. (ed.): Sleeping, Dreaming and Dying: An Exploration of Consciousness with the Dalai Lama. Wisdom Publications, Boston (1997)
61. Varela, F.J., Maturana, H.R., Uribe, R.: Autopoiesis: the organization of living systems, its characterization and a model. BioSystems **5**, 187–196 (1974)
62. Varela, F.J., Thompson, E., Rosch, E.: The Embodied Mind: Cognitive Science and Human Experience. MIT Press, Cambridge, MA (1991)
63. Villalobos, M., Ward, D.: Living systems: autonomy, autopoiesis and enaction. Philos. Technol. **28**(2), 225–239 (2015)
64. Weber, A., Varela, F.J.: Life after Kant: natural purposes and the autopoietic foundations of biological individuality. Phenom. Cognit. Sci. **1**, 97–125 (2002)
65. Young, S.: Designer Evolution: A Transhumanist Manifesto. Prometheus Books, Amherst, NY (2006)
66. Zenil, H., Schmidt, A., Tegnér, J. (in press): Causality, information and biological computation: an algorithmic software approach to life, disease and the immune system. In: Walkers, S.I., Davies, P.C.W., Ellis, G. (eds.) From Matter to Life: Information and Causality. Cambridge University Press, Cambridge

Language Processing, Computational Representational Theory of Mind and Embodiment: Inferences on Verbs

J. Ezquerro and M. Iza

Abstract The computational representational theory of mind conceptualized by Jerry Fodor has dominated the schedule of cognitive science for many years. Fodor defended it as an empirical theory whose plausibility depends on facts about how the mind works. However, current developments in philosophy, psycholinguistics and neurosciences present a challenge to this received view. These developments assume embodied and situated cognition. This paper reviews and assesses the prospects of these new accounts, focusing on processing verbs and on the role of emotions in language processing.

1 Introduction

Sometimes, philosophical theses fix the schedule of empirical research. This is the case of Jerry Fodor. In *The Language of Thought* (1975), Fodor defended the computational representational theory of mind (CRTM) and through the years developed the main conceptual consequences of such a view. Two of these consequences were methodological individualism and methodological solipsism. In *Psychosemantics* (1987), Fodor characterizes these theses in the following way:

> Methodological individualism is the doctrine that psychological states are individuated *with respect to their causal powers*. Methodological solipsism is the doctrine that psychological states are individuated *without respect to their semantic evaluation* (p. 42)

J. Ezquerro (✉)
Institute for Logic Cognition Language and Information (LCLI),
University of the Basque Country, UPV/EHU, Donostia-San Sebastián, Spain
e-mail: jesus.ezquerro@ehu.eus

M. Iza
Department of Psychology, University of Malaga, Málaga, Spain
e-mail: iza@uma.es

© Springer International Publishing AG 2017
G. Dodig-Crnkovic and R. Giovagnoli (eds.), *Representation and Reality in Humans, Other Living Organisms and Intelligent Machines*, Studies in Applied Philosophy, Epistemology and Rational Ethics 28, DOI 10.1007/978-3-319-43784-2_4

These two theses constitute the core of what is usually known as "the received view". They go hand in hand, although Fodor tried to distinguish them carefully. Methodological individualism opposes anti-individualism defended by Tyler Burge:

> Anti-individualism is the view that not all of an individual's mental states and events can be type-individuated independently of the nature of the entities in the individual's environment. There is, on this view, a deep individuative relation between the individual's being in mental states of certain kinds and the nature of the individual's physical or social environments...... ([9], p. 46)

It is clear that this characterization appeals to relational properties (physical and social environments) in order to identify psychological states, and Fodor concedes that it could be so *if and only if* these relational properties can affect causal powers. Thus, Fodor thinks that individualism is a conceptual point: if something lacks causal powers, then it lacks of explanatory force. Therefore, it should be ruled out from psychological explanation.

However, at this point methodological solipsism enters the scene. Far from being a purely conceptual point, Fodor qualifies it as an empirical thesis:

> 'Methodological solipsism' is, in fact, an empirical theory about the mind: it's the theory that mental processes are computational, hence syntactic. I think this theory is defensible; in fact, I think it's true. But this defense can't be conducted on a priori grounds, and its truth depends simply on the facts about how mind works. (1987, p. 43)

In this way, methodological solipsism constrains and sustains methodological individualism: even if the later does not prohibit relational properties, the former does. Given that semantical properties are relational, they fall out of the scope of scientific explanation [23]. The story is well known. Realizing that we cannot do without semantics, Fodor engaged with a program—Psychosemantics—trying to account for meaning in a way that respects the fundamentals of CRTM. However, this program was abandoned because it was unable to respond to some conceptual challenges, especially, twin earth cases.

Since then, time has shown that the computational representational theory of mind is probably false, at least in the way J. Fodor characterized it. On the philosophical side, Burge's anti-individualist position, and A. Clark's extended mind thesis go in the opposite direction and have gained acceptance [10, 12]; Furthermore, inasmuch as Fodor considers methodological solipsism as an empirical theory, our purpose is to revise and assess the current developments in cognitive psychology and cognitive linguistics in order to show that the direction of research has clearly departed from the received view, and its results and findings go against it. Our attention will concentrate on language processing, and within this field, on the case of verbs.

Current research shows that visual, motor and emotional information are active during sentence processing [1, 24, 43]. For this reason, embodied theories of cognition reject the received view. This view affirms that meaning depends only on the structural relations with other symbols, following the assumption of methodological solipsism above mentioned. Embodied accounts suggest, instead, that symbols have to be grounded to their referents in the environment [36, 63].

There are two hypotheses concerning the storage of meaning in cognitive psychology. The received view defends the complexity hypothesis, assumed by Fodor [21], Kinstch [46] and Thorndyke [73]. This hypothesis affirms that a word with many semantic components would require more processing resources, comprehension time and long-term memory space than a word with a smaller number of components, thus interfering more strongly with memory for surrounding words.

The connectivity hypothesis, on the contrary, considers verb semantic structures as frames for sentence representation, claiming that memory strength between two nouns in a sentence increases with the number of underlying verb subpredicates that connect the nouns.

Following the above assumptions, therefore, the complexity hypothesis predicts that a verb with many subpredicates will lead to poorer memory strength between the surrounding nouns than a verb with few subpredicates. The connectivity hypothesis, instead, predicts that verbs with many subpredicates will lead to "greater" memory strength between nouns in cases where the additional subpredicates provide semantic connection between the nouns.

The complexity hypothesis assumes the "bin" view of memory, in which the capacity limitations of various stages of memorial processing form a central theoretical notion. The connectivity hypothesis, instead, holds a structural view of memory, in which the representational assumptions are crucial.

Gentner [25] tested the memory predictions of the above two accounts. In three experiments, subjects recalled subject–verb–object sentences, given subject nouns as cues. General verbs with relatively few subpredicates were compared with more specific verbs whose additional subpredicates either did or did not provide additional connections between the surrounding nouns. The level of recall of the object noun, given the subject noun as cue, was predicted by the relative number of "connecting" subpredicates in the verb, but not by the relative "number" of subpredicates. Results did support the connectivity hypothesis over the complexity hypothesis. Gentner interpreted the results in terms of a model in which the verb conveys a structured set of subpredicates that provides a connective framework for sentence memory.

The proposed model, called the central components model, differs from both the extreme decompositional model and the extreme meaning-postulate model. Here, the verb's representation is aimed to specify the pattern of inferences that is most dependably activated when the verb is comprehended. The model is clearly decompositional: it assumes that one verb leads to several separable (though structurally related) inferences, and that both lexical generalities and psychological phenomena can be stated in terms of connected sets of subpredicates embodying these inferences.

However, these inferences are not intended to embody necessary-and-sufficient conditions for use. Instead, the representation offered for a given verb expresses the central set of inferences (the set most frequently and reliably associated with the verb's use). The representations are not considered as exhaustive; indeed, it is very clear that they are not. For instance, the verb "give" clearly has other possible inferences: that the giver is generous, that she has the means to give away objects and so on. There is no fixed stopping point for this kind of inferential processing. Furthermore, the subpredicates are not required to be atoms belonging to a primitive base. A component is useful in a psychological representation if it functions as a familiar unit at that level of representation. Components at one level of representation may be decomposed at the next level down into a further network of linked components.

These results are compatible with the assumption that inferencing begins on-line, before the syntactic parser is completed. Inferencing would begin with the central set and continue, radiating outward more esoteric inferences, if the central set of inferences is not sufficient for a satisfactory interpretation of the sentence [14], or, even, if nonlinguistic cues provide enough information to spread forward inferences [40].

The importance of perceptual and action simulations has been recurrently stressed in the area of concrete language processing. Researchers emphasize the importance of embodiment by either amplifying evidence for the activation of embodied representations, or by establishing boundary conditions (such as temporal overlap and integratability in the domain of perception, linguistic focus, grammar, affordances and the kind of effectors involved in response to the domain of action).

For example, Glenberg and Kaschak [27] asked participants to view series of sensible sentences that either described transfer toward a reader ("Open the drawer") or away from the reader ("Close the drawer"), as well as series of nonsense sentences ("Boil the air") that did not entail any transfer. The task was to judge whether sentences made sense by pressing one of the vertical keys on a three-button box that required a movement either towards or away from the body. They found that responses were faster when the motion implied by the sentence matched the actual hand motion (action sentence compatibility effect).

In a similar way, Zwaan and Taylor [78] showed that this effect is quite specific: an activation of compatible motor responses was localized on the verb region of the sentence, but not on the preverb, postverb and sentence-final regions. Then, it was tested whether maintaining focus on the action, by following the verb with an adverb implying action, might cause motor resonance to affect both the verb and the adverb that follows it (focus on the action, "When he saw a gas station, he exited slowly", versus on the agent, "When he saw a gas station, he exited eagerly"; see also [70].

From now on, we will concentrate on three issues. First, our review focuses on sentence and discourse processing, whereas most of the previous reviews focus on single-word processing [45]. Our previous research stressed the importance of discussing the role of sensory-motor and affective processes in comprehension of language segments that provide enough context to constrain the interpretation of verb meaning [18, 20, 38]. Second, we will emphasize the trajectory of embodiment research in the domain of verb inferencing, attending to behavioral and neuroscientific data. In our opinion, these data put in jeopardy the received view [13, 71, 76, 8]. Finally, we will discuss a variety of old and new theoretical approaches to understanding of the role of inferences and verbs in language processing.

2 Automatic Processing

Grammatical categories across languages distinguish between nouns and verbs. This distinction tend to rely on lexical-semantic criteria (nouns refer to entities, while verbs refer to actions or events), and syntactic or distributional criteria (nouns and verbs play different roles in sentences and occur in conjunction with different sets of grammatical morphemes). In this sense, all languages have a mechanism for mapping lexical items with particular semantic structures to specific syntactic roles, which in some languages are marked morphologically (e.g. Basque language). These different criteria are engaged specifically in the lexical-semantic processing of verbs and nouns, as semantic priming studies show.

Semantic priming refers to the processing advantage that occurs when a word (the target) is preceded by another word (the prime) when it is related in meaning to the target. The nature of the meaning relationship can be due to shared physical, functional or visual features and membership in the same semantic category. The phenomenon of priming has become ubiquitous within cognitive psychology, and the existence of positive semantic priming has been solidly established [7, 69].

The automatic character of this process is shown by experimental data from a lexical decision task (see, e.g. [48]). In these experiments, subjects were asked to decide as quickly as possible whether each occurrence of a letter string was an English word, and to respond in order to provide reaction times. The left side of the screen was taken up by a patch containing computer-related verbal "garbage", which subjects were instructed to ignore, but sometimes contained a word planted to relate to one of a subject's current concerns. When the target string was indeed a word, the reaction time of reporting this was significantly slower if the distractor patch contained a concern-related word. Thus, concern-related stimuli seem to impose an extra load on cognitive processing even when they are peripheral and subjects are consciously ignoring them. This finding adds a further face to the automaticity of the effect (see also [44]).

Besides, this effect was shown in another cognitive process with a modified Stroop procedure [64]. For instance, MacKay et al. [55] demonstrated three taboo Stroop effects that occur when people name the color of taboo words (e.g. death, war). These effects are (i) longer color-naming times for taboo than for neutral words, an effect that diminishes with word repetition; (ii) superior recall of taboo words in surprise memory tests following color naming; (iii) better recognition memory for colors consistently associated with taboo words rather than with neutral words. They argue that taboo words trigger specific emotional reactions that facilitate the binding of taboo word meaning to salient contextual aspects.

Emotion Stroop experiments found little evidence of automatic vigilance, for instance, slower lexical decision time or naming speed for negative words after controlling for lexical features. Estes and Adelman [17] analyze a set of words, controlling for important lexical features, and find a small but significant effect for word negativity and conclude that this effect is categorical. Larsen et al. [50] analyze the same data set but include the arousal value of each word. They find non-linear and interaction effects in predicting lexical decision time and naming speed. Not all negative words produce the generic slowdown. Only negative words that are moderate to low on arousal produce more lexical decision time slowing than negative words higher on arousal. Similarly, Kahan and Hely [41] showed that the roles of valence and word frequency interact in contributing to the emotional Stroop effect.

Altarriba and Canary [4] also examined the activation of arousal components for emotion-laden words in English (for instance, kiss, death) in two groups of monolingual (English) and bilingual (Spanish–English) subjects. Prime–target word pairs were presented for lexical decisions to English word targets in either high arousal, moderate or unrelated conditions. Results revealed positive priming effects in both arousal conditions for both groups of subjects. Nevertheless, while the baseline conditions were similar across groups, the arousal conditions produced longer latencies for bilinguals than for monolinguals.

3 The Assumption in Cognitive Linguistics Regarding Verb Taxonomy May Be in Error

A widely shared assumption in cognitive linguistics [49, 39] is that verb representations are organized in a taxonomic structure during cognition processes. Also, it is considered that this taxonomy resembles the one of noun prototype structure and taxonomy found by Rosch [66–68]. There is however little or no empirical evidence supporting the hypothesis that verbs are associated through taxonomic structures.

A cross-linguistic (Finland–Spain) study on categorization of lexemes indeed suggests that the psycholinguistic processes relative to the categorization of verbs lack taxonomic structures [37]. The subjects (100 subjects in each country) were unable to associate either the subordinate or the superordinate verbs suggested in cognitive linguistics [49, 39]. The task was formulated in four different wordings for four different groups as well as controlled by tasks using nouns. The subjects performed according to the taxonomy hypothesis on nouns but not on verbs.

A small-scale developmental study with teenagers aged between 13 and 19 years showed that the ability to associate in taxonomic structures with nouns was age related. The younger group made more errors, while the age group 18–19 performed successfully. This suggests that there are questions requiring further study, for example, whether the taxonomy-principle is an inherited property of the mental lexicon or a relatively late-learned association pattern.

In dialogue situations, it is usual to assume that partners know lots of things, so that agents in communication make explicit only those facts they believe that the partner does not know. This shared implicit knowledge is also known as common-sense knowledge. Due to this fact, it is difficult or impossible to analyze a discourse without relying on a large background of knowledge. For instance, this is the case of some computer programs that try to establish a relationship between logic and language. In neurolinguistic programming (NLP) systems, often verbs are considered as predicates and verb valences as arguments. In this approach, new sentences or propositions can be inferred from the discourse. Furthermore, an inference system for verb frames is used and an evaluation process is normally required.

Learning inference relations between verbs and propositions is at the heart of many semantic applications. However, most prior work on learning such rules was focused on a narrow set of information sources, that is to say, mainly distributional similarity and to some extent manually constructed verb co-occurrence patterns (see, e.g. [6]). In this regard, we claim below that it is imperative to use information from various discursive scopes to provide: a much richer set of inferential cues to detect entailment between verbs and combine them as features in an unsupervised classification framework.

4 Self-organizing Maps

Within the emergentist paradigm [16, 56], recent studies [52, 51] interpret the interaction among the categories of "aspect" and "actionality" as the result of the learners' analysis of the probabilities of co-occurrences between aspect morphology and actionality semantics in the linguistic input. Children extract from the input the statistical frequencies of the combinations between aspect forms and actionality classes. Initially, they strengthen the production of the most frequent associations, until prolonged exposure to the input and the increasing amount of data from the input reduce the statistical difference between the most and least frequent combinations.

Li and Shirai [51] tested this hypothesis by means of computational simulations based on self-organizing maps [47]. These maps consist in supervised associative neural networks of "knot receptors" that classify input data, translating relationships of similarity into topological relationships of proximity. Through incremental exposure to an increasing amount of data, the receptors are topologically organized on the network in such a way that associated receptors have the tendency to recognize homogeneous sets of data. These maps are biologically plausible models: human cerebral cortex can be conceived as essentially a multiple feature-map, where all neurons are initially co-activated and the associative strengths between neurons become more focused in parallel with distributional increase of the corresponding co-occurrences in the input.

If the acquisition of the "semantic core" of a construction (and of the semantics of particular actions) and entrenchment (verb frequency) are both probabilistic and incremental processes, then it is easier to imagine how the two processes could work together. For example, perhaps the frequency effect observed is mediated by the effect of verb semantics. Each time a verb is listen in a new context, the learner is given a further opportunity to increase her knowledge of the precise semantics of the verb, and hence the constructions in which the verb can appear. Each time the learner finds a causative construction (and infers the speaker's intended meaning), this constitutes evidence that the construction encodes direct external causation. We think that this result fits very well with the general account proposed by Clark [11] that conceives the brain as a predictive system.

Alishahi and Stevenson [3] present a computer model that makes use of this statistical correlation between semantics and syntax. The model takes frames that are representations of scene–utterance pairings as input (i.e. of the utterance heard by the child and the scene it describes). These representations specify the verb (e.g. take), its semantic primitives (cause, move), argument roles (agent, theme goal), categories (human, concrete, destination-predicate) and syntactic structure (arg1 verb arg2 arg3). Whenever the model encounters a new frame, it stores the semantic and syntactic information in the lexical entry for the verb. It then groups this frame together with the existing class of frames whose members share the greatest number of syntactic and semantic features, or, if none are sufficiently similar, creates a new class (since the classes are not prespecified, this will be the case for the majority of frames in the early stages of learning). Whenever the model is presented with a previously encountered frame, it increases the representational strength of that frame and of its parent class.

To simulate generalization, every five training pairs, the model is presented with a frame from which the syntactic representation has been deleted, but with all other information intact. The task of the model is to choose the most appropriate syntactic structure. In order to choose the most appropriate pattern, the simulation can make use of two sources of information. These sources are, on the one hand, the syntactic structures stored in the lexical entry for that particular verb (item-based knowledge), and, on the other, those stored in the lexical entry of the other verbs in the same class (i.e. the other verbs to which this verb is semantically and syntactically similar; class-based knowledge).

5   Pragmatics and Emotional Inferences

As happens in dialogue contexts, it is usually taken for granted that, in the course of reading a text, world knowledge is often required in order to establish coherent links between sentences [58, 77]. The content grasped from a text turns out to be strongly dependent upon the reader's additional knowledge that allows a coherent interpretation of the text as a whole (see [31]).

The world knowledge directing the inference may be of a distinctive nature. Gygax et al. [33] showed that mental models related to human action may be of a perceptual nature and may include behavioral as well as emotional elements. However, Gygax [32] showed the unspecific nature of emotional inferences and the prevalence of behavioral elements in readers' mental models of emotions. Inferences are performed in both directions; emotional inferences are based on behavior, and vice versa.

Harris et al. [15] and Pons et al. [62] showed that different linguistic skills—in particular lexicon, syntax and semantics—are closely related to emotion understanding. Iza and Konstenius [37] showed that additional knowledge about social norms affects the predictions of participants about what should be inferred as the behavioral or emotional outcome of a given social situation.

Syntactic and lexical abilities are the best predictors of emotion understanding, but making inferences is the only significant predictor of the most complex components (reflective dimension) of emotion comprehension in normal children. Recently, Farina et al. [19] showed in a study that the relation between pragmatics and emotional inferences might not be so straightforward. Children with high functioning autism and Asperger syndrome present similar diagnostic profiles, characterized by satisfactory cognitive development, good phonological, syntactic and semantic competences, but poor pragmatic skills and socio-emotional competences. After training in pragmatics, descriptive analyses showed that the whole group displayed a deficit in emotion comprehension, but high levels of pragmatic competences. Furthermore, this indicates the necessity of studying the relationship between emotion and inference in normal subjects too.

Vanhatalo [75] showed that a group of synonyms of speech act verbs actually had semantically distinctive emotional elements as well as different social norms associated to these lexemes. This semantic knowledge related to inferences was not present either in dictionaries or in current literature, increasing demands for empirical studies directed at native speaker intuitions.

We also suggest that, while behavioral elements may indeed be of perceptual nature and the inference between emotion and behavior less culturally dependent, especially when concerned with basic emotions the inference concerned with social norms may be more complex, requiring elaborative inference. We suggest that, in further studies, a distinction between basic emotions and nonbasic emotions, social settings and non-social settings should be taken into account. The cognitive models concerned with social action may be of a more complex nature, but with recognizable features on lexical and syntactic levels.

An increasing amount of research has been done with the aim of testing whether reenactment of congruent and incongruent emotional states affects language comprehension. For example, Havas et al. [34] asked subjects to read sentences describing emotional or nonemotional events while being in a matching or mismatching emotional state. The major result was that sentences describing pleasant events were processed faster when subjects were smiling (pen in the teeth condition). Contrarily, unpleasant sentences were processed faster when subjects were prevented from smile (pen in the lips condition). Later, Havas et al. [35] injected subjects with botulinum toxin A to temporarily paralyze a facial muscle responsible for frowning. Subjects were instructed to read sad and angry sentences. They found that the reading of sad and angry sentences was slowed after Botox injections, suggesting that being prevented from frowning makes it more difficult to simulate sadness and anger.

Using a different procedure, namely electromyography (EMG) measurement of the zygomatic major and corrugator supercilii muscle regions, Foroni and Semin

[24] found that motor resonance was induced when participants processed adjectives describing emotion (happy, sad), but to a lesser extent than when participants processed action emotional verbs (to smile, to frown). Finally, Lugli et al. [54] asked subjects to judge the sensibility of positively and negatively valenced sentences describing toward transfer ("The object is nice. Bring it toward you".) and away transfer ("The object is ugly. Give it to another person".) while responding with a mouse in a direction that either matched or mismatched the direction described by the sentence. They found that responding to sentences describing positive objects was faster with a "toward the body" hand motion. Likewise, responses to sentences describing negative objects were faster with an "away from the body" hand motion.

To sum up, the results of all these studies suggest that sentence comprehension is facilitated when the suggested mood of the sentence is congruent with the concurrent mood of the comprehender.

6 Discussion

Successful language learning combines generalization and the acquisition of lexical constraints. This conflict is especially clear for verb argument structures, which may generalize to new verbs or resist generalization with certain lexical items. In this sense, we have to capture the emergence of feature biases in word learning. In this work, we have described a model capable of learning about structure variability simultaneously on several levels. This architecture makes inferences about construction variation and applies them when faced with verbs for which it has very little data. Even more, this framework can try to explain the slight divergences between model predictions and human behavior (see, e.g. [28, 63]).

The model action-based language [26] offers an action-based account of language comprehension by making use of neurophysiologic findings on mirror neurons [60], and by adopting controller and predictor models from theories of motor control responsible for computing goal-oriented motor commands and predicting sensori-motor effects of these commands [30]. Here, language comprehension is tantamount to predicting sensorimotor and affective effects of the performed action. For instance, upon hearing the word "walk", a person's speech mirror neurons activate an associated action controller responsible for generating the motor commands necessary for interaction. Then, the predictor (sensory, motor or emotional) of the target word establishes possible consequences of the action to be performed; That is to say, under this approach, the same hierarchical mechanisms that are used in controlling action (control and predictor) are utilized for generating grammatical consequences in language processing.

Although it remains limited in scope, the model presented here lays the basis for an approach to entailment detection that combines a robust semantic representation with inference performance. We have illustrated the potential of the approach by showing how it could handle a limited range of interaction between nominal predication, verbal predication and cognitive inferences. Current work concentrates on extending the model coverage and on evaluating it on a full-size benchmark designed to illustrate a wider range of focused inferences (such as affective or status relations) between basic predications and the various semantic phenomena present in the discourse (see, e.g. [68]).

Several studies have been concerned with the brain regions exhibiting different responses to noun and verb processing tasks. Such differences manifest the location of verb-specific processing and the activity patterns within common verb-specific regions.

Evidence from aphasic patients suggests that noun deficits tend to arise from lesions to the left middle-to-anterior temporal lobe. Verb deficits resulted from lesions in left inferior frontal regions, premotor regions, or in posterior temporal to inferior parietal regions [2]. Many imaging studies have found greater activation for verbs compared with nouns in the left prefrontal cortex [5, 59] and in left posterior temporal regions [53, 74].

Recently, Willms et al. [76] showed greater activity for verbs than for nouns in Spanish–English bilinguals in four regions: left posterior middle temporal gyrus, left middle frontal gyrus, pre-supplementary motor area and right middle occipital gyrus. Their results suggest that the neural substrates underlying verb-specific processing are largely independent of language in bilinguals.

When people become committed to pursue a goal, that event initiates an internal state termed a "current concern". One of the properties of this state is potentiating emotional reactivity to cues associated with the goal pursued. The emotional responses thus emitted begin within about 300 ms after exposure to the cue—early enough to be considered purely central, nonconscious responses at this stage. Because they appear to be incipient emotional responses but lack many of the properties normally associated with emotion, they are called "protoemotional" responses (see also [57]).

These responses go on in parallel with early perceptual and cognitive processing, with which they trade reciprocal influences. The intensity (and other features) of protoemotional responses affects the probability that the stimulus will continue to be processed cognitively. The results of continued cognitive processing in turn modulate the intensity and character of the emerging emotional response. That goal setting affects performance at different phases of skill acquisition and for different levels of ability [42]. Furthermore, there are different distributions of cognitive processing and mental content in different phases of goal-striving sequences [29].

As these experimental data show, activating accessible constructs or attitudes through one set of stimuli can facilitate cognitive processing of other stimuli under certain circumstances, and can interfere with it under other circumstances. Some of the results support and converge on those centered on the constructs of current concern and emotional arousal (see [72]).

Future research should take seriously into account the following question: how to develop models where emotion interacts with cognitive processing. One example could be the work of Pitterman et al. [61] where speech-based emotion recognition is combined with adaptive human–computer modeling. With the robust recognition of emotions from speech signals as their goal, the authors analyze the effectiveness of using a plain emotion recognizer, a speech–emotion recognizer combining speech and emotion recognition, and multiple speech–emotion recognizers at the same time. The semistochastic dialogue model employed relates user emotion management to the corresponding dialogue interaction history and allows the device to adapt itself to the context, including altering the stylistic realization of its speech.

Summing up, both the new directions of research in cognitive linguistics and the empirical results of brain sciences indicate that the received view about how the mind works should be abandoned. Special emphasis must be put on the role of emotions in explaining cognitive processing, given the importance they have in order to explain language comprehension and performance. But this can only be done assuming embodied and situated cognition.

References

1. Borghi, A.M., Di Ferdinando, A., Parisi, D.: Objects, spatial compatibility, and affordances: a connectionist. Study. Cogn. Syst. Res. **12**, 33–44 (2011)
2. Aggujaro, S., Crepaldi, D., Pistarini, C., Taricco, M., Luzzatti, C.: Neuro-anatomical correlates of impaired retrieval of verbs and nouns: interaction of grammatical class, imageability and actionality. J. Neurolinguist. **19**, 174–194 (2006)
3. Alishahi, A., Stevenson, S.: A cognitive model for the representation and acquisition of verb selectional preferences. Proceedings of the ACL-2007 Workshop on Cognitive Aspects of Computational Language Acquisition, pp. 41–48. Czech Republic, Prague (2007)
4. Altarriba, J., Canary, T.M.: The influence of emotional arousal on affective priming in monolingual and bilingual speakers. J. Multiling. Multicult. Dev. **25**(2), 248–265 (2004)
5. Bedny, M., Caramazza, A., Grossman, E., Pascual-Leone, A., Saxe, R.: Concepts are more than percepts: the case of action verbs. J. Neurosci. **28**, 11347–11353 (2008)
6. Berant, J., Dagan, I., Goldberger, J.: Learning entailment relations by global graph structure optimization. Comput. Linguist. **38**(1), 73–111 (2012)
7. Besner, D., Stolz, J.A.: Unintentional reading: Can phonological computation be controlled? Can. J. Exp. Psychol. **52**, 35–42 (1998)
8. Binder, J.R., Desai, R.H.: The Neurobiology of semantic memory. Trends. Cogn. Sci. **15**(11), 527–536 (2011)
9. Burge, T.: Philosophy of language and mind: 1950–1990. Philos. Rev. **101**, 3–52 (1992)
10. Clark, A.: Language, embodiment and the cognitive niche. Trends Cogn. Sci. **10**(8), 370–374 (2006)
11. Clark, A.: Whatever next? Predictive brains, situated agents, and the future of Cognitive Science. Behav. Brain Sci. **36**, 181–253 (2013)
12. Clark, A., Chalmers, D.: The Extended Mind. Analysis **58**, 7–19 (1998)
13. Coello, Y., Bartolo, A. (eds.): Language and Action in Cognitive Neuroscience. Psychology Press, New York (2012)
14. Collins, A.M., Loftus, E.F.: A spreading-activation theory of semantic processing. Psychol. Rev. **82**, 407¬–428 (1975)
15. De Rosnay, M., Harris, P.L.: Individual differences in children's understanding of emotion: the roles of attachment and language. Attach. Hum. Dev. Apr 4(1), 39–54 (2002)

16. Ellis, N.C.: Constructions, chunking and connectionism: the emergence of second language structures. In: Doughty, C.J., Long, M.H. (eds.) The Handbook of Second Language Acquisition, pp. 63–103. Blackwell, Oxford (2003)
17. Estes, Z., Adelman, J.S.: Automatic vigilance for negative words in lexical decision and naming. Emotion **8**(4), 441–444 (2008)
18. Ezquerro, J., Iza, M.: Modeling the interaction of emotion and cognition. Int. J. Res. Rev. Appl. Sci. **17**(3) (2013)
19. Farina, E., Albanese, O., Pons, F.: Making inferences and comprehension of emotions in children of 5–7 years of age. Psychol. Lang. Commun. **11**(2), 3–19 (2007)
20. Fischer, M.H., Zwaan, R.A.: Grounding cognition in perception and action. Q. J. Exp. Psychol. Sect. A—Hum. Exp. **61**, 825–857 (2008)
21. Fodor, J.: The Language of Thought. Harvard University Press, Cambridge, Mass (1975)
22. Fodor, J.: Psychosemantics. The Problem of Meaning in the Philosophy of Mind. MIT Press, Cambridge, Mass (1987)
23. Fodor, J.: The Mind Doesn't Work That Way. The Scope and Limits of Computational Psychology. MIT Press, Cambridge, Mass (2000)
24. Foroni, F., Semin, G.R.: Language that puts you in touch with your bodily feelings: the multimodal responsiveness of affective expressions. Psychol. Sci. **20**(8), 974–980 (2009)
25. Gentner, D.: Verb semantic structures in memory for sentences: evidence for componential representation. Cogn. Psychol. **13**, 56–83 (1981)
26. Glenberg, A.M., Gallese, V.: Action-based language: a theory of language acquisition, comprehension and production. Cortex **48**, 905–922 (2012)
27. Glenberg, A.M., Kaschak, M.P.: Grounding language in action. Psychon. Bull. Rev. **9**, 558–565 (2002)
28. Glenberg, A.M., Webster, B.J., Mouilso, E., Havas, E., Lindeman, L.M.: Gender, emotion and the embodiment of language comprehension. Emot. Rev. **1**, 151–161 (2009)
29. Gollwitzer, P.M.: The volitional benefits of planning. In: Gollwizer, P.M., Bargh, J.A. (eds.) The Psychology of Action: Linking Cognition and Motivation to Behavior, pp. 287–312. Guilford Press, NY (1996)
30. Grush, R.: The emulation theory of representation: motor control, imagery and perception. Behav. Brain Sci. **27**, 377–442 (2004)
31. Guendouzi, J., Loncke, F., Williams, M.J. (eds.): The Handbook of Psycholinguistic and Cognitive Processes. Psychology Press, New York (2011)
32. Gygax, P.M.: L'inférence émotionnelle durant la lecture et sa composante comportementale. [Emotion inference during reading and its behavioral component]. L'Année psychologique **110**, 253–273 (2010)
33. Gygax, P., Tapiero, I., Carruzzo, E.: Emotion inferences during reading comprehension: what evidence can the self-pace reading paradigm provide? Discourse Process. **44**, 33–50 (2007)
34. Havas, D.A., Glernberg, A.M., Rinck, M.: Emotion simulation during language comprehension. Psychon. Bull. Rev. **14**, 436–441 (2007)
35. Havas, D., Glenberg, A., Gutowski, K., et al.: Cosmetic use of botulinum toxin-A affects processing of emotional language. Psychol. Sci. **21**(7), 895–900 (2010)
36. Horchak, O.V., Giger, J.C., Caral, M., Pochwatko, G.: From demonstration to theory in embodied language comprehension: a review. Cogn. Syst. Res. **29–30**, 66–85 (2014)
37. Iza, M., Konstenius, R.A.: Verb representation and modularity. In: Pérez Miranda, L.A., Izagirre, A. (eds.) Advances in Cognitive Science: Learning, Evolution, and Social Action. University of Basque Country Press (2010)
38. Iza, M., Ezquerro, J.: Computational emotions. In: Liljenström, H. (ed.) Advances in Cognitive Neurodynamics (IV). Springer, Dordrecht (2015)
39. Jackendoff, R.: Possible stages in the evolution of language capacity. Trends. Cogn. Sci. **3**(7) (1999)
40. Jackendoff, R.: Semantics and Cognition. MIT Press (1983)
41. Kahan, T.A., Hely, C.D.: The role of valence and frequency in the emotional Stroop task. Psychon. Bull. Rev. **15**(5), 956–960 (2008)

42. Kanfer, R.: Self-regulatory and other non-ability determinants of skill acquisition. In: Gollwizer, P.M., Bargh, J.A. (eds.) The Psychology of Action: Linking Cognition and Motivation to Behavior, pp. 404–423. Guilford Press, NY (1996)

43. Kaschak, M.P., Borreggine, K.L.: Temporal dynamics of the action-sentence compatibility effect. Q. J. Exp. Psychol. **61**, 883–895 (2008)

44. Kauschke, C., Stenneken, P.: Differences in noun and verb processing in lexical decision cannot be attributed to word form and morphological complexity alone. J. Psycholinguist. Res. **37**(6), 443–452 (2008)

45. Kemmerer, D.: How words capture visual experience: the perspective from cognitive neuroscience. In: Malt, B., Wolff, P. (eds.) Words and the Mind: How Words Capture Human Experience, pp. 289–329. Oxford University Press, Oxford (2010)

46. Kinstch, W.: The Representation of Meaning in Memory. Erlbaum, Hillsdale, NJ (1974)

47. Kohonen, T.: Self-Organizing Maps, 3rd edn. Springer (2001)

48. Lancker Van, Sidtis D.: Has neuroimaging solved the problems of neurolinguistics? Brain Lang. **98**, 276–290 (2006)

49. Langacker, R.W.: Foundations of Cognitive Grammar. Stanford University Press (1987)

50. Larsen, R., Mercer, K.A., Balota, D., Strube, M.J.: Not all negative words slow down lexical decision and naming speed: importance of Word arousal. Emotion **8**(4), 445–452 (2008)

51. Li, P., Shirai, Y.: The acquisition of lexical and grammatical aspect. Mouton de Gruyter, Berlin (2000)

52. Li, P.: Language acquisition in a self-organizing neural network model. In: Quinlan, P.T. (ed.) Connectionist Models of Development, pp. 115–149. Psychology Press (2005)

53. Liljeström, M., Tarkiainen, A., Parviainen, T., Kujala, J., Numminen, J., Hiltunen, J., Laine, M., Salmelin, R.: Perceiving and naming actions and objects. Neuroimage **41**, 1132–1141 (2008)

54. Lugli, L., Baroni, G., Gianelli, C., Borghi, A.M., Nicoletti, R.: Self, others and objects: how this triadic interaction modulates our behavior. Mem. Cogn. **40**, 1373–1386 (2012)

55. MacKay, D.G., Shafto, M., Taylor, J.K., Marian, D.E., Abrams, L., Dyer, J.R.: Relations between emotion, memory and attention: evidence from taboo Stroop, lexical decision and immediate memory tasks. Mem. Cogn. **32**(3), 474–488 (2004)

56. MacWhinney, B.: Emergentist approaches to language. In: Bybee, J., Hopper, P. (eds.) Frequency and Emergency of Linguistic Structure, pp. 449–470. Benjamins, Amsterdam (2001)

57. Mathews, A., MacLeod, C.: Cognitive approaches to emotion and emotional disorders. Annu. Rev. Psychol. **45**, 25–50 (1994)

58. McKoon, G., Ratcliff, R.: Inference during reading. Psychol. Rev. **99**, 440–466 (1992)

59. Mestres-Misse, A., Rodriguez-Fornells, A., Munte, T.F.: Neural differences in the mapping of verb and noun concepts onto novel words. Neuroimage **49**, 2826–2835 (2009)

60. Mukamel, R., Ekstrom, A.D., Kaplan, J.T., Iacoboni, M., Fried, I.: Single-neuron responses during execution and observation of actions. Curr. Biol. **20**(8), 750–756 (2010)

61. Pittermann, J., Pittermann, A., Minker, W.: Handling Emotions in Human-Computer Dialogues. Springer, Berlin (2010)

62. Pons, F., Doudin, P.-A., Harris, P.: Enseigner la comprehension es émotions: Quel type de programme pour quel type d'élève?. In Doudin. P.-A., Lafortune, L. (eds.) Enseignement spécialisé: quelle formation pour les enseignants? Paper, Congrès du Réseau d'Échange Francophone. IEEE Signal Processing Society, Geneva (2003)

63. Pulvermüller, F.: Grounding language in the brain. In: de Vega, M., Graesser, A., Glengberg, A.M. (eds.) Symbols, Embodiment and Meaning, pp. 85–116. Oxford University Press, Oxford (2008)

64. Rieman, B.C., McNally, R.J.: Cognitive processing of personally-relevant information. Cogn. Emot. **9**, 325–340 (1995)

65. Rosch, E.H.: On the internal structure of perceptual and semantic categories. In: Moore, T.E., (ed.) Cognitive Development and Acquisition of Language, pp. 111–144. Academic Press, New York (1973)

66. Rosch, E.H.: Linguistic Relativity. In: Silverstein, A. (ed.) Human Communication. Theoretical Explorations. L.E.A, Hillsdale (1974)
67. Rosch, E.H.: Cognitive representations of semantic categories. J. Exp. Psychol. Gen. 104, 192–233 (1975)
68. Rudolph, U., von Hecker: Three principles of explanation: verb schemas, balance and imbalance repair. J. Lang. Soc. Psychol. 25, 377–405 (2006)
69. Rugg, M.D.: The effects of semantic priming and word repetition on event-related potentials. Psychophysiology 22, 642–647 (1985)
70. Taylor, L.J., Zwaan, R.A.: Motor resonance and linguistic focus. Q. J. Exp. Psychol. Sect. A —Hum. Exp. 61, 896–904 (2008)
71. Taylor, L.J., Zwaan, R.A.: Action in cognition: the case of language. Lang. Cogn. 1, 45–58 (2009)
72. Thill, S., Caligiore, D., Borghi, A.M., Ziemke, T., Balsassare, G.: Theories and computational models of affordance and mirror systems: an integrative review. Neurosci. Behav. Rev. 37(3), 491–521 (2013)
73. Thorndyke, P.W.: Conceptual complexity and imagery in comprehension and memory. J. Verbal Learn. Verbal Behav. 14, 359–369 (1975)
74. Tyler, L.K., Randall, B., Stamatakins, E.A.: Cortical differentiation for nouns and verbs depends on grammatical markers. J. Cogn. Neurosci. 20, 1381–1389 (2008)
75. Vanhatalo, U.: The use of questionnaires in exploring synonymy. Lexical knowledge from native speakers to electronic dictionaries. PhD Thesis. Location of document (2005). http://ethesis.helsinki.fi/julkaisut/hum/suoma/vk/vanhatalo
76. Willms, J.L., Shapiro, K.A., Peelen, M.V., Pajtas, P.E., Costa, A., Moo, L.R., Caramazza, A.: Language-invariant verb processing regions in Spanish-English bilinguals. Neuroimage 57, 251–261 (2011)
77. Wray, A., Perkins, M.: The functions of formulaic language: an integrated model. Lang. Commun. 20, 1–28 (2000)
78. Zwaan, R.A., Taylor, L.J.: Seeing, acting, understanding: motor resonance in language comprehension. J. Exp. Psychol.-General 135, 1–11 (2006)

Part II
Computational Perspectives

Knowledge, Representation and the Dynamics of Computation

Jan van Leeuwen and Jiří Wiedermann

Abstract Cognitive processes are often modelled in computational terms. Can this still be done if only minimal assumptions are made about any sort of representation of reality? Is there a purely knowledge-based theory of computation that explains the key phenomena which are deemed to be computational in both living and artificial systems as understood today? We argue that this can be done by means of techniques inspired by the modelling of dynamical systems. In this setting, computations are defined as curves in suitable metaspaces and knowledge is generated by virtue of the operation of the underlying mechanism, whatever it is. Desirable properties such as compositionality will be shown to fit naturally. The framework also enables one to formally characterize the computational behaviour of both knowledge generation and knowledge recognition. The approach may be used in identifying when processes or systems can be viewed as being computational in general. Several further questions pertaining to the philosophy of computing are considered.

1 Introduction

Can cognitive processes be simulated by machines? Can cognitive processes be understood in computational terms at all? Can this be done without making severe assumptions about any sort of representation of the subjective environment and on the nature of the underlying computational mechanisms? These questions not only

J. van Leeuwen (✉)
Department of Information and Computing Sciences, Utrecht University,
Princetonplein 5, 3584 CC Utrecht, The Netherlands
e-mail: J.vanLeeuwen1@uu.nl

J. Wiedermann
Institute of Computer Science, Academy of Sciences of the Czech Republic,
Pod Vodárenskou v. 2, 182 07 Prague 8, Czech Republic

J. Wiedermann
Czech Institute of Informatics, Robotics, and Cybernetics, Czech Technical
University in Prague, Zikova Street 1903/4, 166 36 Prague 6, Czech Republic
e-mail: jiri.wiedermann@cs.cas.cz

© Springer International Publishing AG 2017
G. Dodig-Crnkovic and R. Giovagnoli (eds.), *Representation and Reality in Humans,
Other Living Organisms and Intelligent Machines*, Studies in Applied Philosophy,
Epistemology and Rational Ethics 28, DOI 10.1007/978-3-319-43784-2_5

challenge our deepest understanding of cognition and its computational modelling [1], but also that of computation in itself, being commonly tied to algorithms and discrete-state systems only since the ground-breaking insights of Turing. What are the basic properties needed of a process in the human or animal brain in order for it to be regarded as being computational?

The term "computation" is increasingly being used to describe aspects of natural processes outside the well-formalized domain, including many occurring in cognition but also e.g. in living cells. In all these cases, computation is more than just a metaphor. It appears as a much more general notion than what is implied by current theories. It has already given rise to many alternative views, including many in which computation is seen as a process for transforming information in some way. Is there a notion of computation that is more fundamental? In particular, should one not focus more on *what* computations do rather than on *how* they do what they do?

In previous studies [2, 3] we presented computation as some process that generates *knowledge* instead of one that merely manipulates symbols or transforms information. The question of what constitutes knowledge and how it may be generated in any subjective context clearly depends on the views or theories of the observer. However, if we adopt the Aristotelian view that knowledge, old or new, should be demonstrable from basic premises, it makes sense to assume an underlying process, mental or otherwise, that can acquire, deduce, combine, transform, adapt and create knowledge, using some kind of causality as an ordering principle. A knowledge-generating process is likely to interact with an environment and even evolve over time, using new premises as new knowledge is generated, dispensing with old knowledge that is no longer viable. We take the view that computation is what these knowledge-generating processes do (and vice versa).

Viewing computation as a knowledge-generating process presupposes certain manipulative possibilities and rules for knowledge. For our purposes we assume that knowledge can be specified as *items* and that distinct knowledge items can be recognized or observed. We assume that items can be processed, combined and composed (fused) in some way but we will not be more specific than this. In the interest of generality, we make no further assumptions about the concrete representation of knowledge (items) or about any deductive or generative framework for the knowledge domain (theory) that is considered. In Sect. 2 we introduce *metaspaces* to capture the sets of items we need.

Viewing computation as knowledge generation entails that computations are 'observed' in suitable metaspaces, while the generating mechanisms 'live' in other suitable metaspaces. After defining the types of metaspaces involved, we give a general definition of *computation* in Sect. 3. We will define computations as *curves*, in a suitable topological setting. Clearly, the question arises of whether this approach can explain known or new computational phenomena in cognition or otherwise, in a more satisfying and general way than earlier approaches. We give several examples that aim to show that it does. The approach enables us to formally characterize knowledge generation and knowledge recognition as computational processes. We do so in Sects. 4 and 5.

Having knowledge and knowledge generation as a starting point is not only interesting for understanding computation. By linking the worlds of computation and knowledge generation we bring two domains together which have been remarkably converging to each other in the present information age. By concentrating on *what* computation is and does, we may be losing the finer details of computation and the strength of the mathematical theory as we know it since Turing, but this may be required to achieve the abstraction we need today. This paper explores a theoretical framework that implements some of the ideas and viewpoints of our philosophy. The new focus brings further insight into the essence of computation and its intimate relation to cognitive processes and knowledge generation.

2 Metaspaces

The classical approaches to computation rely on machine models and algorithms, but this severely limits the general interpretation of the notion. In order to explain computation in the broadest possible way, we need a new way to abstract from the underlying mechanisms that effectuate it.

In this section we introduce *metaspaces* as generic sets of items that can arise in capturing mechanisms in some way. We subsequently introduce two types of metaspaces that play a role here: *action spaces* and *knowledge spaces*. In the next section we show how these spaces enable us to give an elegant and general definition of computation.

2.1 *General*

Given our premise that knowledge is generated by a process of some kind, it is implicit that there is some underlying mechanism producing it. We refrain from making any further technical assumptions about such mechanisms, and merely posit that their features can be *captured* at any desired moment in time. The joint features at any time will form the *meta-item* for the mechanism at that time. The collective set of all meta-items corresponding to a mechanism that can arise is called a *metaspace*.

Metaspaces occur in any context where systems or processes are observed. For example, the *configuration spaces* obtained when physical systems are modelled using vectors of parameters are metaspaces. Metaspaces typically have some structure, derived from the way the underlying mechanism is observed or explained. Hence, meta-items will adhere to some descriptive framework or theory for the space we are interested in. A consequence is that meta-items are presentable and distinguishable, even though we do not care how. We note that meta-items do not necessarily characterize a given mechanism completely. Meta-items need not be unique over the lifetime of a process and may repeat even when the observed system is not cyclic.

Metaspaces are *intensionally* defined. Even if meta-items are observable, this does not mean that we know all of them before we observe them nor that we will actually ever observe them in reality. In particular, we do not assume that all items that 'look' like valid meta-items according to some perception of the descriptive framework actually are members of the given metaspace. For example, if meta-items are like theorems of a non-trivial theory, it is clear that we can at best hope to discover some of them in a gradual way. This is also seen e.g. in metaspaces arising in cognition and in Nature.

For every metaspace $\mathbb{M}$ we assume that there is a *core set* $\mathbb{M}_0$ which is 'known' and that there is a process of some sort to *discover* the remaining elements of $\mathbb{M}$, especially when meta-items that contain valuable information ('knowledge') are believed to exist. If no such process is available, we may wish to design it. *Metaspace discovery* will become crucial later on. We will not worry about questions like: are metaspaces sets (we assume they are), are metaspaces enumerable (they probably are) and are meta-items representable (we will discuss this later).

2.2   Action Spaces

Consider any mechanism (process) that is regarded as being computational. The meta-items of the mechanism in action form a metaspace which we will call the *action space* of the mechanism. The notion of action space is dependent on the way meta-items are viewed and thus on the framework used to model the mechanism. Hence, different frameworks could lead to different action spaces for the same mechanism. This happens, for example, when a refined framework or a different theory altogether is used to capture a mechanism. It is similar to the way different 'spectators' may have different views of reality, as in [4].

Action spaces are not arbitrary metaspaces. Observing meta-items while a given mechanism is acting implies a notion of *proximity* among the meta-items as they are occurring in sequence. This is an aspect of action spaces which is intuitively associated with the idea of being computational. Action spaces may be 'continuous' or 'discrete' in this respect, or a combination of both. In order to delineate the action spaces that we need, we resort to mathematical topology and postulate the following:

Action spaces are topological spaces, with a topology consistent with the proximity relation between action items.

The postulate expresses that the topology of an action space 'derives' from the proximity relation observed during the action of the mechanism (i.e. over time or as induced by some other measure). Given the postulate, one can make use of common topological notions and e.g. define continuous mappings over action spaces. The core set of an action space is the collection of meta-items that correspond to valid 'initializations'.

We do not make any further assumptions about action spaces now; in particular, we make no assumptions on how the mechanism that corresponds to it actually *works*. A mechanism may follow any mode of operation, consist of any number of cooperating components, and interact with any environment. This gives action spaces the full generality we wish to preserve.

Example 1 The observable descriptions of a living *cell* form an action space. The meta-items give information about its development over time, a level of abstraction away from the concrete content of the cell. We may also be interested in some special knowledge, e.g. the fluctuation of a chemical compound or a property of the cell, all to be gleaned from the meta-items. (Note that meta-items may be real-valued.) We may also be interested in the metaspace of a family of cells, as in an experiment.

Example 2 The possible 'full information descriptions' of a computer executing a (known or unknown) chain of instructions form an action space. Meta-items display the possible instances of registers and memory in bits. We may read out or interpret any meta-item as knowledge, if indeed it fits the sort of knowledge we are interested in. The meta-items may correspond to any mode of execution (sequential, parallel or distributed) and to any level of abstraction at which we want to observe the mechanism, i.e. the computer system.

2.3 Knowledge Spaces

In viewing computation as knowledge generation, it is implicit that computations generate knowledge that is meaningful in a suitable knowledge domain. In philosophy one distinguishes between many different types of knowledge. We will be mostly concerned with knowledge in a broad Aristotelian sense, as this is most naturally quantized. Knowledge is then basically the collection of 'actualities justified by an understanding', in a sense that may vary over time.

The strong assumption we make is that the knowledge over a given domain can be qualified and described, and bound to a definite 'point of view'. This may be expressed as in some formal theory, but even the use of natural language is not excluded here. The collection of potential 'knowledge items' for a domain will be called the *knowledge space* of the domain. We assume that there is always a descriptive framework or a deductive theory for defining or generating the items of the knowledge spaces that we consider. Hence we postulate that *knowledge spaces are metaspaces*. The core set of a knowledge space consists of the facts that are known by observation or experience, or just by assumption.

Many ways are known by which knowledge can be generated. Knowledge generation has been studied in philosophy ever since the times of Plato and Aristotle. It has given rise to principles such as formal inference, informal reasoning, analogy, knowledge acquisition by communication, cognition, causality and so on. In [2, 3] we argued that computation is a general mechanism for generating knowledge as well. By definition, knowledge generation is merely an instance of the, more general, metaspace generation problem.

Example 3 The theory of a first-order structure 𝔸 as known in mathematical logic forms a knowledge space. The knowledge items are sentences that hold in 𝔸. The core set of 𝔸 consists of the postulates of 𝔸. The mechanism underlying the metaspace is a combination of first-order inference and the evaluation ('invention') of new sentences. Knowledge here follows the standard pattern of a formalized theory.

Example 4 The structures (worlds) that are possible instantiations of a given first-order language 𝕃 over a fixed base set form a knowledge space. The knowledge items represent the way the 'world' could be shaped, using the functions and relations as they are defined in it. The core set of the space consists of the 'initial worlds' one wishes to observe. The mechanisms underlying the metaspace are 'programs' that modify the assigned values of the functions and relations in a stepwise way, with external influences possibly taking place as well. Worlds correspond to 'states' and the mechanisms to *abstract state machines* as defined by Gurevich [5, 6], provided that certain additional restrictions are imposed (notably, the so-called bounded exploration condition).

Knowledge spaces are special because knowledge is. One may well have mechanisms that act on the items of a knowledge space, turning it into an action space itself. Thus, action spaces and knowledge spaces may be viewed as dual structures, even giving rise to formal equivalences between them if the corresponding actions match, in analogy to similar correspondences between formal structures studied in computer science. Alternatively, a knowledge space may serve as the action space for another, higher-level knowledge space, potentially leading to a hierarchy of levels of abstraction [7].

It is an intriguing thought that the (dispositions of the) *brain* may be viewed as a knowledge space. The knowledge items are our possible mindsets (possibly restricted to a certain topic), and the underlying mechanisms are provided by the facilities of thought. The eternal question of whether the brain is a computer or not (cf. [8, 9]) amounts to the very question of whether, and if so how, the corresponding knowledge spaces can be explored by computation.

3   Computation

Our premise is that, in principle, every computation effectuates some knowledge. We need to have a good model in order to design, explain, prove or understand this and highlight the nature of computation. In this section we give a definition of computation from this viewpoint. The definition will be fully machine- and algorithm-free, and uses minimal assumptions on representation.

We start out by assuming that computation is performed by some process and for some purpose, but how do these things connect? The duality we continually noted between underlying mechanism and knowledge generation is similar to the duality between agency and goal in the *philosophy of action*. We will give a possible formalisation of this intuitive setting, while staying as general as possible. The formalization implicitly leads to a possible criterion for the computationality of (cognitive) processes as well.

The approach presented here uses ingredients from the modelling of dynamical systems. It does not necessarily implement all aspects of our philosophy of computation as knowledge generation [2, 3]. For example, we will make some concrete assumptions in cases where normally more options would have to remain open. However, the framework as presented is an excellent testbed for the ideas.

We first introduce the metaspaces we need, and then define the notion of computation in our present setting. In Sect. 4 we will show that the framework allows one to manipulate, viz. to compose computations in a natural way and reflect on the various further aspects of the framework.

3.1 Relevant Spaces

In our view, a computational process will always involve two metaspaces: an action space $\mathbb{A}$, and a knowledge space $\mathbb{E}$. The two spaces reflect the 'two sides' of the process. We use this to explain computation, but one may use it to explain knowledge generation quite generally as well. Let $\mathbb{A}_0$ and $\mathbb{E}_0$ denote the core sets of $\mathbb{A}$ and $\mathbb{E}$, respectively.

The spaces $\mathbb{A}$ and $\mathbb{E}$ are coupled. In particular, (some) action items x with $x \in \mathbb{A}$ will carry information that maps to (some) knowledge items in $\mathbb{E}$. We do not require uniqueness. Thus, a knowledge item may be obtainable from several different action items. We assume that the mapping is achieved by way of a simple readout functionality called a *semantic map* which aims to bring out the knowledge that is contained in an action item, in the terms of the knowledge domain.

Definition 1 A *semantic map* from $\mathbb{A}$ to $\mathbb{E}$ is any partial mapping $\delta : \mathbb{A} \to \mathbb{E}$ with the property that $\delta(\mathbb{A}_0) \subseteq \mathbb{E}_0$.

Given δ and $x \in \mathbb{A}$, we assume that $\delta(x)$ is obtained by only a simple 'extension' of the observational means that produced x to begin with. In other words, no substantial extra effort should be involved that has not already been expended by the underlying mechanism. Note that $\delta(x)$ may be undefined for some items x, reflecting the fact that an action item may not always contain knowledge that is ripe for 'display'. The condition that $\delta(\mathbb{A}_0) \subseteq \mathbb{E}_0$ is required for *consistency*: the knowledge embedded in the (initial) core set of the action space should be part of the core knowledge known at the outset. In particular, it is assumed that $\delta(x)$ is defined for all $x \in \mathbb{A}_0$.

Example 5 Consider any programming language, implemented on a (universal) machine $\mathcal{M}$. Let $\mathbb{A}$ consist of all possible items $\langle \pi, x, J, y \rangle$, where π is a single-input single-output program, x an input value, J the 'full information vector' with the register contents of $\mathcal{M}$ at any moment during π's execution, and y the output or

$\perp$ (undefined) as implied by J. Clearly $\mathbb{A}$ contains the action items of $\mathcal{M}$, seen as a mechanism (cf. Example 2). Let $\mathbb{E}$ consist of the items $\langle f, a, b \rangle$ with $f : \mathbb{N} \to \mathbb{N}$ a partial function, $a \in \mathbb{N}$, $b \in \mathbb{N} \cup \{\perp\}$, and $f(a) = b$. $\mathbb{E}$ is the knowledge space of all single-parameter partial functions. The two spaces can be linked by the semantic map $\delta : \mathbb{A} \to \mathbb{E}$ defined as follows:

$$\delta(\langle \pi, x, J, y \rangle) = \begin{cases} \text{if } J \text{ indicates that the computation is ongoing:} \\ \quad undefined \\ \text{if } J \text{ indicates that the computation has terminated:} \\ \quad \langle f_\pi, x, y \rangle \end{cases}$$

where f_π denotes the function determined by program π and $f_\pi(x) = y$. The sub-space $\delta(\mathbb{A})$ of $\mathbb{E}$ corresponds to the knowledge of the computable functions only. We have $\mathbb{A}_0 = \{\langle \pi, x, J_{\text{init}}, \perp \rangle \mid \pi \text{ a program}, x \in \mathbb{N}, J_{\text{init}} \text{ the initial information vector}\}$ and $\mathbb{E}_0 = \emptyset$.

Instead of a single knowledge space $\mathbb{E}$ it may be desired to use several spaces and have several semantic maps, to capture different facets of the knowledge that may be generated. This is easily reduced to the case of a single knowledge space only.

3.2 Defining Computation

We can now give a definition of computation, in the present setting. We first define single computations, and then focus on so-called bundles.

Let $\mathbb{E}$ be a knowledge space we are interested in, and let $\mathbb{A}$ be the action space of an underlying mechanism. Let $\delta : \mathbb{A} \to \mathbb{E}$ be a semantic map as above. By assumption, $\mathbb{A}$ is a topological space, and thus we can have topological objects in $\mathbb{A}$ such as *curves*. We posit that curves are precisely the sort of 'trajectories' that are traced by computations.

A *curve* is any continuous function $c : S \to \mathbb{A}$, where S is any segment on the real or integer line that is possibly half-open to the right. (The lines are topological spaces by virtue of the standard metric.) We usually identify c and the image $c(S)$ in $\mathbb{A}$. Given a curve c, we let c^{init} be its starting point and, if it is defined, c^{end} its ending point.

Definition 2 A *computation* is any curve $c \subseteq \mathbb{A}$ with the following properties:

- $\delta(c^{\text{init}})$ is defined, and
- if c^{end} is defined, then $\delta(c^{\text{end}})$ is defined as well.

We require that any computation must start with 'some knowledge'. We do *not* insist a priori that $\delta(c^{\text{init}}) \in \mathbb{E}_0$. If we would be perfectly general, a computation might request 'input knowledge' at later points on the curve as well, but we will not elaborate on this in the present setting. Along the curve, δ need not be defined in every

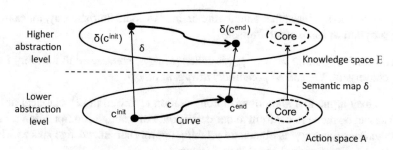

Fig. 1 A schematic diagram of a computation. (Abstraction levels can be iterated)

intermediate action item. However, if the curve ends, δ must be defined in the ending point. A schematic view of a computation is depicted in Fig. 1.

The definition of computation by means of curves is a natural one, fitting the intuition that a computation moves through consecutive action items while respecting the proximity relation in the space. Even 'real-continuous' curves may be needed, for modeling certain natural mechanisms [10].

All information about *how* the computation is effectuated is presumed to be hidden in the details of the action items, which is of no specific concern to us. We may even hide the interactions with other computations in it, checking that their curves 'match' separately. This enables us to concentrate on *what* the computation does. The semantic map will help us read out the 'knowledge' that is generated over the curve (cf. Fig. 1).

A computation c in $\mathbb{A}$ inherits a natural orientation as a curve, progressing from c^{init} towards c^{end} (if it exists). The orientation reflects the (broadly viewed) *serializability* of computations, a notion that is found in many conceptions of computational systems as they operate in any context, regardless of any causal effects whatsoever (but normally consistent with it). A curve can be self-intersecting, without necessarily implying any looping behaviour of the underlying mechanism.

Definition 3 A computation (curve) c is called *convergent* if c^{end} is defined (i.e. as a definite element of $\mathbb{E}$). It is called *divergent* otherwise.

Finally, the term 'computation' is often used to denote not just a single computation but a whole *family of computations* that can be effectuated by the same mechanism or by some conglomerate of mechanisms. We use the term *bundle* here. Let $\mathbb{A}$ be an action space.

Definition 4 A *computation bundle* is any collection of computations $\mathcal{B} = \{c_i\}_{i \in I}$ where I is an index set and for every $i \in I$, c_i is a computation (curve) in $\mathbb{A}$.

Whereas the computations are defined by the underlying mechanism of the action space, it is rather more difficult to define what keeps them together as a bundle. An example of a bundle we are most likely interested in is: $\mathcal{B} = \{c \in F \mid c_{\text{init}} \in \mathbb{A}_0\}$, the bundle of all computations in some feasible set F that begin computing in the

core set. We do not insist that bundles are defined in some finitistic way, for example
by a program or some other artefact.

Definition 5 A bundle $\mathcal{B} = \{c_i\}_{i \in I}$ is called *(always) convergent* if, for every $i \in I$,
c_i is convergent. We call $\mathcal{B}$ *potentially divergent* otherwise.

One may argue that all notions of computation reviewed in [2], classical or otherwise, can be made to conform to our definition. For example, computation seen as
information processing is obtained by taking appropriate knowledge spaces which
just contain 'information' about their domain.

Example 6 Let $\mathbb{I}$ be the internet, M a computer connected to the internet, and π
the client program of a search engine running on M. Answering queries using π is
seen to be computational as follows: Let $\mathbb{E}$ be the collection of all 'facts' that can
be known, and $\mathbb{E}_0$ the subset of currently known facts. Let $\mathbb{A}$ be the set of tuples
$\langle q, a, r \rangle$, where q is a query, a a knowledge item or $\perp$ (undefined), and r any possible
instantaneous description of π as running on M and accessing $\mathbb{I}$. Let semantic map δ
be defined by $\delta(\langle q, a, r \rangle) = a$. Clearly $\mathbb{A}$ can be a very large set, but *we do not need
to know all its elements* as long as the search mechanism can generate the ones we
need. Define the core $\mathbb{A}_0$ as the set of all tuples $\langle q, \perp, r \rangle$ where q is a query and r is
an initial instantaneous description of π. The chain of consecutive action items that
result from initializing a search, moving through items as given by the instantaneous
descriptions, up to and including an item which has an answer to the query (if any)
is a curve in $\mathbb{A}$ and thus a computation. This is easily modified to the case of many
answers. It follows that internet searching, viewed as the collection of all searches
on initial queries, is a computation bundle.

4 Dynamics

We have seen how computation as a knowledge-generation process can be defined
using action and knowledge spaces. However, computations do not stand alone and
their result (knowledge) is often used, and needed, in other computations. We will
show that this feature can be expressed naturally in the given framework. We will
sacrifice some generality in order to show how this can be done. Next we discuss
various further aspects of the framework, from a philosophical viewpoint.

4.1 Composing Computations

The question is: What do computations as defined actually entail. Can a given com-
putation c be effectuated even when $\delta(c^{\text{init}}) \notin \mathbb{E}_0$? If not then, supposing c is part of
a bundle $\mathcal{B}$, may c be effectuated based on knowledge that is generated by another
computation in $\mathcal{B}$? Realistically, many computations will depend on knowledge that

is yet to become available. These computations have to wait for their 'turn' until other computations have produced the lacking knowledge.

Definition 6 A computation c is called *input-enabled* if c can be effectuated fully as soon as c^{init} is available.

There is no reason beforehand to require that computations are input-enabled. It might happen e.g. that a computation needs extra knowledge that is not contained in the core set $\mathbb{E}_0$ and that it cannot compute itself. We will not model all modalities here, and simply assume that each computation is self-contained and 'runnable' whenever c^{init} is 'known', delegating any interactions to the definition of the curve. This will be sufficient to illustrate the principles. Thus:

The computations in all bundles $\mathcal{B} = \{c_i\}_{i \in I}$ we consider are assumed to be input-enabled.

By the assumption, all computations $c \in \mathcal{B}$ with $c^{\text{init}} \in \mathbb{A}_0$ can be effectuated immediately. However, it also makes sense now to define an important further property that is often desired, namely that of *compositionality*.

Definition 7 Let c, d be computations (curves). Let c be convergent and let $c^{\text{end}} = d^{\text{init}}$. Then the curve $c' = c \circ d$ obtained by glueing c and d together at c^{end}, is called the *direct composition* of c and d.

An immediate consequence of this definition is that, if c is enabled and convergent and $c^{\text{end}} = d^{\text{init}}$, then $c \circ d$ is well-defined and enabled as well. The following associativity property is easily verified.

Proposition 1 *Let c, d and e be computations, let c and d be convergent, and let $c^{\text{end}} = d^{\text{init}}$ and $d^{\text{end}} = e^{\text{init}}$. Then $(c \circ d) \circ e = c \circ (d \circ e)$.*

From a computational point of view, direct composition alone is not satisfactory. As a computation c proceeds, we may want to *pre-empt* it at any point that is somehow reasonable, viz. at any point x for which $\delta(x)$ is defined. Any such point might be a valid starting point of a new computation.

Definition 8 Let c, d be computations (curves). Let x be any point on c for which $\delta(x)$ is defined. Let c_x be the curve of c from c^{init} to x, and let $x = d^{\text{init}}$. Then the curve $c' = c_x \circ d$ obtained by glueing c_x and d together at $x = c_x^{\text{end}}$, is called a *grafted composition* of c and d. The set of all possible grafted compositions of c and d is denoted by $c \triangle d$.

Note that the definition does not require either c or d to be convergent. Also note that a point x may occur more than once on a curve. In general $c \triangle d$ may consist of many curves (computations). One easily verifies the following:

Proposition 2 *If c, d and e are computations (curves), then:*
- *If $c \circ d$ is defined, then $(c \circ d) \in c \triangle d$.*
- *$(c \triangle d) \triangle e = c \triangle (d \triangle e)$.*

Definition 9 A bundle $\mathscr{B} = \{c_i\}_{i \in I}$ is said to be *closed under composition* if, for all $i, j \in I$: if $c_i \bigtriangleup c_j$ is defined, then $c_i \bigtriangleup c_j \subseteq \mathscr{B}$.

If a bundle $\mathscr{B}$ is closed under (grafted) composition, we simply call it *compositional*. In the next section we will see what role compositionality plays in the analysis of computations for knowledge generation.

4.2 Reflections

The question of identifying the nature and character of computation has been the subject of many studies and discussions [2]. The idea of viewing computation as a process of some kind seems well accepted [11], but the opinions on what makes processes computational differ considerably. For example, concrete processes may be viewed as being computational if a representative abstraction of them is [12, 13]. The definition of computation we gave here implements our philosophy that computation is a process of *knowledge generation* and that this is its driving characteristic.

4.2.1 Cross Connections

Various connections to other areas should be observed. In particular, the way we view computations here is reminiscent of the way systems are viewed in *control theory*. Any dynamical system that evolves over time may be viewed as computational, provided it is generating knowledge from some perspective in the first place. Connections between computer science and control theory were observed before, e.g. by Arbib [14] in the 1960s. Like [14], we believe that the philosophy of computation 'can gain tremendously' from the ideas in general systems theory.

Another connection can be found in the theory of *concurrent systems*. In particular, Mazurkiewicz [15] already showed in the 1970 s that the behavioural aspects of these systems can be adequately studied using *traces* that represent the possible serializations of the interactions that can take place. Sets of traces are an analogue of what we called bundles. This is where the analogies diverge, as the theory of traces has been elaborated entirely at the symbolic level.

Last but not least, we note that topology has been used extensively in the construction of semantic models of programming systems, notably of the λ-calculus [16]. This has led to powerful approaches to the computability of functions and type theory [17]. In general, *computable topology* has focussed on the computability of 'topological objects', rather than on computational processes as we do here. Again, we believe that much can be gained from the ideas in this domain.

4.2.2 Evaluation

Our definition of computation is sufficiently different from extant notions that a critical evaluation is warranted. For example, there is still considerable flexibility in the underlying notions of knowledge and action spaces. This could give the impression that the definition will allow one to declare just about anything as being computational. We reject this idea, as knowledge and action are sufficiently refined notions to exclude abuse.

Nevertheless, we stress again that the notions are relative to the frameworks and theories in which they are understood, as is the resulting notion of computation. As an example, consider a light switch. The operation of a light switch is not seen as computational, as no notion of knowledge is involved. However, if one declares the signals going around in the switch as being 'knowledge' of its components at suitable times, then one may say that what goes on in the light switch *is* computational. We recall the actor-spectator phenomenon again (cf. Sect. 2.2).

An interesting issue is whether the definition of computation as we gave it is 'free' of representation. It has been claimed that the intensionality of computation requires some form of grounding in a symbolic domain, a view which seems to have been inspired heavily by the classical notion of computing by 'computers'. Fodor [18] has expressed this very eloquently as follows:

> ...it is natural to think of the computer as a mechanism that manipulates symbols. A computation is a causal chain of computer states and the links in the chain are operations on semantically interpreted formulas in a machine code. To think of a system (such as the nervous system) as a computer is to raise questions about the nature of the code in which it computes and the semantic properties of the symbols in the code. In fact, the analogy between minds and computers actually implies the postulation of mental symbols. There is no computation without representation.

In the definition of computation we gave, however, 'symbolic representation' plays no role. Representation is left entirely implicit. This conforms to the view of Piccinini [19], who argues that functional properties of computation may be specified without a need for any semantic properties. It is a great advantage to separate the two notions.

Finally, note that we concentrated on 'what' a computation does and not on 'how' it is effectuated by an underlying mechanism, following [2, 3] and in keeping with the broader views of computation today. Nevertheless, one may still argue that some intuitive machine concept is embodied in the notion of action space. We do not object in principle, as long as the notion of machine is kept as general and open as e.g. in the following definition by Beck [4]:

> A machine [...] is an arrangement of matter devised so that a dependable correspondence is secured between controlled input and usable output.

However, our definition of action spaces does not involve any constraints in terms of input or output or any determined correspondence between them beforehand, leaving room for arbitrary influences from an 'uncontrolled' environment. We reject the idea that computation as a notion requires an analogy to artefacts such as machines, viz. computers. Cases that make use of it are easily subsumed by our definition.

Example 7 It can be argued that the nervous system is computational, using the analogy to complex computing systems. For example, Piccinini and Bahar [20] reason that the nervous system may be seen as "an information-processing, feedback control, functionally organized, input-output system", although they also point out that this may not explain all of the neural processes involved. In particular they argue that the neural processes are neither analog nor digital. Restricting to the computational part, it is easily seen that this follows from our definition without resort to any functional properties of a computing system.

5   Exploring Knowledge Spaces by Computation

The general problem of discovery in metaspaces was introduced in Sect. 2.1. Can one characterize the knowledge that can be discovered by means of some underlying computational mechanism? And, can one turn the question around and 'recognize' the knowledge items that can be computed?

Let $\mathbb{E}$ be a knowledge space we are interested in, and let $\mathbb{E}_0$ be the core set we have for it. In this section we will explore the following key problems in knowledge space exploration:

- *Knowledge generation*: generate all knowledge items of $\mathbb{E}$.
- *Knowledge recognition*: given a knowledge item e, is it an element of $\mathbb{E}$?

Both knowledge generation and knowledge recognition, when viewed as processes embedded in the human or animal brain, are likely to be constrained further in many ways. For example, knowledge is likely to be aggregated in a coded rather than enumerative way only. Also, recognition may be restricted to the knowledge in a 'known' subset of the items that are potentially knowable. We will not address these constraints but aim to characterize the full extent of the knowledge space that has to be mastered.

We immediately note that knowledge generation and knowledge recognition, when viewed as processes without further constraints, may be indefinite, i.e. without any finite bound on their duration or effect. In the case of knowledge generation this is evident, e.g. when items can be generated multiple times or when the knowledge space itself is infinite and only finitely many knowledge items can be discovered at a time. However, the same can be said of knowledge recognition, e.g. when it relies on some kind of searching without a criterion for when to stop, especially for items that are not valid knowledge and thus cannot be found in the knowledge space at all. This is a well-known phenomenon that occurs when these processes are simulated by artefacts such as Turing machines [21].

This leads to the question of how knowledge generation and recognition can be characterized in our framework. We will first show how to characterize the knowledge in $\mathbb{E}$ that can be generated by computation from $\mathbb{E}_0$, from our present perspective. Next we show how knowledge recognition can be characterized as a computational process, in the defined framework.

5.1   Knowledge Generation

Let $\mathbb{E}$ and $\mathbb{E}_0$ be as above. Assume that we have some mechanism for exploring $\mathbb{E}$ that is in essence computational. Let $\mathbb{A}$ be its action space, and let $\delta : \mathbb{A} \to \mathbb{E}$ be the relevant semantic map. Let $\mathscr{B}$ be the bundle of computations in $\mathbb{A}$ that we potentially have at our disposal. How do we go from here?

A crucial question is how 'knowledge' is actually extracted from enabled computations $c \in \mathscr{B}$. If $c^{\mathrm{init}} \in \mathbb{A}_0$ and c^{end} is defined, then we may tacitly assume that $\delta(c^{\mathrm{end}})$ is a 'logical consequence' of $\mathbb{E}_0$ and thus 'knowledge' of the sort we are after. However, any knowledge computed 'on the way' may be regarded as available too (assuming it is accessible). Thus, if c is enabled, the entire set $\delta(c) \subseteq \mathbb{E}$ may be seen as generated knowledge.

Making this more precise, we first define when a computation is regarded as being enabled (runnable), cf. Definition 6. We do this recursively as follows:

Definition 10 A computation $c \in \mathscr{B}$ is called *enabled* when either $c^{\mathrm{init}} \in \mathbb{A}_0$ or some enabled computation $d \in \mathscr{B}$ contains c^{init}.

The knowledge-generation process in $\mathbb{E}$ now proceeds as follows: We begin with $\mathbb{E}_0$ and all computations $c \in \mathscr{B}$ with $c^{\mathrm{init}} \in \mathbb{A}_0$, and see what we get. Whenever any new computation gets enabled in the process, we allow it to perform as well. Iterating this ad infinitum, we obtain all knowledge in $\mathbb{E}$ that can possibly become 'known' or, at least, generated (i.e. by this mechanism).

Definition 11 Let $e \in \mathbb{E}$ be a knowledge item. We say that e is *producible* and that computations $c_1, \ldots, c_n$ with $c_i \in \mathscr{B}$ $(1 \le i \le n)$ constitute a *production* for e (denoted by $c_1, \ldots, c_n \vDash e$) if and only if the following properties hold:

- $c_1^{\mathrm{init}} \in \mathbb{A}_0$, and
- $e \in \delta(c)$ for some $c \in c_1 \triangle \cdots \triangle c_n$.

Recalling that $c_1^{\mathrm{init}} \in \mathbb{A}_0$ expresses that c_1 is enabled as a first 'step' in the computational argument, the definition captures precisely what it means for an item to be knowable (by computation).

Let $K_{\mathscr{B}} \subseteq \mathbb{E}$ be the set of all producible knowledge items. We will show that $K_{\mathscr{B}}$ is indeed a well-defined set. To this end, we first define the function $g : \mathbb{E} \to 2^{\mathbb{E}}$ as follows, for all $e \in \mathbb{E}$:

$$g(e) = \{e' \mid \text{there are computations } c_1, \ldots, c_n, c_{n+1} \in \mathscr{B} \ (n \ge 0) \text{ such that}$$
$$c_1, \ldots, c_n \vDash e \text{ and } c_1, \ldots, c_n, c_{n+1} \vDash e'\}.$$

Notice that $g(e) = \emptyset$ for any e that is not producible or in case it is and $c_1, \ldots, c_n \vDash e$, if no computation exists in $\mathscr{B}$ that can still be grafted onto c_n. The effect of g is that it extends the knowledge obtainable through productions of some length n to all knowledge producible by one computation more. Now consider the following set-theoretic operator $G : 2^{\mathbb{E}} \to 2^{\mathbb{E}}$:

$$G(K) = K \cup \bigcup_{e \in K} g(e).$$

One observes that the iterative procedure for generating all knowledge in $\mathbb{E}$ that can possibly be produced implies that $K_{\mathscr{B}} = \mathbb{E}_0 \cup G(\mathbb{E}_0) \cup G^2(\mathbb{E}_0) \cup \cdots$.

Theorem 1 $K_{\mathscr{B}}$ *is the least fixpoint of G that includes the core set* $\mathbb{E}_0$. *In particular,* $K_{\mathscr{B}}$ *is well defined.*

Proof Clearly $2^{\mathbb{E}}$ is a complete partially ordered set (cpo) under inclusion. By its very definition G is a monotone operator, but a stronger property can be proved:

Claim G is *chain-continuous*, i.e. if $K_1 \subseteq K_2 \subseteq \cdots$ and $\bigcup_{i \geq 1} K_i = K$, then $G(K_1) \subseteq G(K_2) \subseteq \cdots$ and $\bigcup_{i \geq 1} G(K_i) = G(K)$.

Proof By monotonicity one has $G(K_1) \subseteq G(K_2) \subseteq \cdots$ and $\bigcup_{i \geq 1} G(K_i) \subseteq G(K)$. To prove that $G(K) \subseteq \bigcup_{i \geq 1} G(K_i)$ as well, consider any e with $e \in G(K)$. As $G(K) = K \cup \bigcup_{e \in K} g(e)$, we can distinguish the following cases:

- $e \in K$: then there is an $i \geq 1$ such that $e \in K_i$. By monotonicity we obtain that $e \in G(K_i)$ and thus that $e \in \bigcup_{i \geq 1} G(K_i)$.
- $e = g(e')$ *for some* $e' \in K$: then there is an $i \geq 1$ such that $e' \in K_i$. It follows by definition that $e \in G(K_i)$ and again that $e \in \bigcup_{i \geq 1} G(K_i)$.

We conclude that $\bigcup_{i \geq 1} G(K_i) = G(K)$.   □

It now follows from the Tarski–Kantorovich fixed point theorem[1] that $K_{\mathscr{B}}$ is indeed the least fixpoint of G in the collection of all sets K with $K \supseteq \mathbb{E}_0$.   □

If a bundle is closed under (grafted) composition, then the characterization of $K_{\mathscr{B}}$ reduces to a much simpler form.

Corollary 1 *Let* $\mathscr{B}$ *be compositional. Then* $K_{\mathscr{B}} = G(\mathbb{E}_0)$.

Proof Let G be the operator as defined above. One easily verifies that the compositionality of $\mathscr{B}$ implies that $G^2(\mathbb{E}') = G(\mathbb{E}')$, for any $\mathbb{E}' \subseteq \mathbb{E}$. Hence, we obtain that $K_{\mathscr{B}} = \mathbb{E}_0 \cup G(\mathbb{E}_0) \cup G^2(\mathbb{E}_0) \cup \cdots = \mathbb{E}_0 \cup G(\mathbb{E}_0) = G(\mathbb{E}_0)$.   □

Corollary 1 shows that, if bundles are compositional, all knowledge that can be generated from $\mathbb{E}_0$ can be generated using at most *one* computation from the bundle. This may also be seen from the definition of compositionality directly. Compositionality is a strong property, but it can be expected to hold for all knowledge spaces that are based on a deductive theory.

Finally, the characterization of $K_{\mathscr{B}}$ allows us to define another important notion for knowledge generation by computation, namely *universality*. The concept is of key

---

[1] The Tarski–Kantorovich fixed point theorem states the following: Let $\langle X, \leq \rangle$ be a cpo and let $H : X \to X$ be chain-continuous. If there is an $x \in X$ such that $x \leq H(x)$, then $x' = \sup_n H^n(x)$ is a fixpoint and in fact the least fixpoint of H among all y with $y \geq x$. For a proof see e.g. [22]. Chain-continuity is also known as Scott-continuity.

importance in many branches of science and philosophy. In our approach here, we may use it to signify that the underlying mechanism is powerful enough to generate the entire knowledge space.

Definition 12 A bundle $\mathscr{B}$ is *universal* for $\mathbb{E}$ if and only if $K_{\mathscr{B}} = \mathbb{E}$.

In classical computability theory, universality refers to the property that all Turing machine programs can be simulated on one single (universal) Turing machine. However, in the approach here, the notion of simulation is completely avoided. This may lead to a possible answer to the quest for a *clear-cut notion of universality* as expressed by Abramsky [23].

5.2 Knowledge Recognition

Now consider the recognition problem, i.e. the problem of determining whether a given knowledge item $e \in \mathbb{E}$ is 'obtainable' from the core knowledge. Our aim will be to define recognition as a process, and show that this process is computational.

Before we get into this question, it should be noted that 'recognition' of knowledge can be of greater concern than generation. For example, recognition processes take place in natural systems such as found on the surfaces of cells and in cognition. One may argue that in recognition there is as much generation of knowledge going on as there is in any computation, except that the *usage scenario* differs. Let us make this more precise.

We want to think of recognition as a concrete computational process, working on an input datum e from some 'interesting' subdomain $D \subseteq \mathbb{E}$ and 'flagging' it as soon as the process finds that e is recognized. Typically, D will consist of items that have the right form but have to be tested for being valid knowledge, i.e. for being in $\mathbb{E}$. Clearly, when a recognition process is brought to bear on an item e with $e \notin K_{\mathscr{B}}$, one should allow for the indefinite behaviour alluded to before, notably when computational criteria are lacking for items not in $K_{\mathscr{B}}$.

It is well-known from classical automata and formal language theory [21] that the processes of recognition and generation are closely related. We show that this phenomenon emerges at the present, very general level as well. In order to make this concrete, we will show how to define recognition as a computational process in our framework, in a way that it is dual to generation.

The following definition expresses exactly what we expect from the recognition process, hiding all specificities of how the computations in a bundle work. For every $d \in D$, let d^+ be a (new) knowledge item expressing its positive recognition. Let $D^+ = \{d^+ \mid d \in D\}$.

Definition 13 A *recognizer* $\mathscr{R}$ for some domain $D \subseteq \mathbb{E}$ consists of the following components:

- An action space $\mathbb{B}$ and a knowledge space $\mathbb{F} \supseteq D \cup D^+$
- A semantic mapping $\mu : \mathbb{B} \to \mathbb{F}$

- Core sets $\mathbb{B}_0$ and $\mathbb{F}_0$ such that $\{\mu(x) \mid x \in \mathbb{B}_0\} \subseteq \mathbb{F}_0 = D \cup D^+$
- A computation bundle $\mathscr{S}$

$\mathscr{R}$ is said to *recognize* item $e \in D$ if there are computations $s_1, \ldots, s_n \in \mathscr{S}$ with $\delta(s_1^{\text{init}}) \in \{e, e^+\}$ and $s_1, \ldots, s_n \vDash e^+$. The set of all knowledge items from D recognized by $\mathscr{R}$ is denoted by $D_{\mathscr{R}}$.

We now show how a recognizer can be constructed from the computational, generative process that underlies $\mathbb{E}$. We assume that the items in $\mathbb{E}$ have a known form so they can be identified as reasonable inputs. Let $\mathbb{A}$ and $\delta : \mathbb{A} \to \mathbb{E}$ correspond to the computational mechanism for $\mathbb{E}$. Let $\mathbb{B}$ be the bundle we have available for it. Assume that $\{\delta(x) \mid x \in \mathbb{A}_0\} = \mathbb{E}_0$.

Theorem 2 *With the given conventions, a recognizer for the full set $K_{\mathscr{B}}$ can be constructed from the computational mechanism underlying $\mathbb{E}$.*

Proof We define the components of a recognizer $\mathscr{R}$ with $D \equiv \mathbb{E}$, using the generative process as follows:

- Let $\mathbb{B} = \mathbb{A} \times \mathbb{E}$, and let $\mathbb{F} \supseteq D \cup D^+$. Note that, if we supply $\mathbb{E}$ with the discrete (pointwise) topology and take the product with the topology of $\mathbb{A}$, then $\mathbb{B}$ is a topological space again (as required).
- Define the semantic map $\mu : \mathbb{B} \to \mathbb{F}$ in terms of δ as follows:

$$\mu([x, d]) = \text{' if } \delta(x) = d \text{ then } d^+ \text{ else } d\text{'}.$$

The map reflects the intention that, whenever an action item contains evidence that an item d is recognized, it is flagged.
- Let $\mathbb{B}_0 = \mathbb{A}_0 \times \mathbb{E}_0$ and $\mathbb{F}_0 = D \cup D^+$.
- In order to define $\mathscr{S}$ we do the following: For each $c \in \mathscr{B}$ and $d \in D$, let c_d be the curve $c \times \{d\}$, which is a curve in the product topology on $\mathbb{B}$. Let $\mathscr{S} = \{c_d \mid c \in \mathscr{B} \text{ and } d \in D\}$.

The construction specifies a recognizer for $D \equiv \mathbb{E}$, as desired. Moreover, by the assumption that $\{\delta(x) \mid x \in \mathbb{A}_0\} = \mathbb{E}_0$, it follows that the items d that can be recognized 'at the start', are precisely those of $\mathbb{E}_0$. Definition 13 implies that the further items that can be recognized are precisely those that can be generated. It follows that $D_{\mathscr{R}} = K_{\mathscr{B}}$. $\qquad\qquad\square$

By a similar argument one can show that a recognizer for a domain $D \subseteq \mathbb{E}$ which satisfies the specifications of Definition 13 can be 'moulded' into a generator for D. This would prove the *functional equivalence* of knowledge generation and recognition, now resulting from the philosophy of computation that we followed.

6 Conclusion

The question of how to characterize computation as an intrinsic notion is a complex one. While analogies to classical models of computation have proved quite satisfactory in the past, the spreading of the computing metaphor to natural systems has made those analogies far less convincing and productive. The question of defining computation adequately therefore remains an intriguing one. Can one capture computation in such a way that the forms of *computationality* as understood today are covered. Can new, so far unfathomed forms of computationality be identified?

In this paper we have followed up on the philosophy developed in [2, 3], in which computation is viewed as a process of generating knowledge. We have presented a theoretical framework in which computations are viewed as processes operating against the backdrop of suitable spaces of knowledge and actions. The framework is widely applicable and allows for a theory of computation which covers the wide variety of processes that are all regarded as computational, without any assumptions on how they work but focussing solely on *what* they do.

Computation is, in our theory, the generation of knowledge in action, with the help of a suitable underlying mechanism. The framework we developed does not require any concrete assumptions on representation, except that there is a 'natural' topology in the relevant action space so computations can be characterized as being *continuous* over the course of their existence. In the resulting framework, knowledge generation can be shown to be a well-defined process. Also knowledge recognition can be captured computationally, from a logico-epistemic point of view.

While the notion of computation has wide reach as intended, it will be of interest to test it on more cases than the current ones we used from conventional and unconventional computing. Philosophically intriguing boundary cases are plenty and can be found e.g. in cognition [24], the more general *computational theory of mind* [25] and in the even more general realm of *pancomputationalism* [26]. As an example one might consider the presumed computationality of the *Universe*. It could be seen as a system which evolves dynamically, producing (implicit) knowledge in the form of life, and life eventually produces explicit knowledge. See also [27] for an expansion on this theme. Hence one may view the meaning of life as being to compute, to produce knowledge and, eventually, wisdom.

Whether a phenomenon can be meaningfully viewed as computational depends on the frameworks and theories through which it is viewed. We posit that, if a process or system is to be viewed as computational, one should be able to characterize it as a knowledge-generating process in some perspective. Then, of course, the decision whether a process is computational becomes observer dependent. Nevertheless, in this way we have provided a 'test' for computationality with wider applicability than the previous tests based on analogies to classical computing systems.

Acknowledgements The work of the second author was partially supported by ICS AS CR fund RVO 67985807 and the Czech National Foundation Grant No. 15-04960S.

References

1. Farkaš, I.: Indispensability of computational modeling in cognitive science. J. Cognit. Sci. **13**, 401–435 (2012)
2. Wiedermann, J., van Leeuwen, J.: Rethinking computations. In: 6th AISB Symp. on Computing and Philosophy: The Scandal of Computation—What is Computation? AISB Convention 2013 Proceedings, pp. 6–10. AISB, Exeter, UK (2013)
3. Wiedermann, J., van Leeuwen, J.: Computation as knowledge generation, with application to the observer-relativity problem. In: 7th AISB Symp. on Computing and Philosophy: Is Computation Observer-Relative? AISB Convention 2014 Proceedings, AISB, Goldsmiths, University of London (2014)
4. Beck, L.W.: The actor and the spectator—foundations of the theory of human action. Yale University Press (1975) (Reprinted: Key Texts, Thoemmes Press, 1998)
5. Blass, A., Gurevich, Y.: Algorithms: a quest for absolute definitions. Bulletin EATCS **81**, 195–225 (2003)
6. Gurevich, Y.: Foundational analyses of computation. In: Cooper, S.B., Dawar, A., Löwe, B. (eds.), How the World Computes, Proc. CiE 2012. Lecture Notes in Computer Science, vol. 7318, pp. 264–275. Springer (2012)
7. Floridi, L.: The Philosophy of Information. Oxford University Press, Oxford (2011)
8. Searle, J.R.: Minds, brains, and programs. Behavioral Brain Sci. **3**, 417–457 (1980)
9. Searle, J.R.: Is the brain a digital computer? Proceedings and Addresses of the American Philosophical Association **64**(3), 21–37 (1990)
10. Tong, D.: The unquantum quantum. Sci. Am. **307**, 46–49 (2012)
11. Frailey, D.J.: Computation is process. In: Ubiquity Symposium 'What is Computation?', ACM Magazine Ubiquity, November issue, Article No 5 (2010)
12. Horsman, C., Stepney, S., Wagner, R.C., Kendon, V.: When does a physical system compute? Proc. Royal Soc. A **470**(2169), 20140182 (2014)
13. Horsman, C., Kendon, V., Stepney, S., Young, J.P.W.: Abstraction and representation in living organisms: when does a biological system compute? In: Representation and Reality: Humans, Animals and Machines. Springer, Heidelberg (2017)
14. Arbib, M.A.: Automata theory and control theory—a rapprochement. Automatica **3**, 161–189 (1966)
15. Mazurkiewicz, A.: Concurrent program schemes and their interpretation. Technical Report No. PB-17, DAIMI, Datalogisk Afdeling, Aarhus University, Aarhus (1977)
16. Scott, D.S.: Continuous lattices. In: Lawvere, F. (ed.), Toposes, Algebraic Geometry and Logic. Lecture Notes in Mathematics, vol. 274, pp. 97–136. Springer (1972)
17. Longo, G.: Some topologies for computations, invited lecture. In: Géométrie au XX siècle, 1930–2000, Paris. http://www.di.ens.fr/users/longo/files/topol-comp.pdf (2001)
18. Fodor, J.A.: The mind-body problem. Sci. Am. **244**, 124–132 (1981)
19. Piccinini, G.: Computation without representation. Philos. Stud. **137**(2), 205–241 (2008)
20. Piccinini, G., Bahar, S.: Neural computation and the computational theory of cognition. Cognit. Sci. **37**(3), 453–488 (2013)
21. Hopcroft, J.E., Ullman, J.D.: Formal Languages and their Relation to Automata. Addison-Wesley, Reading, MA (1968)
22. Ok, E.A.: Elements of order theory. Ch 6: Order-theoretic fixed point theory. https://sites.google.com/a/nyu.edu/efeok/books
23. Abramsky, S.: Two puzzles about computation. In: Cooper, S.B., van Leeuwen, J. (eds.) Alan Turing—His Work and Impact, pp. 53–57. Elsevier, Amsterdam (2013)
24. Pylyshyn, Z.W.: Computation and cognition: toward a foundation for cognitive science. MIT Press, Cambridge MA (1984)
25. Putnam, H.: Brains and behavior. Presented to the American Association for the Advancement of Science, section L (History and Philosophy of Science), 27 Dec 1961

26. Piccinini, G.: Computational modelling vs computational explanation: is everything a Turing machine, and does it matter to the philosophy of mind? Aust. J. Philoso. **85**(1), 93–115 (2007)
27. Dodig-Crnkovic, G.: Modeling life as cognitive info-computation. In: Beckman, A., Csuhaj-Varjú, E., Meer, K. (eds.), Language, Life, Limits, Proc. CiE 2014. Lecture Notes in Computer Science, vol. 8493, pp. 153–162. Springer (2014)

Abstraction and Representation in Living Organisms: When Does a Biological System Compute?

Dominic Horsman, Viv Kendon, Susan Stepney and J. P. W. Young

Abstract Even the simplest known living organisms are complex chemical processing systems. But how sophisticated is the behaviour that arises from this? We present a framework in which even bacteria can be identified as capable of representing information in arbitrary signal molecules, to facilitate altering their behaviour to optimise their food supplies, for example. Known as Abstraction/Representation theory (AR theory), this framework makes precise the relationship between physical systems and abstract concepts. Originally developed to answer the question of when a physical system is computing, AR theory naturally extends to the realm of biological systems to bring clarity to questions of computation at the cellular level.

1 Introduction

The language of information processing is widespread in biology. From DNA replication to nerve impulses to brain activity, systems are frequently talked of as storing and processing data, and even as performing intrinsic computation. It has previously been difficult, however, to argue that this is more than an analogy: is there, in fact, computation happening in biological systems? Is it possible to model such systems

DH published previously as Clare Horsman.

D. Horsman
Department of Computer Science, University of Oxford, Oxford, UK

D. Horsman · V. Kendon
Department of Physics, Durham University, Durham, UK

S. Stepney (✉)
Department of Computer Science and York Centre for Complex Systems Analysis,
University of York, York, UK
e-mail: susan.stepney@york.ac.uk

J.P.W. Young
Department of Biology, University of York, York, UK

© Springer International Publishing AG 2017   91
G. Dodig-Crnkovic and R. Giovagnoli (eds.), *Representation and Reality in Humans,
Other Living Organisms and Intelligent Machines*, Studies in Applied Philosophy,
Epistemology and Rational Ethics 28, DOI 10.1007/978-3-319-43784-2_6

as computations? Is it even the case that the ability to compute is so basic to living organisms that we can use it as a definition of life?

Biological and computational processes share many similarities, such as: the encoding of process data in proteins; signal transduction from input to processed output [3]; cells mimicking computers [22]; similarities between computer networks and biological distributed systems and viruses [18].

Computational biology aims to use these similarities to model biological systems computationally, with the dual aims of better understanding their basic processes, and of producing biological systems artificially. Computer simulations of biological systems have inspired key areas in machine intelligence [23, Chap. 8]. Organisms such as bacteria [26], slime moulds [1], and hybrid biological/silicon devices [8] have been closely studied as potential non-standard computational devices.

Here we make precise this previously informal relationship between biological and computational processes. We locate computing within the broader category of *representational activity*, and give conditions for when a biological system is making fundamental use of representation to perform a range of tasks including engineering, communication and signalling and computing. We use the formal framework of the recently developed *Abstraction/Representation Theory* (AR theory), introduced in [13] and extended in [11]. AR theory was introduced to give a rigorous characterisation of the relationship between abstract representation and physical system, primarily in the context of determining when a physical system is being used as a computer. It was developed with non-standard human-designed computational devices in mind, and has already been put to good use determining whether representational activity, including computation, is occurring in unconventional computing substrates [12, 14].

AR theory's ability to deal with representation as a whole, and computing outside standard silicon-based digital models, gives it the capacity to extend further to considerations of the computing/representational activities of organic systems. There are, however, challenges to using AR theory with respect to biological systems that do not occur when considering devices that have been deliberately engineered.

At the centre of AR theory is the *representation relation*, mediating between physical and abstract objects. This permits the encoding and decoding of abstract information in physical systems, as is necessary for communication and computing. Physical states are represented abstractly, and in certain tightly-defined situations this representation relation from physical to abstract can effectively be 'reversed' (by engineering the system) to an instantiation relation from abstract to physical. By this means abstract information can be instantiated in a physical system.

In the human-user context, the representation relation is both determined by and located in easily-identifiable intelligent and conscious *representational entities*, namely the human users taking part in scientific, technological or computational activities. A representational entity is required for any representational activity to take place, and is required to be a physical entity that is part of the representational system. This distinguishes, in AR theory, a system being used as part of representational activity, and one that is 'going about its own business'. This also allows for the distinction between a system being used as a computer, and one that is *post hoc* represented as computing: in the first case the representational entity is part of the system under consideration, and in the second it is not [13].

In natural biological systems, the challenge in identifying representational activity is both to identify the representational entity present, and to determine that representation is occurring within the system in the absence of a conscious or intelligent user who can inform us of this fact. The type of representational activity (engineering, communication, computing) is then a further property of the system to be determined.

Here we investigate representational activity (including signalling and computing) in low-level biological systems. We give methods for determining the presence of representational activity in the absence of high-level representational entities, and pose the question: how simple can a representational entity be? We find evidence for representation in systems far removed from conscious entities, and show that representation does not require structures as complex as a brain or collections of neurons. We see that key to determining representational activity (as opposed to 'manipulation of stuff') is the identification of *arbitrariness* within representation. That is, that the instantiation of information occurs in a one-to-many mapping between abstract and physical, so that the *same* outcome could have occurred using a *different* physical material or process. From the point of view of the biological process, it is the abstract process that is key in determining the correct physical outcome, not its particular instantiation.

We analyse three specific biological examples using AR theory to illustrate the presence and type of representational activity. By a close consideration of three biological processes—bacterial chemotaxis, the genetic code and photosynthesis—we show how the arbitrariness of information representation allows us to determine whether these systems are engaged in representational activity or not—yes for the first and second, no for the third—and we identify specific examples of computing happening in biological systems. We find that low-level biological systems use representation as an integral part of their behaviour and their interaction with their environment, and use this ability in certain situations to store and manipulate information in an equivalent process to that used by human-designed computers. Abstraction and representation can now be seen to be fundamental processes engaged in by most, maybe all, living organisms.

2 The Framework

AR theory was introduced in [13] and extended and formalised in [11]. These works should be consulted for the full physical, philosophical and formal background to the framework. It is a framework in which science, engineering/technology, computing and communication/signalling are all defined as *representational activity* requiring the fundamental use of the representation relation in order to define their operation.

AR theory was developed to answer the specific questions of when a physical system is computing [13]. This turns out to be a question about the relationship between an abstract object (a computation) and a physical object (a computer). What is needed is a formal language of relations, not from mathematical objects to mathematical objects (as is usual in mathematics and theoretical computer science), but between physical objects and those in the abstract domain. The core of AR theory is the representation relation, mapping from physical objects to abstract objects. Experimental science, engineering and computing all require the interplay of abstract and physical objects via representation in such a way that formal descriptive diagrams commute: the same result can be gained through either physical or abstract evolutions. The key result of [13] defined computing as *the use of a physical system to predict the outcome of an abstract evolution*.

2.1 Formalising Representation

AR theory identifies objects in the domain of physical systems, abstract objects (including mathematical and logical entities) and the representation relation which mediates between the two. The distinction between the two spaces, abstract and physical, is fundamental in the theory, as is their connection *only* by the (directed) representation relation. An intuitive example is given in Fig. 1: a physical switch is represented by an abstract bit, where in this case it is zero for up, one for down.

An example of an object in the domain of physical entities is a *computer*. It has, usually, internal degrees of freedom, and a physical evolution that connects initial input and final output states. An example of an abstract object is a *computation*, which is a set of objects and relations as described in the logical formalisms of theoretical computer science. Likewise, an object such as a bacterium is a physical entity, and its theoretical representation within biology is an object in the domain of abstract entities. In what follows, we use bold font to indicate where an object **p** or evolution **H** is physical; and italic font for abstract objects represented within equations, for example, in giving the abstract object $m_\mathbf{p}$.

The elementary *representation relation* is the directed map from physical to abstract objects, $\mathcal{R} : \mathbf{P} \to M$, where $\mathbf{P}$ is the set of physical objects, and M is the set of abstract objects. When two objects are connected by $\mathcal{R}$ we write them as $\mathcal{R} : \mathbf{p} \to m_\mathbf{p}$. The abstract object $m_\mathbf{p}$ is then said to be the *abstract representation* of

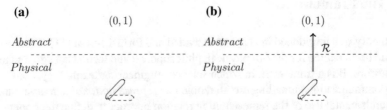

Fig. 1 Basic representation. **a** Spaces of abstract and physical objects (here, a switch with two settings and a binary digit). **b** The directed representation relation $\mathcal{R}$ mediating between the spaces

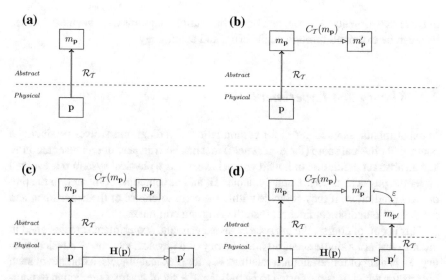

Fig. 2 Parallel evolution of abstract evolution (e.g. an algorithm) and potential physical computing device. **a** The basic representational triple, $\langle \mathbf{p}, \mathcal{R}, m_{\mathbf{p}} \rangle$: physical system $\mathbf{p}$ is represented abstractly by $m_{\mathbf{p}}$ using the modelling representation relation $\mathcal{R}_{\mathcal{T}}$ of theory $\mathcal{T}$. **b** Abstract dynamics $C_{\mathcal{T}}(m_{\mathbf{p}})$ give the evolved abstract state $m'_{\mathbf{p}}$. **c** Physical dynamics $\mathbf{H}(\mathbf{p})$ give the final physical state $\mathbf{p}'$. **d** $\mathcal{R}_{\mathcal{T}}$ is used again to represent $\mathbf{p}'$ as the abstract output $m_{\mathbf{p}'}$, $|m_{\mathbf{p}} - m_{\mathbf{p}'}| = \varepsilon$. (Adapted from [13])

the physical object $\mathbf{p}$, and together they form one of the basic composites of AR theory, the *representational triple* $\langle \mathbf{p}, \mathcal{R}, m_{\mathbf{p}} \rangle$. The basic representational triple is shown in Fig. 2a.

Similarly, abstract evolution takes abstract objects to abstract objects, which we write as $C : M \to M$. An individual example is shown in Fig. 2b, for the mapping $C(m_{\mathbf{p}})$ taking $m_{\mathbf{p}} \to m'_{\mathbf{p}}$. The corresponding physical evolution map is given by $\mathbf{H} : \mathbf{P} \to \mathbf{P}$. For individual elements in Fig. 2c this is $\mathbf{H}(\mathbf{p})$, which takes $\mathbf{p} \to \mathbf{p}'$.

In order to reach the next key concept in AR theory, we now apply the representation relation to the outcome state of the physical evolution to give its abstract representation $m_{\mathbf{p}'}$, Fig. 2d. We now have two abstract objects, $m'_{\mathbf{p}}$ and $m_{\mathbf{p}'}$. For some (problem-dependent) error quantity ε and norm $|.|$, if $|m_{\mathbf{p}'} - m'_{\mathbf{p}}| \leq \varepsilon$ then the diagram (Fig. 2d) *commutes*. Commuting diagrams are fundamental to the use of AR theory. If a set of abstract and physical objects form a commuting diagram under representation, then $m_{\mathbf{p}}$ is a *faithful abstract representation* of physical system $\mathbf{p}$ for the evolutions $C(m_{\mathbf{p}})$ and $\mathbf{H}(\mathbf{p})$.

The main reason why commuting diagrams are important, along with faithful abstract representations for physical systems, is that the final state of a physical object undergoing evolution can be known either by tracking the physical evolution and then representing the output abstractly, or by theoretically evolving the representation of the system. In the first case, the 'lower path' of a commuting diagram is followed; in the latter, the 'upper path'. Finding out which diagrams commute is the business of basic experimental science; and once commuting diagrams have been established they can be exploited through engineering and technology.

2.2 Theory and Experiment

In experimental science, a test for commutation of a diagram involves producing a controlled physical setup (the experiment) that has both an abstract representation $\mathcal{R}$ and an abstract prediction of how it will behave, C. The physical system $\mathbf{p}$ is evolved under the physical experimental dynamics $\mathbf{H}$, and the outcome compared to the theoretical prediction. If they coincide within the error tolerance of the experiment and the desired outcome confidence, then the diagram commutes.

This is not, of course, the *purpose* of an experiment. Experiments are designed in order to test not a single scenario but a *theory* of a physical system. A physical **theory**, $\mathcal{T}$, is a set of representation relations $\mathcal{R}_\mathcal{T}$ for physical objects, a domain of such objects for which it is purported to be valid, and a set of abstract predictive dynamics for the output of the representations, $m_\mathbf{p}$, $C(m_\mathbf{p})$. If a theory supports commuting diagrams for all scenarios in which it has been both defined and tested, then it is a *valid* theory. A physical system or device that is both well tested and well understood will in general have a large number of commuting diagrams supporting it.

2.3 Engineering

The representation relation defined so far is directed, from physical to abstract objects. This is *modelling*: giving an abstract representation of a physical object. The question can now be posed: is it possible to give a reversed representation relation, an *instantiation* relation? This will not be a basic relation in the same way as the ordinary (modelling) representation relation is basic: abstract representation can be given for any physical object (this is language), but there are plenty of abstract objects that do not have a physical instantiation ('unicorn', 'free lunch', etc.). Only in very specific circumstances can an instantiation relation $\widetilde{\mathcal{R}}_\mathcal{T}$ be given for a theory $\mathcal{T}$.

To find these circumstances, consider again the 'upper' and 'lower' paths of a commuting diagram, $(\mathbf{p}_0 \to m_{\mathbf{p}_0} \to m'_{\mathbf{p}_0})$ and $(\mathbf{p}_0 \to \mathbf{p} \to [m_\mathbf{p} = m'_{\mathbf{p}_0}])$ respectively. Between them, these paths describe the process of finding some $\mathbf{p}_0$ such that when it is subjected to the physical process $\mathbf{H} : \mathbf{p}_0 \to \mathbf{p}$ it becomes the physical system $\mathbf{p}$ whose abstract representation is $m_\mathbf{p}$. In other words, if both paths are present and form a commuting diagram, the theory $\mathcal{T}$ can be used to *engineer* system $\mathbf{p}$ from system $\mathbf{p}_0$ given a desired abstract specification $m_\mathbf{p}$: this is the *instantiation relation* $\widetilde{\mathcal{R}}_\mathcal{T}$, Fig. 3.

Fig. 3 Engineering the system **p**, using the instantiation relation $\widetilde{\mathcal{R}}_T$

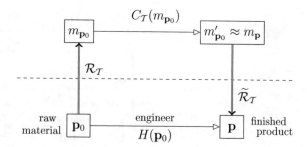

A use of the instantiation relation can be seen as a counterfactual use of the representation relation: which physical system, when represented abstractly, would give the abstract representation that we are trying to instantiate? The method by which it is achieved will vary considerably given different scenarios: trial and error, abstract reasoning, numerical simulation, etc. What connects these methods is that they are not straightforward: it is generally a skilful and cumulative process to reverse a representation relation.

2.4 Computation

A commuting diagram in the context of *computation* connects the physical computing device, **p**, and its abstract representation $m_{\mathbf{p}}$. It makes integral use of the instantiation relation: a computer is an engineered device. $m_{\mathbf{p}}$ can be a number of different abstract representations; a common one draws from the set of binary strings. The abstract evolution is then the (binary) program to be run on the computer, and the physical evolution is how the state of the computer changes during the program (change of voltages, etc.). The full commuting diagram describes the parallel evolution of physical computer and abstract algorithm, connected via the representation given by the theory of the computing device, $\mathcal{R}_T$.

The AR description of physical computing is not simply the parallel evolution of physical and abstract. The *compute cycle* starts from a set of abstract objects: the program and initial state that are to be computed. The most important use of a computing system is when the abstract outcome $m'_{\mathbf{p}}$ is unknown: when computers are used to solve problems. Consider as an example the use of a computer to perform the binary arithmetical problem $01 + 10$. If the outcome were unknown, and the computing device being used to compute it, the final abstract state, $m'_{\mathbf{p}} = (11)$, would not be evolved abstractly. Instead, confidence in the technological capabilities of the computer would enable the user to reach the final, abstract, output state $m_{\mathbf{p}'} = m'_{\mathbf{p}}$ using the physical evolution of the computing device alone.

This use of a physical computer is the compute cycle, Fig. 4: the use of a physical system (the computer) to predict the outcome of a computation (an abstract evolution).

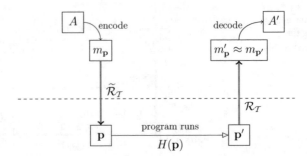

Fig. 4 Physical computing in which an abstract problem A is encoded into the model $m_{\mathbf{p}}$ of the computer $\mathbf{p}$, then instantiated into $\mathbf{p}$; the computer calculates via $\mathbf{H}(\mathbf{p})$, evolving into $\mathbf{p}'$, from which the representation relation is used to obtain the model $m_{\mathbf{p}'} \simeq m'_{\mathbf{p}}$, from which the output of the computation A' can be decoded

2.5 Encoding and Representation

An important element in the AR analysis of computing is the integral use of *encoding* and *decoding* in the compute cycle. These bear a close resemblance to the representation and instantiation relations, and we will see that in certain circumstances that they can be composed. However, unlike representation, encoding and decoding maps live entirely above the abstract/physical dividing line, and take abstract objects to abstract objects.

Encoding as the first step in the compute cycle embeds the computation to be performed into the abstract specification of the physical computer. This stage, frequently overlooked when analysing computation, is necessary in order to translate between the language of the problem specification (for human users, frequently linguistic) and that of the input interface of the device. Similarly, the decoding step is fundamentally necessary in order to translate the answer from the abstract representation of the end-state of the physical computer.

Encoding and decoding are, in general, fully composable: multiple encodings can be used within a system, and we can always consider a combination of encodings to be itself a single encoding (likewise for decodings). In other words, given a series of encodings γ_i, we can define the result of applying all of them in turn to be a single encoding, $\gamma = \gamma_1 \circ \gamma_2 \circ \dots$. Encodings can also be composed with representation: encoding the abstract representation of a physical system is equivalent to representing it in a different manner, $\mathcal{R}_{\mathcal{T}} \circ \gamma = \mathcal{R}'_{\mathcal{T}}$. A single representation can also generally be decomposed in this way into another representation (often a simpler one) and an encoding. Encodings can in this way often be dispensed with *notationally* by rolling them in to the definition of a representation; care must be taken, however, to ensure that important elements of diagrams are not thereby obscured, because encoding and decoding can come with significant computational overheads. The converse is not possible: representation cannot fully be replaced with encoding or decoding. At some stage in between the physical system and the abstract problem, a representation relation *must* be used to cross the line between abstract and physical.

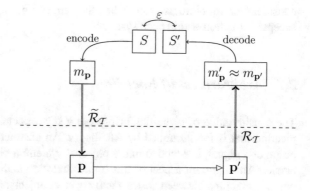

Fig. 5 A signal S instantiated into a physical carrier **p**

2.6 Signalling

Science, engineering and computing all build on each other within AR theory: faithful abstract scientific representations and good device theories are needed for engineering, and the instantiation relation and commuting engineering cycles are required in order to have a functioning computer. There is another important category of representational activity that sits alongside computing at the top of this stack: communication or, alternatively, signalling.

Signalling entails the encoding of a signal into a carrier, and then its decoding back to the original signal, within some ε error tolerance, Fig. 5. This demonstrates how it can be considered as a specific type of computation, where the abstract evolution to be 'predicted' is the identity operation (that is, the abstract output is equal to the abstract input); hence the diagram ends at the same abstract output, the signal $S \approx S'$, as it begins. The physical carrier starts in state **p** and ends in state **p′**; these states are typically separated either spatially or temporally. Furthermore, a given signal may be carried by several different physical substrates in the course of a single communication (for example, electrons in copper, and photons in fibre); what is important is that the correct *abstract* signal can be decoded at the end.

Signalling can therefore be viewed as a simpler form of representational activity than full computing, but requiring the same key elements of AR theory. Engineering is necessary to allow instantiation of an abstract signal in the carrier, and so good theories of the signalling device (and commuting 'science' diagrams) are required. These good theories may be discovered through the process of science, in the case of human-engineered signalling, or through evolution, in the case of *intrinsic* computation (see Sect. 3).

The blurred lines between communication and computation are familiar within Computer Science (where communication tasks between 'Alice' and 'Bob' can be used to perform some even quite complex computations). With AR theory, it becomes clearer that this boundary is similarly blurred when considering when biological system are performing signalling operations (a relatively common consideration for biologists) and when they are computing (a rather less universally accepted situation within such systems).

2.7 Arbitrariness of Encoding

The consideration of signalling brings out a key aspect of computing and communication that is foregrounded by AR theory. An abstract object, such as a computation or a signal, is encoded into a physical system: a physical computer or signal carrier. An important aspect of this implementation is that there is not a necessary one:one mapping between abstract and physical objects. A given signal may be carried by many different physical systems, a situation familiar to human language users (a word can be written, transmitted through speech, or through physical sign language, to name just a few possibilities). This also occurs frequently in the biological domain. For example some plants coevolve a communication language with other organisms; a particularly sophisticated case of this is where plants use chemical carriers to implement a signal to their predators' predators [24]; different plants use different chemicals.

The converse can also be true: a single physical system can be represented by multiple different abstract representations. These abstractions may be related in a particular manner that occurs frequently in Computer Science, where an abstract computation or communication is encoded into a physical computing/signalling device. There can be a number of models at different levels of abstraction, each representing the same physical computer/signal carrier, connected through a *refinement relation* [5, 10]. That is, the same abstract computational object has several more concrete encodings, each progressively more 'refined' or 'reified'. These are all abstract as opposed to physical, but some of them, the more concrete ones, are somehow deemed 'closer' to the physical system. That is, the eventual instantiation relation, that maps down from these abstract states to a physical implementation, is somehow 'simpler' or 'more natural' for these concrete representations.

In refinement theory, an abstract object can be refined in many different ways. For example, an abstract set may be refined to a (still abstract, but more concrete) list, array, linked list, or other data structure. A data word may be encoded in a string of bytes in big-endian or little-endian order in memory (most significant byte first, or last). Another example is how an alphabetical character may be encoded as a string of eight bits in ASCII or EBCDIC formats. When larger character sets that require more than eight bits for encoding are considered, many more possibilities exist.

It is refinement theory that allows us to connect the two seemingly different situations within computing/signalling of a single abstract object encoded in multiple physical implementations, and a single physical implementation supporting multiple abstract representations. It is simplest to consider these within a signalling scenario, Fig. 6.

Figure 6a shows a refinement stack: a single physical carrier $\mathbf{p}$ has abstract representation $m_{\mathbf{p}}$, which is a refinement through the function γ_2 of the abstract object n,

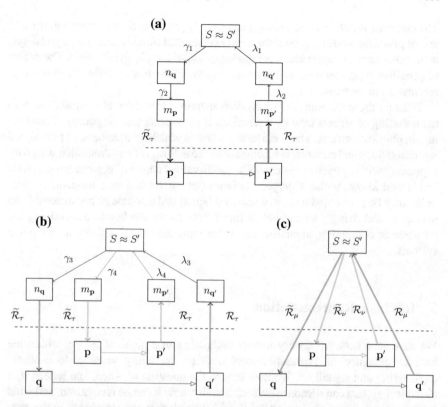

Fig. 6 Multi-valued representation and instantiation of a signal S: **a** one signal carrier with two different abstract representations connected by a refinement; **b** two different refinements of the same signal instantiated in two different physical carriers (the different colours indicate the distinct signalling cycles); **c** (**b**) redrawn as two representation relations connecting the signal carriers

which in turn is an encoding of the signal S through the function γ_1 (and similarly for the decoding stage). This is a single carrier being represented by different abstract models, both of which represent the same 'refined' signal S. Note that it may require considerable computation to implement these steps, including calculation and compilation (for encoding), and rendering (for decoding).

In comparison, Fig. 6b has the same signal, S, encoded in two different abstract models n (here given a subscript to form n_q), via refinement γ_3, and m_p, via the refinement γ_4. Each of these in turn is instantiated in a separate physical carrier, **p** and **q**. This is a single signal being transmitted by two different carriers. They are equivalent if $\gamma_3 \equiv \gamma_1$ and there is a composition of the refinements such that $\gamma_4 \equiv \gamma_1 \circ \gamma_2$. It is worth noting that there are situations where this equivalence does not follow, most notably when considering the heterotic systems analysed in [11].

Figure 6b is also equivalent within AR theory to Fig. 6c, where there is only representation between the signal and the abstract layer. As noted in Sect. 2.5, encodings and representation relations can always be combined to form a new representation.

The converse is not, however, always necessarily possible. Some systems may make use of *primitive representation*: the signal is encoded directly into the physical system, rather than via some abstract representation of the physical system. The notion of primitive representation will become important when we consider intrinsic representation in systems.

What all these different situations demonstrate is that there is an arbitrariness to the encoding of signals (and by equivalent, if more complex, diagrams, of computing) in physical carriers. There is a strong sense in which arbitrariness, of both signal representations and carriers, is a *hallmark* of these forms of representational activity happening within physical systems. The multi-valued nature of abstract *versus* physical is well-known within Computer Science (where, for instance, the same computation may be performed on both a standard laptop and a computer constructed from beer cans and string); we see below that it also forms the key to determining the presence of computing, signalling and other representational activity in biological systems.

3   Intrinsic Representation

We have seen how AR theory locates computing in a physical system within the broader category of *representational activity*; including science, technology/engineering and signalling. We now turn to the question of when, and indeed if, a biological system can demonstrate such activity, how it can be recognised, and what considerations are needed to extend the framework that was developed in the context of human-centric representational activities to systems without such organisms present. This requires us to extend the range of key concepts within the framework. In this section we give the theoretical extensions and insights necessary for the identification of representational activity in biological systems, giving the framework within which discussion of specific examples then takes place in section Sect. 4.

3.1   Representational Entities

The first consideration when analysing computing in biological systems is the ability of the system to be performing any representational activity at all. In AR theory, the ability to represent depends on there being a representational entity. Originally termed a 'computational entity' in [13], this is an entity capable of establishing a representation relation, and capable of encoding and decoding abstract information in physical systems. It physically locates the representation relation, and is the entity that performs abstraction, and uses the output of representational activity (including computation). If there is representational activity then there is always *something* that is performing it.

When considering standard computing, the representational entity is almost always a human being: the computer designer, or programmer, or user, given the representational cycle being determined. It locates and generates $\mathcal{R}_\mathcal{T}$, bridging the gap between abstract and physical. We now introduce some new terminology that is necessary for discussing computing outside this scenario. If a physical object **p** participates in a representational cycle (science, technology, computing, signalling) with a representation $\mathcal{R}_\mathcal{T}$ given by representational entity **e** (bold font as the representational entity must be physical), then we say that the system comprising $\{\mathbf{p}, \mathbf{e}, \mathcal{R}_\mathcal{T}\}$ forms a *closed representational system*. If the cycle is a compute cycle, then the set forms a *closed computational system* (and similarly for signalling, etc.). If the computational system $\{\mathbf{p}, \mathcal{R}_\mathcal{T}\}$ does *not* include the physical representational entity **e**, we say that it is *open under representation*.

In standard computing scenarios, the computer (e.g. a laptop) does not form a closed representational system: the representational entity is separate from the computing device, and not even necessarily co-located with it. This is an example of *designed* computing. Biological systems can compute/signal in exactly this way when they are used by a human (or otherwise) entity to perform a computation. Examples include DNA computing [2] and the use of slime molds to compute shortest paths [1]. In these cases, **p** is the biological system, **e** the human experimenter/programmer, and $\mathcal{R}_\mathcal{T}$ the representation they have predetermined. This is another example of designed computing, fitting entirely within the field of non-standard or unconventional computing devices, and whose analysis will exactly mirror those given for, e.g. chemical and quantum computers [13].

The core of the present investigation, however, is whether, and if so *how*, a biological system can be said to be computing *intrinsically*. Is there a meaningful way within AR theory to describe a biological process in the absence of any human computer users or experimenters as computing, or indeed performing any other representational activity? AR theory gives us a straightforward way to phrase this question. We require a closed computational system to be present in order for computing to be occurring, $\{\mathbf{p}, \mathbf{e}, \mathcal{R}_\mathcal{T}\}$. In designed computing, **p** and **e** are separate physical objects; for example, a biological system is being used as a computer. If, however, the biological system itself is closed under representation then the system is performing *intrinsic* computation. In such a case, the physical system **p** (for example 'a leaf' or 'a bacterium') includes within it the representational entity, and the system $\{\mathbf{p}, \mathbf{e} \subseteq \mathbf{p}, \mathcal{R}_\mathcal{T}\}$ forms a closed computational system. The first step in identifying intrinsic computing is then to determine whether a biological system under consideration is acting as its own representational entity.

It is important to note immediately that AR theory itself gives no requirements as to the level of complexity that a representational entity must have; only that it is a physical and not an abstract object. Importantly, there is no requirement that representation activity needs to take place in the presence of conscious, sentient, or intelligent agents. This is crucial if we are to investigate the presence of computing (or other representational activity) in low-level biological systems. The presence of representational entities, and of representation itself, is the key element in AR theory's ability to give a meaningful answer to the question of when a biological system is computing. Without an understanding of the crucial role of representation, answers given previously have been of only two types, neither particularly interesting: either that computation is an activity purely of conscious beings, and therefore any resemblance to biological processes is entirely misleading, or else that everything is computing at all times, not only human beings and bacteria, but rocks and subatomic particles. By focussing on the presence or otherwise of representation and representational entities, AR theory enables us to make important and meaningful distinctions between when a system is computing (or otherwise representing) and when it is merely 'going about its business'.

Non-intelligent representational entities do, however, pose a particular problem that rarely arises in human-computer interactions. When the representational activity is being performed by humans (engineering, technology, computing, etc.), the representational entity is generally both obvious and articulate. It is therefore usually straightforward to see that representation is happening (for example, a computer user can say that they are using a computer to perform a certain calculation), and to determine the representation relation (for example, the computer user tells us how they are encoding data on their laptop). In the absence of the ability to interrogate the representational entity, however, some other method is needed to determine whether representation is occurring within the processes under consideration and, if so, in what way it is being put to use. This requires careful analysis to avoid the most obvious pitfall: it is always possible for an external observer (e.g. us) to impose a *post hoc* abstract representation on the physical system that is not itself part of how the system is operating. We must avoid smuggling ourselves in as representational entities 'through the back door'; this results in a situation where everything, including rocks, compute. Computation then becomes a word without meaningful content. Instead, we must look at the representation happening within the system itself: do we as scientists and external observers represent that system itself as participating in representational activity intrinsic to itself, forming its own representational entity; and, if so, what is being represented?

3.2 Signatures of Representation

The key to determining the presence of intrinsic representational activity is the understanding that when a system is 'going about its business' as normal, its dynamics are given by the physical system itself. The physical system $\mathbf{p}$ is the important element. If, however, it includes representation, then at some point the important element will be an abstract object $m_{\mathbf{p}}$. When considering intrinsic computation or signalling, abstract representation will always be an intermediary process: it always results in *physical* behavioural changes. What is seen from the outside is always some physical process: all data and information are embodied. In the absence of entities (such as conscious humans) that can use an abstract result (as in the human use of computers), the use of representation in biological systems will be to produce physical output. For example (see Sect. 4.1), consider the system comprising a bacterium in a chemical gradient. The output of interest might be the observed part of this system that describes the bacterium's movement: swimming versus tumbling. As another example (see Sect. 4.2), consider a system comprising a cell containing DNA and other molecules. The output of interest might be the observed part of this system that is the protein expressed from a gene. The question is whether that output has been produced via the irreducible use of compute/signal cycles along the way.

Both representation (from physical to abstract) and instantiation (abstract to physical) are one:many maps. That is, a given physical system has many possible abstract representations (under different relations), and a given abstract object has many possible physical instantiations. If a physical object *qua* physical object is important in a biological process, then no other object will do: if $\mathbf{p}$ itself is needed in a process, then some other $\mathbf{p}_1$ will not do. However, if it is the *abstract* object and evolution that matters, then it is $m_{\mathbf{p}}$ that is needed; and there will in general be multiple ways in which this abstract object can be instantiated in a physical system.

It is precisely this multiplicity of mappings that is described in Sect. 2.7 in the context of computing and signalling. They are complex types of representational activity, but the arbitrariness of encoding and decoding identified there has its roots in the arbitrariness of representation, and the more broad type of one:many mapping that is present in all representational activity. In computing, it allows for the possibility of forming multiple alternative compute cycles. More generally, it gives many implementations of a single abstract process.

This is the key to analysing when a biological system is computing intrinsically: if a certain process can or could occur in multiple different ways, all of which instantiate a single abstract object or evolution, then this is the signature that it is the abstract and not the physical operation that is important. Can an arbitrariness of encoding and decoding within a given process be identified; and, from this, can it be seen that the given cycle could be implemented in multiple different physical systems with the same physical outcome at the end?

In order to make this precise, we here concentrate on the situation of intrinsic signalling within systems. As noted above, signalling can be viewed as a simplified form of computing; and within AR theory they share almost all their key elements. An ability to identify signalling and communication within biological systems is an important stepping-stone to an eventual understanding of when such systems compute. If signalling is present in a biological organism, then according to AR theory it has almost all the relevant components in order also to compute intrinsically.

Consider again the arbitrariness of encoding in signalling in designed representational systems, Fig. 6. This figure starts and ends in the abstract domain with the abstract signal S. By extending this to the intrinsic case, we give the equivalent signatures of signalling happening where the representational entity is not available for comment. In this situation, what is observed external to the system is an evolution of the whole system (e.g. a plant or a bacterium) from state $\mathbf{p}$ to $\mathbf{p}'$. By itself there is no way to determine if abstract objects or processes are necessary to this process. However, if there is a part of this evolution that is being implemented by the use of a signal, then there will be an identifiable use of representation within the system, as shown in Fig. 7. Let $\mathbf{q}$ be some physical subsystem of the overall physical system $\mathbf{p}$;

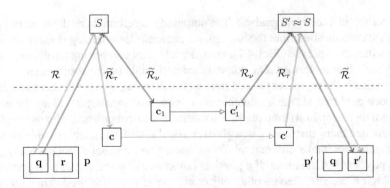

Fig. 7 An intrinsic use of signalling, with two possible signal carriers $\mathbf{c}$ or $\mathbf{c}_1$ as an intermediary in the physical evolution $\mathbf{p} \rightarrow \mathbf{p}'$. The physical elements $\mathbf{q}$ and $\mathbf{r}$ are subsystems of the system $\mathbf{p}$. The different colours denote the two separate signalling pathways

this is the part of the system that is responsible for initiating the signal. The intrinsic use of signalling involves an initial representation $\mathcal{R}$ of $\mathbf{q} \subset \mathbf{p}$ as the abstract signal to be sent, S. This is an instance of what is noted in Sect. 2.7 as *primitive representation*: the organism itself has no 'abstract model' of itself, so there is no intermediate step between the physical system and the abstract signal that is being encoded.

In order for the signal S to be transmitted, it is instantiated via $\widetilde{\mathcal{R}}_\tau$ in the physical signal carrier $\mathbf{c}$. This physical carrier evolves to $\mathbf{c}'$: standardly, it is transmitted with no change in its state other than time or space coordinates. $\mathbf{c}'$ is represented as signal S' via $\mathcal{R}_\tau$. If the transmission is faithful, $S' \approx S$. Now consider $\mathbf{r}$, the part of the system $\mathbf{p}$ that is receiving the signal. By instantiating S' in the new state $\mathbf{r}'$, the final state of the full system $\mathbf{p}'$ is produced.

Figure 7 also shows an alternative pathway for the signal; it is the identifiable possibility of these alternatives that gives us the signature of signalling happening in these systems. The signal carrier $\mathbf{c}$ is not the only possible transmission system: precisely as in Fig. 6, the signal S could also be instantiated using $\widetilde{\mathcal{R}}_\nu$ into a different physical system: here, $\mathbf{c}_1$. By decoding at the other end using $\mathcal{R}_\nu$, the same change to subsystem $\mathbf{r}$ is instantiated, and the same physical evolution $\mathbf{p} \rightarrow \mathbf{p}'$ is effected. The system has evolved this way through the fundamental use of representation: the different ways in which signal has been transmitted have in common only that they are instantiations of a given abstract object S.

The key then is that the semantics of the physical process that uses signalling as an intermediary are given by the abstract signal S, not the specific physical carrier $\mathbf{c}$, which is just one of many that could have been used to perform the critical abstract operation. The presence of other signal carriers is a strong sign that representation is occurring fundamentally. This is not however an immediately sufficient criterion. In any given situation a close analysis of the actual signalling pathways will need to be done to back up this hypothesis. This requires a detailed understanding of the biology of the systems under consideration, and a close interrogation of the best scientific understanding of each individual system in order to identify all the separate levels of encoding, decoding and representation that must be present for such representational activity. We turn to examples of such a detailed analysis in the next section.

This is then our challenge for identifying representation in biological systems: determining where there is an arbitrariness in the physical process (another type of physical system could perform the same operation), but where all the possible processes are physical instantiations of a single abstract object or process. In such a case, the physical objects and processes are functioning to instantiate abstract processes, and we argue that the biological system is engaging in signalling; a stepping-stone towards a further analysis of systems where, in certain cases, the biological system is *computing*.

4 Representation in Biological Systems

We now consider in detail three example biological systems. We will look specifically at them as candidates for the presence of the intrinsic use of signalling in biological organisms. This is a first stage towards, and proof-of-principle of, how to consider computation (the most complex of the representation activities considered here) in biological systems. As noted in the introduction, we focus on relatively low-level systems far removed from the degree of organisation and complexity needed for a neural/conscious organism. We consider specific processes within the candidate systems, and interrogate them for the presence of representation via the multiple realisability of information. We identify both representation and computation present even at these low levels and contrast it with an example where representational activity is not present.

4.1 Bacteria

Our first example is of bacteria that possess a motor-driven propellor called a flagellum that enables them to move around in a watery environment. The biological detail in this section is taken from the review article [21].

Such bacteria will swim towards food using their flagellum. The control of the flagellar motor provides a clear example of computation by a biological system, in which signals are processed and integrated through a series of interactions between proteins. Receptor proteins that protrude through the cell membrane detect the level of nutrient outside the cell. This information is passed via the linker protein CheW to CheA. If the nutrient level is declining, CheA transfers a phosphate group to CheY, which moves through the cell and binds to a component of the flagellar motor. If enough CheY is bound, the motor will switch direction from anticlockwise, which drives the cell forward, to clockwise, which causes the cell to tumble, changing its orientation (See Fig. 8).

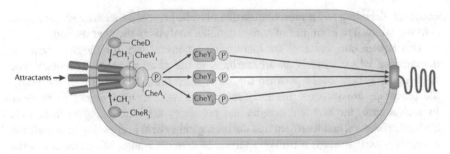

Fig. 8 Some signalling pathways in *E. coli*. Taken from [21]

In order to determine whether nutrient levels are declining, the system requires a memory, so CheA also adds a phosphate to another protein CheB, which takes methyl groups off the receptor proteins, reducing their ability to activate CheA. This happens on a slower timescale, providing a 'memory' of the average of recent conditions. The net result of all this protein chemistry is that a bacterium that experiences steady or increasing nutrient concentrations keeps swimming in the same direction, but if it is heading into a region with less nutrient, it stops, spins round and tries a different direction at random.

The system just described has been investigated in detail in the well-known bacterium *Escherichia coli*, but many other bacteria have variants of the same system. In *Rhodobacter sphaeroides*, there is an additional set of internal receptors that monitor the level of nutrients inside the cell and modulate the system appropriately. In other words, before chasing food, it asks itself 'am I hungry?' and integrates that information with the external nutrient situation to determine its behaviour. Although these two bacteria use related systems to process the information, the actual nature of the nutrients detected is different: *E. coli* senses amino acids and sugars, whereas *R. sphaeroides* is attracted to acids such as acetate. Hence, the same internal protein-based process represents different external realities. The arbitrariness of the representation is further emphasised by a consideration of more distantly related bacteria such as *Bacillus subtilis*, in which the meaning of the pathway is inverted because it has a different 'memory' mechanism, so that the phosphate-bound form of CheY promotes smooth swimming, rather than tumbling. Furthermore, the same basic information processing system has been adapted to handle different kinds of input and output. Besides attractants and internal nutrient status, inputs can include repellants, oxygen and light, while the system can be used to control other complex behaviours such as developmental gene expression, cyst or biofilm formation.

This bacterial control system makes complex and integral use of representation through signalling, and even some signal processing. The receptor proteins serve as transducers that convert various external inputs into an arbitrary internal representation. This representation is then manipulated, integrating information (including a memory of past states) to generate an output signal that then leads to action by the flagellum or other transducers. Each bacterium has multiple receptors—sometimes

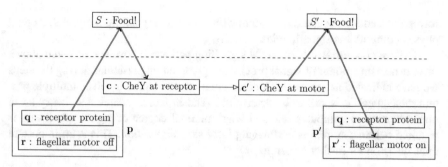

Fig. 9 Signalling in the *E. Coli* bacterium

dozens—and may have several pathways operating in parallel with different inputs and possibly different outputs. By comparing different signalling pathways across different bacteria, as well as within a single instance, we can see the arbitrariness of the signal carrying substrate in action.

To see in detail how this system satisfies the AR category for signalling we consider again the specific example of *E. coli*, comparing the schematic diagram of its signalling pathways, Fig. 8, with that for intrinsic signalling in AR theory, Fig. 7. The bacterium itself is the full system **p**, and we isolate the three subsystems needed for an AR description of signalling. The receptor proteins form the subsystem initiated the signal, **q**. The CheY proteins are the signal carriers **c**, and finally the flagellar motor forms the subsystem receiving the signal, **r**.

The AR signalling schematic for the system is shown in Fig. 9. The part of the full bacterial system **p** that senses the external environment is **q**, the receptor. This is represented as a signal ("nutrient level rising", or, more simply, "food!"). The signal to be sent is instantiated as the CheY proteins, sent from the receptor to the flagellum. The protein at the flagellum represents "food!", which is then instantiated as the required action of another subsystem of the bacterium: the flagellar motor **r**.

We have drawn Fig. 9 using single instantiation and representation steps for simplicity (see Fig. 6b, c). However, in this system these steps are non-trivial. First, a single signal *S*, indicating rising nutrient levels, is instantiated into multiple carrier protein molecules: it is the number of CheY which determines the signal. Then representation back to the signal *S'* combines the multiple molecules of the signal carriers into a single signal.

The multi-instantiation within this system is seen in the different ways in which the signal can be instantiated in different carriers. Different proteins could be used, as seen in different bacteria, and the essential role of the instantiation and representation steps make plain the central role of the CheY proteins as carrying the signal rather than anything else intrinsic to their physical state. The proteins themselves do not 'carry information' in isolation; it is their concentration that determines whether a signal is transmitted and received. Only in the context of the flagellar motor receptors does the number of CheY proteins form a representation of a signal. Without that representational context, the signal would not be transmitted. As it is, the bacterium

D. Horsman et al.

forms a closed representational system with the use of signalling integral to its final physical output: it is signalling intrinsically.

We have concentrated here on the signalling pathways; however, as noted above, there is also an amount of signal processing going on, in particular during the steps we have indicated as representation/instantiation (the signal involves multiple protein molecules). It is still to be determined (and in fact pre-dates AR theory as an open question in computer science) what minimal degree of signal processing is required before a system is performing intrinsic *computation*. This is an important open question for further research.

4.2 DNA

We now turn to the area of biology where the language of information, encoding and decoding is arguably the most used: DNA systems as they store and transmit the data used for the construction of cellular life. There are a number of redundancies in the way information is coded in DNA, demonstrating a key instance of the use of intrinsic representation in biological systems.

DNA is a heteropolymer, a sequence comprised of four different nucleotide bases: adenine, cytosine, guanine and thymine. Triplets of bases (called codons) encode, or represent, particular amino acids, and sequences of triplets represent sequences of amino acids, or proteins. The genetic code is the particular mapping between base triplets and the corresponding amino acids; see Fig. 10.

UTT	Phe	TCT	Ser	TAT	Tyr	TGT	Cys
TTC	Phe	TCC	Ser	TAC	Tyr	TGC	Cys
TTA	Leu	TCA	Ser	TAA	Stop	TGA*	Stop
TTG	Leu	TCG	Ser	TAG	Stop	TGG	Trp
CTT	Leu	CCT	Pro	CAT	His	CGT	Arg
CTC	Leu	CCC	Pro	CAC	His	CGC	Arg
CTA	Leu	CCA	Pro	CAA	Gln	CGA	Arg
CTG	Leu	CCG	Pro	CAG	Gln	CGG	Arg
ATT	Ile	ACT	Thr	AAT	Asn	AGT	Ser
ATC	Ile	ACC	Thr	AAC	Asn	AGC	Ser
ATA*	Ile	ACA	Thr	AAA	Lys	AGA*	Arg
ATG	Met	ACG	Thr	AAG	Lys	AGG*	Arg
GTT	Val	GCT	Ala	GAT	Asp	GGT	Gly
GTC	Val	GCC	Ala	GAC	Asp	GGC	Gly
GTA	Val	GCA	Ala	GAA	Glu	GGA	Gly
GTG	Val	GCG	Ala	GAG	Glu	GGG	Gly

Fig. 10 The genetic code. How the 64 DNA codons map to 20 amino acids and a stop signal. This code is almost universal. But in vertebrate mitochondria, the starred codons code differently

Fig. 11 Production of amino acids through translation from mRNA codons via tRNA within a ribosome. Taken from [17]

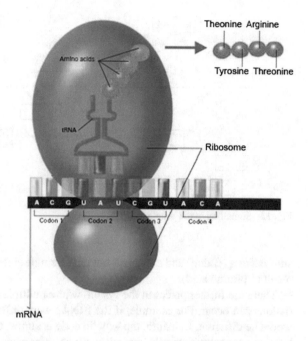

The transfer of information from DNA to protein happens in two stages. First, a copy of the DNA is made in RNA, a molecule that is chemically different from DNA but has the same structure. This messenger RNA (mRNA) represents the DNA sequence base for base. It is the same text, just in a different font, so the process is 'transcription'. The translation from codons to amino acids is realised by a small molecular machine called a tRNA (transfer RNA). This has two 'ends', one of which binds to a codon of the mRNA, and the other to an amino acid, Fig. 11. The tRNAs are the dictionary entries that map the 'words' in the language of DNA into the protein language, so this is aptly called 'translation'. This translation exhibits the first of the levels of redundancy: in most cases, multiple codons are translated to a given amino acid.

As with the case of bacteria, we can analyse this system from within the framework of AR theory. As in the previous example, the output of the AR cycle is a physical system: the amino acid produced. This is comparable with the bacterial output, which was the specific motion of the flagellum in the direction of food. While this physical output is specific, much of the intermediate stage includes the arbitrariness that we see in the use of intrinsic representation. The system begins with the "physical encoding" step, where the structure of the codon is instantiated in the tRNA. This is similar to the first stage of signalling within a bacterium.

Figure 12 illustrates this situation in terms of signalling in AR theory. The signal S is "make a given amino acid". A codon represents an amino acid; this representation is instantiated in a tRNA. Different codons, with their associated tRNAs can represent and instantiate the *same* amino acid. In this system, there are multiple pairs of

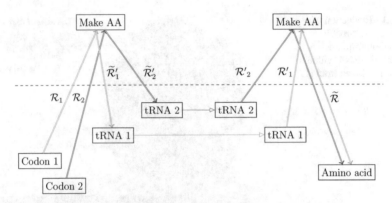

Fig. 12 Signalling in the genetic code

subsystem **q** (codon) and **c** (tRNA), but only a single signal S and the physical output result **r′** (amino acid).

There are further places in the system where multiple representations and instantiations can occur. For example, if the tRNAs were different, then the genetic code would be different. In nature, the genetic code is almost universal across organisms, but there are some naturally occurring minor differences in the genetic code. In vertebrate mitochondria, TGA is not a stop codon, but codes for tryptophan (Trp); ATA does not code for isoleucine (Ile), but for methionine (Met); AGA and AGG become stop codons [19]. That is, these codons represent different information in different contexts. Other organisms have reassigned codons in many other ways. Similar styles of changes have also been engineered within synthetic biology into the genetic code, for example in order to add an extra amino acid [25]. One could also, in principle, have a very different code, by altering all the tRNAs, and then altering the DNA so that these new tRNAs still produce the original proteins.

Even without changing the genetic code itself, there is a certain arbitrariness to the representation. For example, the DNA has to be read three bases at a time to form codons, but where to start? This is defined by the *reading frame*. Some DNA in viruses and mitochondria supports multiple reading frames [6]. So the same piece of DNA can represent different proteins, corresponding to different reading frames.

The underlying nucleotide bases structure of DNA has been extended with further 'unnatural' base pairs, extending the A-T C-G pairs [27]. Such extended DNA has been included in a plasmid, inserted into *E. coli*, and successfully replicated [15]. Potentially, new nucleotides can expand the codon dictionary and be translated into novel amino acids [4].

4.3 Photosynthesis

As a final example, we consider the energy transport processes in photosynthesis. It is a complex process involving specialised cell structures in a cascading sequence of excitations until the electromagnetic energy of the incoming light has been converted into chemical energy in molecules in the cells. This is accomplished by using an incoming photon (a quantum unit of light) to create an exciton (a quantum quasi-particle) in the light-harvesting molecules. Roughly speaking, an exciton is what you get when several electrons are excited into higher energy levels in a collective way. This then needs to be transported to the molecular reaction centre, where the chemical reactions can take place to store the energy. During transport, the exciton moves through the molecules, losing a fraction of its energy to vibrations in the molecules. This energy loss allows the dynamic process of energy transfer to proceed. A lower energy exciton will fit into neighbouring electron excitations better than the current ones, so the excitations transfer. This clearly requires exquisitely tuned molecular structures, and these have been intensively studied to understand the process in detail [7, 9].

Photosynthesis has caught the attention of the quantum information community, who have contributed insights to our understanding of how exactly this process works so well [16, 20]. As anyone who has ever suffered sunburn will know, photons from the sun can be dangerous. They are capable of breaking chemical bonds and thus damaging essential cellular machinery. It is crucial that the captured incoming photon is safely transferred to the reaction centre without causing unwanted rearrangement of the cell's molecules; energy is conserved, so it has to go somewhere once inside the cell. Achieving this high transfer efficiency is best explained using quantum mechanics to describe how the collective excitations maintain *coherence* to transfer the energy step by step. These processes are well-studied in the quantum information context, enabling fragile quantum information to be transported and stored, or processed in quantum computers.

Within the context of photosynthesis, however, these collective excitations are not being *used* by a leaf (or other photosynthesising system) to store or process quantum information. There has been an element of confusion present in the quantum information community since the discovery that this transmission can be modelled as a quantum walk [16], which is a process that can be used in quantum computing, and which can find the shortest path between two points faster than a classical random walk. Problems arise when these results are mistakenly seen to demonstrate that an exciton is 'quantum-computing' the shortest path to the reaction centre.

We can see within AR theory why it is not the case that a light-harvesting complex is computing, or indeed engaged in any other representational activity. The aim of photosynthesis is the transmission of the exciton: the semantics of this process can be understood entirely 'below the line' as a physical process transmitting energy, without needing to describe the system as intrinsically representing anything. There are multiple different ways that this process could be implemented, with different energy carriers, but they all involve the carrying of the energy in the exciton, not storing or

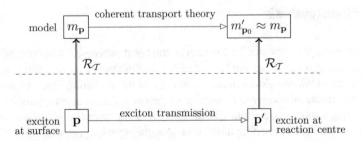

Fig. 13 A commuting 'science' diagram (c.f. Fig. 2d). Note we have a representation relation at both the beginning and end of the physical evolution. There is no instantiation relation, and hence no computing occurring here

processing any information. There is no multiplicity of instantiation of an abstract object (a signal or other information) created intrinsically by the light-harvesting complex itself. We can draw a 'science' diagram of our external representation of the process (Fig. 13), but not a signalling diagram involving intrinsic representation and instantiation of a signal.

As always, the key is to look to our best theories of the process under consideration: do they represent the system as itself representing information, or is that a description that can be given only *post hoc*, in the presence of external representational entities? In photosynthesis, a full description of the process of energy transmission requires only the physical elements of the exciton. If there is some information being computed that is essential for the process (such as a shortest path), then we would expect multiple ways this could be instantiated for a given exciton. However that is not the case: there is no mechanism for a given exciton to get the 'information' about the shortest path to the reaction centre by using a different physical mechanism. The specific exciton being transported has to undergo quantum coherent transmission, and nothing else. There is no signature of representational activity present intrinsically.

Photosynthesis is an example of the problems that arise with a failure to distinguish between intrinsic and imposed representational activity in natural systems. A light-harvesting complex plus photon/exciton considered in terms of computing a shortest path is not a closed representational system: $\mathcal{R}_T$ for this system would need to be set up by external representational entities (e.g. physicists or biologists looking at the system in this way). There is no evidence of the system itself using this representation of the process in order to complete the energy transport. In fact, even with an external entity imposing a representation in terms of information processing, a single instance of photosynthesis is still not computing: the representation is imposed *post hoc* as a parallel description of the physical process, but the physical system is not being used to predict the outcome of any abstract evolution. In the absence of a prediction element, the system does not satisfy the conditions for computation. Only if a photosynthesising organism were engineered into a larger computing system, and the path of the exciton used to encode some other abstract object, would the process of photosynthesis be computing. As it stands, it is a purely physical process, with no evidence for any intrinsic representational activity.

5 Summary and Conclusions

We have described in some detail our abstraction-representation framework in which representational activity such as signalling or computation can be identified in simple living systems that are far below the sophistication of nervous systems or consciousness. Crucial to identifying representation is the arbitrariness of the systems used to represent the information concerned, information such as 'hungry' or 'food molecules detected'. However, since in many cases the same molecules can be processed both as a source of energy (food) and as triggers for behaviour modification (swim towards food), identifying such systems as representational requires a detailed analysis of the processes. Often this means studying a whole class of organisms to draw out common features that are united by the representational activity. That life uses arbitrary representations at a fundamental level is most clearly illustrated by the variations in the encoding of amino acids by DNA. Conversely, there are many systems that are dedicated to processing energy, to power the cellular machinery, and these do not *a priori* need to include any representational activity. There is clearly much further work to be done to elucidate the conditions that identify representation in biological systems, including determination of the simplest possible representational system, and the ubiquity of representation in living organisms. Our framework is designed to bring clarity to this endeavour, by clearly defining what it means to be representational and how to identify when it is, and is not, occurring.

References

1. Adamatzky, A.: Physarum Machines: Computers from Slime Mould. World Scientific (2010)
2. Amos, M.: Theoretical and Experimental DNA Computation. Springer (2005)
3. Brent, R., Bruck, J.: 2020 computing: can computers help to explain biology? Nature **440**, 416–417 (2006)
4. Bain, J.D., Switzer, C., Chamberlin, A.R., Benner, S.A.: Ribosome-mediated incorporation of a non-standard amino acid into a peptide through expansion of the genetic code. Nature **356**(6369), 537–539 (1992)
5. Cousot, P., Cousot, R.: Abstract interpretation: a unified lattice model for static analysis of programs by construction or approximation of fixpoints. In: Proceedings of the 4th ACM SIGACT-SIGPLAN Symposium on Principles of Programming Languages, pp. 238–252. ACM (1977)
6. Chirico, N., Vianelli, A., Belshaw, R.: Why genes overlap in viruses. Proc. R. Soc. B: Biol. Sci. **277**(1701), 17–3809 (2010)
7. Cho, M., Vaswani, H.M., Stenger, J., Brixner, T., Fleming, G.R.: Exciton analysis in 2D electronic spectroscopy. J. Phys. Chem. B **109**(21), 10542–10556 (2005)
8. De Marse, T.B., Dockendorf, K.P.: Adaptive flight control with living neuronal networks on microelectrode arrays. In: Proceedings of the 2005 IEEE International Joint Conference on Neural Networks. IJCNN'05, vol. 3, pp. 1548–1551. IEEE (2005)

9. Engel, G.S., Calhoun, T.R., Read, E.L., Ahn, T.-K., Cheng, Y.-C., Mancal, T., Blankenship, R.E., Fleming, G.R.: Evidence for wavelike energy transfer through quantum coherence in photosynthetic systems. Nature **446**, 782–786 (2007)

10. He, J., Hoare, C.A.R., Sanders, J.W.: Data refinement refined resume. In: ESOP 86, pp. 187–196. Springer (1986)

11. Horsman, D.C.: Abstraction/representation theory for heterotic physical computing. Philos. Trans. R. Soc. A **373**, 20140224 (2015)

12. Horsman, C., Stepney, S., Kendon, V.: When does an unconventional substrate compute? In: UCNC 2014 Poster Proceedings, University of Western Ontario Technical report #758 (2014)

13. Horsman, C., Stepney, S., Wagner, R.C., Kendon, V.: When does a physical system compute? Proce. R. Soc. A **470**(2169), 20140182 (2014)

14. Kendon, V., Sebald, A., Stepney, S.: Heterotic computing: past, present and future. Philos. Trans. R. Soci. A **373**, 20140225 (2015)

15. Malyshev, D.A., Dhami, K., Lavergne, T., Chen, T., Dai, N., Foster, J.M., Correa, I.R., Romesberg, F.E.: A semi-synthetic organism with an expanded genetic alphabet. Nature **509**(7500), 385–388 (2014)

16. Mohseni, M., Rebentrost, P., Lloyd, S., Aspuru-Guzik, A.: Environment-assisted quantum walks in photosynthetic energy transfer. J. Chem. Phys. **129**, 174106 (2008)

17. National Institute of General Medical Sciences. The New Genetics. NIH Publication No.10-662 (2010). http://www.nigms.nih.gov

18. Navlakha, S., Bar-Joseph, Z.: Distributed information processing in biological and computational systems. Commun. ACM **58**(1), 94–102 (2015)

19. Osawa, S., Jukes, T.H., Watanabe, K., Muto, A.: Recent evidence for evolution of the genetic code. Microbiol. Rev. **56**(1), 229–264 (1992)

20. Plenio, M.B., Huelga, S.F.: Dephasing assisted transport: quantum networks and biomolecules. New J. Phys. **10**, 113019 (2008)

21. Porter, S.L., Wadhams, G.H., Armitage, J.P.: Signal processing in complex chemotaxis pathways. Nat. Rev. Microbiol. **9**(3), 153–165 (2011)

22. Regev, A., Shapiro, E.: Cellular abstractions: cells as computation. Nature **419**, 343 (2002)

23. Wooley, J., Lin, H. (eds.): Catalyzing Inquiry at the Interface of Computing and Biology. National Academies Press (2005)

24. War, A.R., Paulraj, M.G., Ahmad, T., Buhroo, A.A., Hussain, B., Ignacimuthu, S., Sharma, H.C.: Mechanisms of plant defense against insect herbivores. Plant Signal. Behav. **7**(10), 1306–1320 (2012)

25. Wang, Q., Parrish, A.R., Wang, L.: Expanding the genetic code for biological studies. Chem. Biol. **16**(3), 323–336 (2009)

26. Wang, B., Kitney, R, Joly, N., Buck, M.: Engineering modular orthogonal genetic logic gates for robust digital-like synthetic biology. Nat. Commun. **2**(508) (2011)

27. Yang, Z., Hutter, D., Sheng, P., Sismour, A.M., Benner, S.A.: Artificially expanded genetic information system: a new base pair with an alternative hydrogen bonding pattern. Nucleic Acids Res. **34**(21), 6095–6101 (2006)

The Information-Theoretic and Algorithmic Approach to Human, Animal, and Artificial Cognition

Nicolas Gauvrit, Hector Zenil and Jesper Tegnér

Abstract We survey concepts at the frontier of research connecting artificial, animal, and human cognition to computation and information processing—from the Turing test to Searle's Chinese room argument, from integrated information theory to computational and algorithmic complexity. We start by arguing that passing the Turing test is a trivial computational problem and that its pragmatic difficulty sheds light on the computational nature of the human mind more than it does on the challenge of artificial intelligence. We then review our proposed algorithmic information-theoretic measures for quantifying and characterizing cognition in various forms. These are capable of accounting for known biases in human behavior, thus vindicating a computational algorithmic view of cognition as first suggested by Turing, but this time rooted in the concept of algorithmic probability, which in turn is based on computational universality while being independent of computational model, and which has the virtue of being predictive and testable as a model theory of cognitive behavior.

Nicolas Gauvrit and Hector Zenil authors contributed equally

N. Gauvrit (✉)
Human and Artificial Cognition Lab, Université Paris 8 & EPHE, Paris, France
e-mail: ngauvrit@me.com

H. Zenil · J. Tegnér
Unit of Computational Medicine, Department of Medicine Solna,
Centre for Molecular Medicine & SciLifeLab, Karolinska Institutet,
Stockholm, Sweden
e-mail: hector.zenil@algorithmicnaturelab.org

J. Tegnér
e-mail: jesper.tegner@ki.se

H. Zenil
Department of Computer Science, University of Oxford, Oxford, UK

© Springer International Publishing AG 2017
G. Dodig-Crnkovic and R. Giovagnoli (eds.), *Representation and Reality in Humans,
Other Living Organisms and Intelligent Machines*, Studies in Applied Philosophy,
Epistemology and Rational Ethics 28, DOI 10.1007/978-3-319-43784-2_7

1 The Algorithmic Model of Mind

Judea Pearl, a leading theorist of causality, believes that every computer scientist is in a sense a frustrated psychologist, because computer scientists learn about themselves (and about others) by writing computer programs that are emulators of intelligent human behavior [31]. As Pearl suggests, computer programs are in effect an enhanced version of an important intellectual component of ourselves: they perform calculations for us, albeit not always as we ourselves would perform them. Pearl claims that by emulating ourselves we externalize a part of our behavior; we mirror our inner selves, allowing them to become objects of investigation. We are able to monitor the consequences of changing our minds by changing the code of a computer program, which is effectively a reflection of our minds.

Perhaps Alan Turing was the first such frustrated psychologist, attempting to explain human behavior by means of mechanical processes, first in his seminal work on universal computation [66], but also in his later landmark paper connecting intelligence and computation through an imitation game [67]. On the one hand, artificial intelligence has sought to automatize aspects of behavior that would in the past have been considered intelligent, in the spirit of Turing's first paper on universal computation. It has driven the evolution of computer programs from mere arithmetic calculators to machines capable of playing chess at levels beyond human capacity and answering questions at near—and in some domains beyond—human performance, to machines capable of linguistic translation and face recognition at human standards. On the other hand, the philosophical discussion epitomized by Turing's later paper [67] on human and machine intelligence prompted an early tendency to conflate intelligence and consciousness, generating interesting responses from scholars such as John Searle. Searle's Chinese room argument [60] (CRA) is not an objection against Turing's main argument, which has its own virtues despite its many limitations, but a call to distinguish human consciousness from intelligent behavior in general. Additionally, the philosophy of mind has transitioned from materialism to functionalism to computationalism [19], but until very recently little had been done by way of formally—conceptually and technically—connecting computation to a model of consciousness and cognition.

Despite concerns about the so-called *integrated information theory* (IIT)—which by no means are devastating or final even if valid[1]—there exists now a contending formal theory of consciousness, which may be right or wrong but has the virtue of being precise and welldefined, even though it has evolved in a short period of time. Integrated information theory [51] lays the groundwork for an interesting computational and information-theoretic account of the necessary conditions for consciousness. It proposes a measure of the integration of information between the interacting elements that account for what is essential to consciousness, viz. the feeling of an internal experience, and therefore the generation of information within the system

---

[1] See, e.g. http://www.scottaaronson.com/blog/?p=1799, as accessed on December 23, 2015, where Tononi himself provided acceptable, even if not definite, answers.

in excess of information received from the external environment and independent of what the system retrieves, if anything.

Moreover, there now exists a formal theory capable of accounting for biases in human perception that classical probability theory could only quantify but not explain or generate a working hypothesis about [25, 26]. This algorithmic model provides a strong formal connection between cognition and computation by means of recursion, which seems to be innate or else developed over the course of our lifetimes [26]. The theory is also not immune to arguments of super-Turing computation [73], suggesting that, while the actual mechanisms of cognition and the mind may be very different from the operation of a Turing machine, its computational power may be that of the Turing model. But we do not yet know with certainty what conditions would be necessary or sufficient to algorithmically account for the same mental biases with more or less computational power, and this is an interesting direction for further investigation. More precisely, we need to ascertain whether theorems such as the algorithmic coding theorem are valid under conditions of super- or sub-universality. Here it is irrelevant, however, whether the brain may look like a Turing machine, which is a trivialization of the question of its computational power because nobody would think the brain operates like a Turing machine.

The connections suggested between computation and life go well beyond cognition. For example, Sydney Brenner, one of the fathers of molecular biology, argues that Turing machines and cells have much in common [9] even if these connections were made in a rather top-level fashion reminiscence of old attempts to trivialize biology. More recently, it has been shown in systematic experiments with yeast that evolutionary paths taken by completely different mutations in equivalent environments reach the same evolutionary outcome [39], which may establish another computational property known as *confluence* in abstract rewriting systems, also known as the *Church–Rosser property* (where the Turing machine model is but one type of rewriting system). This particular kind of contingency makes evolution more predictable than expected, but it also means that finding paths leading to a disease, such as a neurodegenerative disease, can be more difficult because of the great diversity of possible causes leading to the same undesirable outcome. Strong algorithmic connections between animal behavior, molecular biology, and Turing computation have been explored in [79, 83]. But the connections between computation and cognition can be traced back to the very beginning of the field, which will help us lay the groundwork for what we believe is an important contribution towards a better understanding of cognition, particularly human cognition, through algorithmic lenses. Here, we take the first steps towards revealing a stronger non-trivial connection between computation (or information processing) on one side and cognition in the other side by means of the theory of algorithmic information.

1.1 The Turing Test Is Trivial, Ergo the Mind Is Algorithmic

In principle, passing the Turing test is trivially achievable by brute force. This can be demonstrated using a simple Chinese room argument-type thought experiment. Take the number of (comprehensible) sentences in a conversation. This number is finite because the set of understandable finite sentences is bounded and is therefore finite. Write a *lookup table* with all possible answers to any possible question. A thorough introduction to these ideas is offered in [48]. Passing the Turing test is then trivially achievable in finite time and space by brute force; it is just a combinatorial problem, but nobody would suspect the brain of operating in this way. Lookup tables run in $O(1)$ computational time, but if a machine is to pass the Turing test by bruteforce, their size would grow exponentially for sentences that only grow linearly, given the wide range of possible answers. Passing the Turing test, even for conservative sentence and conversation lengths, would require more space than the observable universe.

One can construct the case of certain sentences of some infinite nature that may be understood by the human mind. For example, we can build a nested sentence using the word *"meaning"*. *"The meaning of meaning"* is still a relatively easy sentence to grasp, the third level of nesting, however, may already be a challenge, and if it is not, then one can nest it n times until the sentence appears beyond human understanding. At some point, one can think of a large n for which one can still make the case that some human understanding is possible. One can make the case for an unbounded, if not infinite n, from which the human mind can still draw some meaning, making it impossible for a lookup table implementation to deal with, because of the unbounded, ever increasing n. This goes along the lines of Hofstadter self-referential loops [20]: Think of the sentence *"What is the meaning of this sentence"*, which one can again nest several times, making it a conundrum but, from which an, for arbitrary number of "nestings", for the human mind may still be able to grasp something, even if a false understanding of its meaning, and maybe even collapsing on itself, making some sort of *strange loop* where, the larger the value of n, the less absolute information content there is in the sentence, not only relative to its size, hence approaching zero additional meaning for ever increasing n, hence finite meaning out of an asymptotically infinite nested sentence. Understanding these sentences in the described way seems to require "true consciousness" of the nature of the context in which these sentences are constructed, similar to arguments in favor of "intuiton" [54] spanning different levels of understanding (inside and outside the theory) leading to and based upon Gödel's [29] type arguments.

While it is true that not all combinations of words form valid grammatical sentences, passing the Turing test by brute force may also actually require the capacity to recognize invalid sentences in order either to avoid them or to find suitable answers for each of them. This conservative number of combinations also assumes that for the same sentence the lookup table produces the same answer, because the same sentence will have assigned to it the same index produced by a hash function. This amounts to implementing a lookup table with no additional information—such as the likelihood that a given word will be next to another—thus effectively reducing the number of combinations. But assuming a raw lookup table with no other algorithmic implementation, in order to implement some sort of memory, it would need to work at the level of conversations and not sentences or even paragraphs. Then the questioner can have a reasonable answer to the question *"Do you remember what my first question was?"*, as they must in order to be able to pass a well-designed Turing test. This means that, in order to answer any possible question related to previous questions, the lookup table has to be simply astronomically even larger, but not that a lookup table approach cannot pass the Turing test. While the final number of combinations of possible conversations, even using the most conservative numbers, is many orders of magnitude larger than the largest astronomical magnitudes, this is still a finite number. This means that, on the one hand, the Turing test is computationally trivially achievable because one only needs to build a large enough lookup table to have an answer for each possible sentence in a conversation. On the other hand, given both that the Turing test is in practice impossible to pass by brute force using a lookup table, and that passing the test is in fact achievable by the human mind, it cannot be the case that the mind implements a lookup table [37, 56]. The respective "additional mechanisms" are the key to cognitive abilities.

And this is indeed what Searle helped emphasize. Objections to the Turing test may be of a metaphysical stripe, or adhere to Searle and introduce resource constraints. But this means that either the mind has certain metaphysical properties that cannot be represented and reproduced by science, or that the Turing test and therefore the computational operation of the mind can only make sense if resources are taken into account [21, 22, 52]. Which is to say that the Turing test must be passed using a certain limited amount of space and in a certain limited time, and if not, then machine intelligence is unrelated to the Turing test (and to computing), and must be understood as a form of rule/data compression and decompression. Searle is right in that the brain is unlikely to operate as a computer program working on a colossal lookup table, and the answer can be summarized by Chaitin's dictum "compression is comprehension" [12]. We believe in fact, just as Bennett does [7], that decompression, i.e. the calculation to arrive at the decompressed data, is also a key element, but one that falls into the algorithmic realm that we are defending.

One may have the feeling that Searle's point was related to grounding and semantics versus syntax, and thereby that algorithms are still of syntactic nature, but Searle recognizes that he is not claiming anything metaphysical. What we are claiming is that it follows from the impossibility of certain computer programs to explain understanding of the human mind that human understanding must therefore be related to features of highly algorithmic nature (e.g. compression/decompression) and thus that not all computer programs are the same, particularly when resources are involved. And this constraint of resources comes from the brain optimization of all sorts of cost functions achieved by striving for optimal learning [79] such as minimizing energy consumption, to mention but one example, and in this optimal behavior *algorithmic probability* [63] must have an essential role.

In light of the theoretical triviality of passing the Turing test, it has been stressed [1, 52] that one must consider the question of resources and therefore of *computational complexity*. This means that the mind harnesses mechanisms to compress large amounts of information in efficient ways. An interesting connection to *integrated information* can be established by way of compression. Imagine one is given two files. One is uncompressible and therefore any change to the uncompressed file can be simply replicated in the compressed file, the files being effectively the same. This is because any given bit is independent of every other in both versions, compressed and uncompressed. But reproducing any change in a compressible file leading to an effect in an uncompressed file requires a cryptic change in the former, because the compression algorithm takes advantage of short- and long-range regularities, i.e. non-independency of the bits, encoded in a region of the compressed file that may be very far afield in the uncompressed version. This means that uncompressible files have little integrated information but compressible files have a greater degree of integrated information. Similar ideas are explored in [42], making a case against integrated information theory by arguing that it does not conform to the intuitive concept of consciousness. For what people mean by the use of 'consciousness' is that a system cannot be broken down [51]; if it could it would not amount to consciousness.

In practice, search engines and the *web* constitute a sort of lookup table of exactly the type attacked by Searle's CRA. Indeed, the probability of finding a webpage containing a permutation of words representing a short sentence is high and tends to increase over time, even though the probability grows exponentially slowly due to the combinatorial explosion. The web contains about 4.73 billion pages (http://www.worldwidewebsize.com/ as estimated and accessed on Wednesday, 07 January, 2015) with text mostly written by humans. But that the web seen through a search engine is a lookup table of sorts is an argument against the idea that the web is a sort of "global mind" and therefore consonant with Searle's argument. Indeed, there is very little evidence that anything in the web is in compressed form (this is slightly different from the Internet in general, where many transactions and protocols implement compression in one way or another, e.g. encryption). We are therefore proposing that compression is a necessary condition for minds and consciousness, but clearly not a sufficient one (cf. compressed files, such as PNG images on a computer hard drive).

It has been found in field experiments that animals display a strong behavioral bias. For example, in Reznikova's communication experiments with ants, simpler instructions were communicated faster and more effectively than complex ones by scout ants searching for patches of food randomly located in a maze. The sequences consist of right and left movements encoded in binary (R and L) (see http://grahamshawcross.com/2014/07/26/counting-ants/ accessed on Dec 23, 2014) [57–59] that are communicated by the scout ants returning to their foraging team in the colony to communicate instructions for reaching the food. We have quantified some of these animal behavior experiments, confirming the author's suspicion that algorithmic information theory could account for the biases with positive results [74].

Humans too display biases in the same *algorithmic* direction, from their motion trajectories [53] to their perception of reality [13]. Indeed, we have shown that cognition, including visual perception and the generation of subjective randomness, shows a bias that can be accounted for with the seminal concept of algorithmic probability [24–26, 33, 44]. Using a computer to look at human behavior in a novel fashion, specifically by using a reverse Turing test where what is assessed is the human mind and an "average" Turing machine or computer program implementing any possible compression algorithm, we will show that the human mind behaves more like a machine. We will in effect reverse the original question Turing posed via his imitation game as to whether machines behave like us.

2 Algorithmic Complexity as Model of the Mind

Since the emergence of the Bayesian paradigm in cognitive science, researchers have expressed the need for a formal account of complexity based on a sound complexity measure. They have also struggled to find a way of giving a formal normative account of the probability that a deterministic algorithm produces a given sequence, such as the heads-or-tails string "HTHTHTHT", which intuitively looks like the result of a deterministic process even if it has the same probability of occurrence as any other sequence of the same length according to classical probability theory, which assigns a probability of $1/2^8$ to all sequences of size 8.

2.1 From Bias to Bayes

Among the diverse areas of cognitive science that have expressed a need for a new complexity measure, the most obvious is the field of probabilistic reasoning. The famous work of Kahneman et al. [32] aimed at understanding how people reason and make decisions in the face of uncertain and noisy information sources. They showed that humans were prone to many errors about randomness and probability. For instance, people tend to claim that the sequence of heads or tails "HTTHTH-HHTT" is more likely to appear when a coin is tossed than the series "HHHH-HTTTTT".

In the "heuristics and bias" approach advocated by Kahneman and Tversky, these "systematic" errors were interpreted as biases inhering in human psychology, or else as the result of using faulty heuristics. For instance, it was believed that people tend to say that "HHHHHTTTTT" is less random than "HTTHTHHHTT" because they are influenced by a so-called representativeness heuristic, according to which a sequence is more random the better it conforms to prototypical examples of random sequences. Human reasoning, it has been argued, works like a faulty computer. Although many papers have been published about these biases, not much is known about their causes.

Another example of a widespread error is the so-called equiprobability bias [40], a tendency to believe that any random variable should be uniform (with equal probability for every possible outcome). In the same vein as the seminal investigations by

Kahneman et al. [32], this bias too, viz. the erroneous assumption that randomness implies uniformity, has long been interpreted as a fundamental flaw of the human mind.

During the last decades, a paradigm shift has occurred in cognitive science. The "new paradigm"—or Bayesian approach—suggests that the human (or animal) mind is not a faulty machine, but a probabilistic machine of a certain type. According to this understanding of human cognition, we all estimate and constantly revise probabilities of events in the world, taking into account any new pieces of information, and more or less following probabilistic (including Bayesian) rules.

Studies along these lines often try to explain our probabilistic errors in terms of a sound intuition about randomness or probability applied in an inappropriate context. For instance, a mathematical and psychological reanalysis of the equiprobability bias was recently published [27]. The mathematical theory of randomness, based on algorithmic complexity (or on entropy, as it happens), does in fact imply uniformity. Thus, claiming that the intuition that randomness implies uniformity is a bias does not fit with mathematical theory. On the other hand, if one follows the mathematical theory of randomness, one must admit that a combination of random events is, in general, not random anymore. Thus, the equiprobability bias (which indeed is a bias, since it yields frequent faulty answers in the probability class) is not, we argue, the result of a misconception regarding randomness, but a consequence of the incorrect intuition that random events can be combined without affecting their property of randomness.

We now believe that, when we have to compare the probability that a fair coin produces either "HHHHHTTTTT" or any other 10-item long series, we do not really do so. One reason is that the question is unnatural: our brain is built to estimate the probabilities of the causes of observed events, not the a priori probability of such events. Therefore, researchers say, when we have participants rate the probability of the string s = "HHHHHTTTTT" for instance (or any other), they do not actually estimate the probability that such a string will appear on tossing a fair coin, which we could write as $P(s|R)$ where R stands for "random process", but the reverse probability $P(R|s)$, that is, the probability that the coin is fair (or that the string is genuinely random), given that it produced s. Such a probability can be estimated using Bayes' theorem:

$$P(R|s) = \frac{P(s|R)P(R)}{P(s|R)P(R) + P(s|D)P(D)},$$

where D stands for "not random" (or deterministic).

In this equation, the only problematic element is $P(s|D)$, the probability that an undetermined but deterministic algorithm will produce s. It was long believed that no normative measure of this value could be assumed, although some authors had the intuition that it was linked to the complexity of s: simple strings are more likely to appear as a result of an algorithm than complex ones.

The algorithmic theory of information actually provides a formal framework for this intuition. The algorithmic probability of a string s is the probability that a randomly chosen program running on a universal (prefix-free) Turing machine will produce s and then halt. It therefore serves as a natural formal definition of $P(s|D)$. As we will see below, the algorithmic probability of a string s is inversely linked to its (Kolmogorov–Chaitin) algorithmic complexity, defined as the length of the shortest program that produces s and then halts: simpler strings have a higher algorithmic probability.

One important drawback of algorithmic complexity is that it is not computable. However, there now exist methods to approximate the probability, and thus the complexity, of any string, even short ones (see below), giving rise to a renewed interest in complexity in the cognitive sciences.

Using these methods [24], we can compute that, with a prior of 0.5, the probability that the string HHHHHTTTTT is random amounts to 0.58, whereas the probability that HTTHTHHHTT is random amounts to 0.83, thus confirming the common intuition that the latter is "more random" than the former.

2.2 The Coding Theorem Method

One method for assessing Kolmogorov–Chaitin complexity, namely lossless compression, as epitomized by the Lempel–Ziv algorithm, has been long and widely used. Such a tool, together with classical Shannon entropy [68], has been used recently in neuropsychology to investigate the complexity of EEG or fMRI data [11, 42]. Indeed, the size of a compressed file gives an indication of its algorithmic complexity. The size of a compressed file is, in fact, an upper bound of true algorithmic complexity. On the one hand, compression methods have a basic flaw; they can only recognize statistical regularities and are therefore implementations of variations of entropic measures, only assessing the rate of entropic convergence based on repetitions of strings of fixed sliding-window size. If lossless compression algorithms work to approximate Kolmogorov complexity, it is only because compression is a sufficient test for non-randomness, but they clearly fail in the other direction, when it is a matter of ascertaining whether something is the result of or is produced by an algorithmic process (e.g. the digits of the mathematical constant π). That is, they cannot find structure that takes other forms than simple repetition. On the other hand, compression methods are inappropriate for short strings (of, say, less than a few hundred symbols). For short strings, lossless compression algorithms often yield files that are longer than the strings themselves, hence providing results that are very unstable and difficult, if not impossible, to interpret.

In cognitive and behavioral science, researchers usually deal with short strings of at most a few tens of symbols, for which compression methods are thus useless. This is the reason they have long relied on tailor-made measures instead.

The coding theorem method [18, 62] (CTM) has been specifically designed to address this challenge. By using CTM, researchers have provided values for the "algorithmic complexity for short strings" (which we will abbreviate as "ACSS").

ACSS is available freely as an R-package (named *ACSS*), or through an online (http://www.complexitycalculator.com Accessed on 26 Dec, 2014) complexity calculator [25].

At the root of *ACSS* is the idea that algorithmic probability may be used to capture algorithmic complexity. The algorithmic probability of a string *s* is defined as the probability that a universal prefix-free Turing machine *U* will produce *s* and then halt. Formally,

$$m(s) = \sum_{U(p)=s} 1/2^{-|p|}.$$

The algorithmic complexity [12, 38] of a string *s* is defined as the length of the shortest program that, running on a universal prefix-free [41] Turing machine *U*, will produce *s* and then halt. Formally,

$$K(s) = \min\{|p| \,:\, U(p) = s\}.$$

$K_U(s)$ and $m_U(s)$ both depend on the choice of the Turing machine *U*. Thus, the expression "the algorithmic complexity of *s*" is, in itself, a shortcut. For long strings, this dependency is relatively small. Indeed, the *invariance theorem* [12, 38, 63] states that, for any *U* and *U'*, two universal prefix-free Turing machines, there exists a constant *c* independent of *s* such that

$$|K_U(s) - K_{U'}(s)| < c.$$

The constant *c* can be arbitrarily large. If one wishes to approximate the algorithmic complexity of short strings, the choice of *U* is thus determinant.

To overcome this inconvenience, we can take advantage of a formal link established between algorithmic probability and algorithmic complexity. The *algorithmic coding theorem* [41] states that

$$K_U(s) = -\log_2(m_U(s)) + O(1).$$

This theorem can be used to approximate $K_U(s)$ through an estimation of $m_U(s)$.

Instead of choosing a particular arbitrary universal Turing machine and feeding it with random programs, Delahaye and Zenil [18] had the idea (equivalent in a formal way) of using a huge sample of Turing machines running on blank tapes. By doing so, they built a "natural" (a quasi-lexicographical enumeration) experimental distribution approaching *m(s)*, conceived of as an average $m_U(s)$.

They then defined *ACSS*(s) as $-\log_2(m(s))$. *ACSS*(s) approximates an average $K_U(s)$. To validate the method, we studied how *ACSS* varies under different conditions. It has been found that *ACSS* as computed with different huge samples of small Turing machines remained stable [25]. Also, several descriptions of Turing machines did not alter *ACSS* [72]. Using cellular automata instead of Turing machines, they showed that *ACSS* remained relatively stable. On a more practical level, *ACSS* is also validated by experimental results. For instance, as we will see below, *ACSS* is linked to human complexity perception. And it has found applications in graph theory and network biology [80].

2.3   The Block Decomposition Method

ACSS provides researchers with a user-friendly tool [24] for assessing the algorithmic complexity of very short strings. Despite the huge sample (billions) of Turing machines used to build $m(s)$ and $K(s)$ using the coding theorem method, two strings of length 12 were never produced, yet *ACSS* can safely be used for strings up to length 12 by assigning the missing two the greatest complexity in the set plus 1.

The same method used with two-dimensional Turing machines produced a sample of binary patterns on grids, but here again of limited range. Not every 5×5 grid is present in the output, but one can approximate any square grid $n \times n$ by decomposing it into 4×4 squares for which there is a known estimation approximated by a two-dimensional coding theorem method. The method simply involved running Turing machines on lattices rather than tapes, and then all the theory around the coding theorem could be applied and a two-dimensional Coding theorem method conceived.

This idea of decomposition led to filling in the gap between the scope of classical lossless compression methods (large strings) and the scope of *ACSS* (short strings). To this end, a new method based on *ACSS* was developed: the block decomposition method [80].

The basic idea at the root of the block decomposition method is to decompose strings into shorter strings (with possible overlapping) for which we have exact approximations of the Kolmogorov complexity. For instance, the string "123123456" can be decomposed into six-symbol subsequences with a three-symbol overlap "123123" and "123456". Then, the block decomposition method takes advantage of known information-theoretic properties by penalizing n repetitions that can be transmitted by using only $\log_2(n)$ bits through the formula

$$C(s) = \sum_p \log_2(n_p) + K(p),$$

where p denotes the different types of substrings, n_p the number of occurrences of each p, and $K(p)$ the complexity of p as approximated by *ACSS*. As the formula shows, the block decomposition method takes into account both the local complexity of the string and the (possibly long distance) repetitions.

With the coding theorem method leading to *ACSS*, the block decomposition method, and compression algorithms, the whole range of string lengths can now be covered with a family of formal methods. Within the field of cognitive and behavioral science, these new tools are really relevant. In the next section, we will describe three recent areas of cognitive science in which algorithmic complexity plays an important role. In these areas, *ACSS* and the block decomposition method have been key in the approximation of the algorithmic complexity of short (2–12) and medium (10–100) length strings and arrays.

3 Cognition and Complexity

An early example of an application to cognition of the coding theorem and the block decomposition methods was able to quantify the short sequence complexity in Reznikova's ant communication experiments, validating their results [76]. Indeed, it was found that ants communicate simpler strings (related to instructions for getting to food in a maze) in a shorter time, thus establishing an experimental connection between animal behavior, algorithmic complexity, and time complexity.

The idea was taken further to establish a relation between information content and energy (spent in both foraging and communicating) to establish a fitness landscape based on the thresholds between these currencies as reported in [79]. As shown in [79], if the environment is too predictable, the cost of information processing is very low. In contrast, if the environment is random, the cost is at a maximum. The results obtained by these information-processing methods suggest that organisms with better learning capabilities save more energy and therefore have an evolutionary advantage.

In many ways, animal behavior (including, notably, human behavior) suggests that the brain acts as a data compression device. For example, despite our very limited memory, it is clear we can retain long strings if they have low algorithmic complexity (e.g. the sequence 123456… vs the digits of π, see below). Cognitive and behavioral science deals for the most part with small series, barely exceeding a few tens of values. For such short sequences, estimating the algorithmic complexity is a challenge. Indeed, until now, behavioral science has largely relied on a subjective and intuitive account of complexity (e.g. Reznikova's ant communication experiments). Through the coding theorem method, however, it is now possible to obtain a reliable approximation of the algorithmic complexity for any string length [25] and the methods have been put to the test and applied to validate several intuitions put forward by prominent researchers in cognitive science over the last few decades. We will describe some of these intuitions and explain how they have been confirmed by experimental studies.

3.1 Working Memory

Researchers in psychology have identified different types of memory in humans and other animals. Two related kinds of memory have been particularly well studied: short-term memory and working memory. Short-term memory refers to a kind of cognitive resource that allows the storage of information over a few seconds or minutes. According to a recent account of short-term memory [6], items stored in short-term memory quickly and automatically decay as time passes, as a mere result of the passage of time, unless attention is devoted to the reactivation of the to-be-remembered item. Thus, constant reactivation is necessary to keep information stored for minutes. This is done by means of an internal language loop (one can continuously repeat the to-be-memorized items) or by using mental imagery (one can produce mental images of the items). On the other hand, some researchers believe that the decay observed in memory is entirely due to interference by new information [50]. Researchers thus disagree as to the reasons, while they all acknowledge that there is indeed a quick decay of memory traces. The span of short-term memory is estimated at around seven items [23, 49]. To arrive at this estimation, psychologists use simple memory span tasks in which participants are required to repeat series of items of increasing length. For instance, the experimenter says "3, 5" and the participants must repeat "3, 5". Then the experimenter says "6, 2, 9" out loud, which is repeated by the participant. Short-term memory span is defined as the length of the longest series one can recall. Strikingly, this span is barely dependent on the type of item to be memorized (letters, digits, words, etc.).

Two observations, however, shed new light on the concept of short-term memory. The first is the detrimental effect of cognitive load on short-term memory span. For instance, if humans usually recall up to seven items in a basic situation as described above, the number of items correctly recalled drops if an individual must perform another task at the same time. For instance, if a subject had to repeat "baba" while memorizing the digits, s/he would probably not store as many as seven items [5]. More demanding interfering tasks (such as checking whether equations are true or generating random letters) lead to even lower spans. The second observation is that, when there is a structure in the to-be-recalled list, one can retain more items. For instance, it is unlikely that anyone can memorize, in a one-shot experiment, a series such as "3, 5, 4, 8, 2, 1, 1, 9, 4, 5". However, longer series such as "1, 2, 3, 4, 5, 6, 5, 4, 3, 2, 1" will be easily memorized and recalled.

The first observation led to the notion of working memory. Working memory is a hypothetical cognitive resource used for both storing and processing information. When participants must recall series while performing a dual task, they assign part of their working memory to the dual task, which reduces the part of working memory allocated to storage, i.e. short-term memory [3].

The second observation led to the notion of chunking. It is now believed that humans can divide to-be-recalled lists into simple sublists. For instance, when they have to store "1, 2, 3, 4, 5, 3, 3, 3, 1, 1, 1" they can build three "chunks": "1, 2, 3, 4, 5", "3, 3, 3", and "1, 1, 1". As this example illustrates, chunks are not arbitrary: they are conceived to minimize complexity, and thus act as compression algorithms [45]. An objective factor showing that chunking does occur is that, while recalling the above series, people make longer pauses between the hypothetical chunks than within them. Taking these new concepts into account, it is now believed that the short-term memory span is not roughly seven items, as established using unstructured lists, but more accurately around four chunks [15].

The short-term memory span is thus dependent on the type of list to be memorized. With structured series, one can recall more items, but fewer chunks. This apparent contradiction can be overcome by challenging the clear-cut opposition between short-term memory span and working memory span. To overcome the limitations of short-term memory, humans can take advantage of the structure present in some series. Doing so requires using the "processing" component of working memory to analyze the data [45]. Thus, even simple tasks where subjects must only store and recall series of items do tax the processing part of working memory.

According to this hypothesis, working memory works as a compression algorithm, where part of the resource is allocated to data storage while another part is dedicated to the "compression/decompression" program. Recent studies are perfectly in line with this idea. Mathy et al. [44] used a Simon®, a popular 1980s game in which one must reproduce series of colors (chosen from among a set of four colors) of increasing length, echoing classical short-term memory tasks. They show that the algorithmic complexity of a series is a better predictor of correct recall than the length of the string. Moreover, when participants make mistakes, they generate, on average, an (incorrect) string of lower algorithmic complexity than the string to be remembered.

All these experimental results suggest that working memory acts as a compression algorithm. In normal conditions, compression is lossless, but when the items to be remembered exceed working memory capacity, it may be that lossy compression is used instead.

3.2 Randomness Perception

Humans share some intuitive and seemingly innate (or at least precocious) concepts concerning mathematical facts, biology, physics, and naïve psychology. This core knowledge [64] is found in babies as young as a few months. In the field of naïve mathematics, for instance, it has been shown that people share some basic knowledge or intuitions about numbers and geometry. Humans and other animals both manifest the ability to discriminate quantities based on an "approximate number sense," thanks to which one can immediately feel that 3 objects are more than 2, or 12 than 6 [17]. Because this number sense is only approximate, it does not allow us to discriminate 12 from 15, for instance. There are many indications that we do have an innate sense of quantity. For instance, both children and adults are able to "count" quantities not greater than four almost immediately, whereas larger quantities require a counting strategy. When faced with one, two or three objects, we immediately perceive how many there are without counting. This phenomenon is known as "subitizing" [43]. Babies as young as 6 months [69], apes [8], and even birds [55] are able to perform basic comparisons between small numbers. However, for quantities above four, we are bound to resort to learned strategies such as counting.

The same results have been found in the field of naïve geometry. In a study of an Amazonian indigenous group, Dehaene et al. [16] found that people without any formal mathematical culture naturally reason in terms of points and lines. They also perceive symmetry and share intuitive notions of topology, such as connexity or simple connexity (holes).

Recent findings in the psychology of subjective randomness suggest that probabilistic knowledge may well figure on the list of core human knowledges. Téglás et al. [65] exposed 1-year-old infants to images of jars containing various proportions of colored balls. When a ball is randomly taken from the jar, the duration of the infant's gaze is recorded. Also, one can easily compute the probability of the observed event. For instance, if a white ball is chosen, the corresponding probability is the proportion of white balls in the jar. It turned out that gazing time is correlated to the probability of the observed event: the more probable the event, the shorter the gaze. This is interpreted as proof that 1-year-old children already have an intuitive and approximate probability sense, in the same way they have an approximate number sense.

This probability sense is even more complex and rich than the previous example suggests. Xu and Garcia [70] proved that infants even younger (8 months old) could exhibit behaviors showing a basic intuition of Bayesian probability. In their experiment, children saw an experimenter taking white or red balls out of an opaque jar. The resulting sample could exhibit an excess of red or white, depending on the trial. After the sample was removed, the jar was uncovered, revealing either a large proportion of red balls or a large proportion of white balls. It was observed that children stared longer at the scene if what transpired was less probable according to a Bayesian account of sampling. For instance, if the sample had a large majority of red balls but the jar, once uncovered, showed an excess of white ones, the infant's gaze would point to the scene longer.

In the same vein, Ma and Xu [47] presented 9- and 10-month-old children with samples from a jar containing as many yellow as red balls. However, the sampling could be done either by a human or a robot. Ma and Xu show that children expect any regularity to come from a human hand. This experiment was meant to trace our intuition of intentionality behind any regularity. However, it also shows that humans have a precocious intuition of regularity and randomness.

Adults are able to discriminate more finely between strings according to their complexity. For example, using classical randomness perception tasks, Matthews [46] had people decide whether binary strings of length 21 were more likely to have been produced by a random process or a deterministic one. Matthews was primarily interested in the contextual effects associated with such a task. However, he also shows that adults usually agree about which strings are more random. A re-analysis of Mathews' data [24] showed that algorithmic complexity is actually a good predictor of subjective randomness—that is, the probability that a human would consider that the string looks random.

In attempting to uncover the sources of our perception of randomness, Hsu et al. [30] have suggested that it could be learned from the world. More precisely, according to the authors, our two-dimensional randomness perception [82] could be based on the probability that a given pattern appears in visual natural scenes. They presented a series of patterns, which were 4×4 grids with 8 black cells and 8 white cells.

Subjects had to decide whether these grids were "random" or not. The proportion of subjects answering "random" was used as a measure of subjective randomness. The authors also scanned a series of still nature shots. The pictures were binarized to black and white using the median as threshold, and every 4×4 grid was extracted from them. From this dataset, a distribution was computed. Hsu et al. [30] found that the probability that a grid appears in random photographs of natural scenes was correlated to the human estimate: the more frequent the grid, the less random it is rated. The authors interpret these results as evidence that we learn to perceive randomness through our eyes. An extension of the Hsu et al. study (2010) confirmed the correlation, and found that both subjective randomness and frequency in natural scenes were correlated to the algorithmic complexity of the patterns [26]. It was found that natural scene statistics explain in part how we perceive randomness/complexity.

In contradiction with the aforementioned experiments, according to which even children under 1 year displayed the ability to detect regularities, these results suggest that the perception of randomness is not innate. Our sense of randomness could deploy very early in life, based on visual scenes we see and an innate sense of statistics, but evolve over time.

Our results [26] suggest that the mind is intrinsically wired to "believe" that the world is algorithmic in nature, that what happens in it is likely the output of a random computer program and not of a process producing uniform classical randomness. To know if this is the result of biological design or a developed "skill" one can look at whether and how people develop this view during their lifetime. Preliminary results suggest that this algorithmic view or filter about the world we are equipped with is constructed over time, reaching a peak of algorithmic randomness production and perception at about age 25 years [28]. This means that the mind adapts and learns from experience before declining. We would not have developed such a worldview peaking at reproductive age had it no evolutionary advantage, as it seems unlikely to be a neutral feature of the mind about the world, since it affects the way it perceives events and therefore how it learns and behaves. And this is not exclusive to the human mind but, appears in other animals' behavior, as shown in [74]. All this evidence points towards the same direction, that the world is or appears to us highly algorithmic in nature, at least transitionally.

3.3   Culture and Structure

In different areas of social science and cognitive science, it has been found that regularities may arise from the transmission of cultural items [61]. For instance, in the study of rumors, it has been found that some categories of rumor disseminate easily, while others simply disappear. The "memes" that are easily remembered, believed, and transmitted share some basic properties, such as a balance between expected and unexpected elements, as well as simplicity [10]. Too-complex stories, it is believed, are too hard to remember and are doomed to disappear. Moreover, in spreading rumors the transmitters make mistakes, in a partially predictable way. As a rule, errors make the message simpler and thus easier to retain.

For instance, Barrett and Nihoff [4] had participants learn fabricated stories including some intuitive and some counterintuitive elements. They found that slightly counterintuitive elements were better recalled than intuitive elements. In research along the same lines, Atran and Norenzayan [2] found an inverse U-shaped curve between the proportion of counterintuitive elements and the correct recall of stories.

Two complementary concepts are prominent in the experimental psychology of cultural transmission. *Learnability* measures how easily an item (story, word, pattern) is remembered by someone. More learnable elements are more readily disseminated. They are selected in a "darwinian" manner, since learnability is a measure of how adapted the item is to the human community in which it will live or die.

Complexity is another important property of cultural items. In general, complex items are less learnable, but there are exceptions. For instance, a story without any interest will not be learned, even if highly simple. Interest, unexpectedness or humoristic level also play a role in determining how learnable and also how "buzzy" an item is.

When researchers study complex cultural items such as stories, arguments, or even theories or religions, they are bound to measure complexity and learnability by using more or less tailor-made tools. Typically, researchers use the number of elements in a story, as well as their interrelations to rate its complexity. The number of elements and relations someone usually retains is an index of learnability.

The paradigm of iterative learning is an interesting way to study cultural transmission and investigate how it yields structural transformations of the items to be transmitted. In the iterative learning design, a first participant has to learn a message (a story, a picture, a pseudo-language, etc.) and then reproduce or describe it to a second participant. The second participant will then describe the message to a third, etc. A chain of transmission can comprise tens of participants.

Certainly, algorithmic complexity is not always usable in such situations. However, in fundamental research in cognitive psychology, it is appropriate. Instead of investigating complex cultural elements such as rumors or pseudo-sciences, we turned to simpler items, in the hope of achieving a better understanding of how structure emerges in language. Some pseudo-languages have been used [35] and showed a decrease in intuitive complexity along the transmission chains.

In a more recent study, researchers used two-dimensional patterns of points [33]. Twelve tokens were randomly placed on 10×10 grids. The first participant had to memorize the pattern in 10 seconds, and then reproduce it on a blank grid, using new tokens. The second participant then had to reproduce the first participant's response, and so on. Participants in a chain were sometimes all adults, and sometimes all children. Learnability can be defined as the number of correctly placed tokens. As expected, learnability continuously increases within each transmission chain, at the same rate for children and adults. As a consequence of increasing learnability, each chain converges toward a highly learnable pattern (which also is simple), different from one chain to the other.

The fact that children's and adults' chains share the same rate of increase in learnability is striking, since they have different cognitive abilities, but may be explained by complexity. Indeed, the algorithmic complexity of the patterns within each chain decreases continuously, but at a faster rate in children's chains than adult's chains.

The less-is-more hypothesis [34] in psycholinguistics designates the idea that babies, with low cognitive abilities, are better than adults at learning language. Thus, less cognitive ability could actually be better when it comes to learning language. One possible explanation of the paradox relies on complexity: as a consequence of the reduced cognitive power of babies, they could be bound to "compress the data" more—build a simpler though possibly faulty representation of the available data they have access to. The recent study of Kempe et al. (2015) supports this idea (albeit remotely), by showing that children achieve the same increase in learnability as adults by a much quicker decrease in algorithmic complexity, which can be a consequence of reduced cognitive power.

4 Concluding Remarks

We have surveyed most of what has formally been done to connect and explain cognition through computation, from Turing to Searle to Tononi to our own work. It may seem that a mathematical/computational framework is taken too far, for example in interpreting how humans may remember long sequences and that remembering such sequences may probably never be useful for any organism. But it is not the object in question (string) at the core of the argument but the mechanism. Learning by systematic incorporation of information from the environment is what ultimately a DNA sequence is, and what the evolution of brains ultimately allowed to speed up the understanding of an organism's environment. Of course, the theoretical implications of a Turing machine model—such as halting, having an unbounded tape, etc.,—are only irrelevant if taken trivially as an analogy for biological processes, but this is missing the point. The point is not that organisms or biological processes are or may look like Turing machines, but that Turing machines are mechanistic explanations of behavior, and as shown by algorithmic probability, they are an optimal model for hypothesis testing of, in our opinion, the uttermost relevance for biological reality and what we call the algorithmic cognition approach to cognitive sciences at the forefront of pattern recognition.

It may also be believed that computational and algorithmic complexity focus only on classifying algorithmic problems according to their inherent difficulty or randomness and as such cannot be made equal to models of cognition. Moreover, that algorithmic analysis can give meaningful results only for a relatively narrow class of algorithms that on one hand are not too general and on the other hand are not too complex and detailed and reflect the main ingredients of cognitive processes. These beliefs would only hold if the power and universality of the results of the field of algorithmic probability (the other side of the same coin of algorithmic complexity) as introduced by Solomonoff [63] and Levin [41] were not understood. An introduction giving it proper credit, as the optimal theory of formal induction/inference, and therefore learning, can be found in [36]. The fact that the theory is based on the Turing machine model is irrelevant and cannot be used as an objection, as shown, for example, by the invariance theorem [41, 63], which not only applies to algorithmic probability, as Levin showed, that disregarding the computational model, the algorithmic probability converges to the only universal semi-measure [36]. We have adopted this powerful theory and used it to advance the algorithmic cognition approach. We have devised specific tools with a wide range of applications [75, 78] that also conform to animal and human behavior [24–26, 33, 44, 74, 76].

Just as Tononi et al. made substantial progress in discussing an otherwise more difficult topic by connecting the concept of consciousness to information theory, we have offered what we think is an essential and what appears a necessary connection between the concept of cognition and algorithmic information theory. Indeed, within cognitive science, the study of working memory, probabilistic reasoning, the emergence of structure in language, strategic response, and navigational behavior is cutting-edge research. In all these areas we have made contributions [24–26, 33, 44, 74, 76] based upon algorithmic complexity as a useful normative tool, shedding light on mechanisms of cognitive processes. We have proceeded by taking the mind as a form of algorithmic compressor, for the reasons provided in this survey emerging from the simple fact that the mind operates otherwise than through a lookup table and the ways in which the mind manifests biases that seem not to be accounted for but by algorithmic probability based purely on computability theory. The research promises to generate even more insights into the fundamentals of human and animal cognition [74], with cross-fertilization taking place from the artificial to the natural realms and vice versa, as algorithmic measures of the properties characterizing human and animal behavior can be of use in artificial systems such as robotics [71], which amounts to a sort of reverse Turing test as described in [81] and [77].

Acknowledgements The authors are indebted to the anonymous referees and to the hard work of the members of the Algorithmic Nature Group, LABORES (http://www.algorithmicnaturelab.org).

References

1. Aaronson, S.: Why philosophers should care about computational complexity. In: Copeland, B.J., Posy, C., Shagrir, O. (eds.) Computability: Turing, Gödel, Church, and Beyond. MIT Press, pp. 261–328 (2013)
2. Atran, S., Norenzayan, A.: Religion's evolutionary landscape: counterintuition, commitment, compassion, communion. Behav. Brain Sci. **27**, 713–770 (2004)
3. Baddeley, A.: Working memory. Science **255**(5044), 556–559 (1992)
4. Barrett, J.L., Nyhof, M.A.: Spreading non-natural concepts: the role of intuitive conceptual structures in memory and transmission of cultural materials. J. Cogn. Culture **1**(1), 69–100 (2001)
5. Barrouillet, P., Bernardin, S., Camos, V.: Time constraints and resource sharing in adults' working memory spans. J. Exp. Psychol. Gen. **133**(1), 83 (2004)

6. Barrouillet, P., Gavens, N., Vergauwe, E., et al.: Working memory span development: a time-based resource-sharing model account. Dev. Psychol. **45**(2), 477 (2009)
7. Bennett, C.H.: Logical depth and physical complexity. In: Herken, R. (ed.) The Universal Turing Machine. A Half-Century Survey. pp. 227–257. Oxford University Press, Oxford (1988)
8. Boysen, S.T., Hallberg, K.I.: Primate numerical competence: contributions toward understanding nonhuman cognition. Cogn. Sci. **24**(3), 423–443 (2000)
9. Brenner, S.: Turing centenary: life's code script. Nature **482**, 461 (2012)
10. Bronner, G.: Le succès d'une croyance. Ann. Soc. **60**(1), 137–160 (2010)
11. Casali, A.G., Gosseries, O., Rosanova, M., Boly, M., Sarasso, S., Casali, K.R., Casarotto, S., Bruno, M.-A., Laureys, S., Tononi, G., Massimini, M.: A theoretically based index of consciousness independent of sensory processing and behaviour. Sci. Transl. Med. **5**(198) (2013)
12. Chaitin, G.J.: On the length of programs for computing finite binary sequences. J. ACM **13**(4), 547–569
13. Chater, N.: The search for simplicity: A fundamental cognitive principle? The Q. J. Exp. Psychol. **52**(A), 273–302 (1999)
14. Church, A., Rosser, J.B.: Some properties of conversion. Trans. Am. Math. Soc. **39**(3), 472–482 (1936)
15. Cowan, N. The magical number 4 in short-term memory: a reconsideration of mental storage capacity. Behav. Brain Sci. **24**(1), 87–114 (2001)
16. Dehaene, S., Izard, V., Pica, P., Spelke, E.: Core knowledge of geometry in an Amazonian indigene group. Science **311**(5759), 381–384 (2006)
17. Dehaene, S.: The Number Sense: How the Mind Creates Mathematics. Oxford University Press, Oxford (2011)
18. Delahaye, J.-P., Zenil, H.: Numerical evaluation of algorithmic complexity for short strings: a glance into the innermost structure of randomness. Appl. Math. Comput. **219**(1), 63–77 (2012)
19. Dodig-Crnkovic, G.: Where do new ideas come from? how do they emerge? epistemology as computation (information processing). In: Calude, C. (ed.) Randomness & Complexity, from Leibniz to Chaitin (2007)
20. Douglas, H.: I am a strange Loop. In: Basic Books (2008)
21. Dowe, D.L., Hájek, A.R.: A computational extension to the Turing test. Technical Report 97/322, Department of Computer Science, Monash University (1997)
22. Dowe, D.L, Hájek, A.R.: A non-behavioural, computational extension to the Turing Test. In: Proceedings of the International Conference on Computational Intelligence and Multimedia Applications, pp. 101–106, Gippsland, Australia (1998)
23. Edin, F., Klingberg, T., Johansson, P., McNab, F., Tegnér, J., Compte, A.: Mechanism for top-down control of working memory capacity. Proc. Nat. Acad. Sci. USA **106**(16), 6802–6807 (2009)
24. Gauvrit, N., Singmann, H., Soler-Toscano, F., Zenil, H.: Algorithmic complexity for psychology: a user-friendly implementation of the coding theorem method. Behav. Res. Methods **148**(1), 314–329 (2014b)
25. Gauvrit, N., Zenil, H., Delahaye, J.-P., et al.: Algorithmic complexity for short binary strings applied to psychology: a primer. Behav. Res. Methods **46**(3), 732–744 (2014a)
26. Gauvrit, N., Soler-Toscano, F., Zenil, H.: Natural scene statistics mediate the perception of image complexity. Vis. Cogn. **22**(8), 1084–1091 (2014c)
27. Gauvrit, N., Morsanyi, K.: The equiprobability bias from a mathematical and psychological perspective. Adv Cogn. Psychol. **10**(4), 119–130 (2014)
28. Gauvrit, N., Zenil, H., Soler-Toscano, F., Delahaye, J. P., Brugger, P.: Human behavioral complexity peaks at age 25. PLoS Comp. Biol. **13**(4), e1005408 (2017)
29. Gödel, K.: Über formal unentscheidbare Sätze der Principia Mathematica und verwandter Systeme, I; On formally undecidable propositions of Principia Mathematica and related systems I in Solomon Feferman, ed., 1986. Kurt Gödel Collected works, vol. I, pp. 144–195. Oxford University Press (1931)
30. Hsu, A.S., Griffiths, T.L., Schreiber, E.: Subjective randomness and natural scene statistics. Psychon. B. Rev. **17**(5), 624–629 (2010)

31. http://blogs.wsj.com/digits/2012/03/15/work-on-causality-causes-judea-pearl-to-win-prize/ Accessed 27 Dec 2014
32. Kahneman, D., Slovic, P., Tversky, A.: Judgment under uncertainty: Heuristics and biases. Cambridge University Press, Cambridge (1982)
33. Kempe, V., Gauvrit, N., Forsyth, D.: Structure emerges faster during cultural transmission in children than in adults. Cognition **136**, 247–254 (2015)
34. Kersten, A.W., Earles, J.L.: Less really is more for adults learning a miniature artificial language. J. Mem. Lang. **44**(2), 250–273 (2001)
35. Kirby, S., Cornish, H., Smith, K.: Cumulative cultural evolution in the laboratory: an experimental approach to the origins of structure in human language. Proc. Nat. Acad. Sci. USA **105**(31), 10681–10686 (2008)
36. Kirchherr, W., Li, M., Vitányi, P.: The miraculous universal distribution. Math. Intell. **19**(4), 7–15 (1997)
37. Kirk, R.: How is consciousness possible? In: Metzinger, T. (ed.) Conscious Experience, Ferdinand Schoningh (English edition published by Imprint Academic), pp. 391–408 (1995)
38. Kolmogorov, A.N.: Three approaches to the quantitative definition of information. Prob. Inform. Transm. **1**(1), 1–7 (1965)
39. Kryazhimskiy, S., Rice, D.P., Jerison, E.R., Desai, M.M.: Microbial evolution. Global epistasis makes adaptation predictable despite sequence-level stochasticity. Science **344**(6191), 1519–22 (2014)
40. Lecoutre, M.P.: Cognitive models and problem spaces in "purely random" situations. Educ. Stud. Math. **23**(6), 557–568 (1992)
41. Levin, L.A.: Laws of information conservation (non-growth) and aspects of the foundation of probability theory. Probl. Inf. Transm. **10**(3), 206–210 (1974)
42. Maguire, P., Moser, P., Maguire, R., Griffith, V.: Is consciousness computable? Quantifying integrated information using algorithmic information theory. In: Bello, P., Guarini, M., McShane, M., Scassellati, B. (eds.) Proceedings of the 36th Annual Conference of the Cognitive Science Society. Cognitive Science Society, Austin, TX (2014)
43. Mandler, G., Shebo, B.J.: Subitizing: an analysis of its component processes. J. Exp. Psychol. Gen. **111**(1), 1 (1982)
44. Mathy, F., Chekaf, M., Gauvrit, N.: Chunking on the fly in working memory and its relationship to intelligence. In: Abstracts of the 55th Annual meeting of the Psychonomic Society. Abstract #148 (p. 32), University of California, Long Beach (2014), pp. 20–23 Nov 2014
45. Mathy, F., Feldman, J.: What's magic about magic numbers? Chunking and data compression in short-term memory. Cognition **122**(3), 346–362 (2012)
46. Matthews, W.J.: Relatively random: context effects on perceived randomness and predicted outcomes. J. Exp. Psychol. Learn. **39**(5), 1642 (2013)
47. Ma, L., Xu, F.: Preverbal infants infer intentional agents from the perception of regularity. Dev. Psychol. **49**(7), 1330 (2013)
48. McDermott, D.: On the claim that a table-lookup program could pass the turing test. Minds Mach. **24**(2), 143–188 (2014)
49. Miller, G.A.: The magical number seven, plus or minus two: some limits on our capacity for processing information. Psychol. Rev. **63**(2), 81 (1956)
50. Oberauer, K., Lange, E., Engle, R.W.: Working memory capacity and resistance to interference. J. Mem. Lang. **51**(1), 80–96 (2004)
51. Oizumi, M., Albantakis, L., Tononi, G.: From the Phenomenology to the Mechanisms of Consciousness: Integrated Information Theory 3.0. PLoS Computational Biology **10**(5), (2014)
52. Parberry, I.: Knowledge, Understanding, and computational complexity. In: Levine, D.S., Elsberry, W.R. (eds.) Optimality in Biological and Artificial Networks?, chapter 8, pp. 125–144, Lawrence Erlbaum Associates (1997)
53. Peng, Z., Genewein, T., Braun, D.A.: Assessing randomness and complexity in human motion trajectories through analysis of symbolic sequences. Front. Hum. Neurosci. **8**, 168 (2014)
54. Penrose, R.: The Emperor's New Mind: Concerning Computers, Minds and the Laws of Physics. Vintage, London (1990)

55. Pepperberg, I.M.: Grey parrot numerical competence: a review. Anim. Cogn. **9**(4), 377–391 (2006)
56. Perlis, D.: Hawkins on intelligence: fascination and frustration. Artif. Intell. **169**, 184–191 (2005)
57. Reznikova, Z., Ryabko, B.: Ants and Bits. IEEE Inf. Theor. Soc. Newsl. (2012)
58. Reznikova, Z., Ryabko, B.: Numerical competence in animals, with an insight from ants. Behaviour **148**, 405–434 (2011)
59. Ryabko, B., Reznikova, Z.: The use of ideas of information theory for studying "language" and intelligence in ants. Entropy **11**, 836–853 (2009). doi:10.3390/e1104083
60. Searle, J.: Minds. Brains Progr. Behav. Brain Sci. **3**, 417–457 (1980)
61. Smith, K., Wonnacott, E.: Eliminating unpredictable variation through iterated learning. Cognition **116**(3), 444–449 (2010)
62. Soler-Toscano, F., Zenil, H., Delahaye, J.-P., Gauvrit, N.: Calculating kolmogorov complexity from the output frequency distributions of small turing machines. PLoS ONE **9**(5), e96223 (2014)
63. Solomonoff, R.J.: A formal theory of inductive inference: Parts 1 and 2. Inf. Control **7**, 1–22 and 224–254, (1964)
64. Spelke, E.S., Kinzler, K.D.: Core knowledge. Dev. Sci. **10**(1), 89–96 (2007)
65. Téglás, E., Vul, E., Girotto, V., et al.: Pure reasoning in 12-month-old infants as probabilistic inference. Science **332**(6033), 1054–1059 (2011)
66. Turing, A.M.: On Computable numbers, with an application to the entscheidungsproblem: a correction. Proc. Lon. Math. Soc. **2**, **43**(6), 544–6 (1937)
67. Turing, A.M.: Computing machinery and intelligence. Mind LIX **236**, 433–460 (1950)
68. Wang, Z., Li, Y., Childress, A.R., Detre, J.A.: Brain entropy mapping using fMRI. PLoS ONE **9**(3), e89948 (2014)
69. Xu, F., Spelke, E.S., Goddard, S.: Number sense in human infants. Dev. Sci. **8**(1), 88–101 (2005)
70. Xu, F., Garcia, V.: Intuitive statistics by 8-month-old infants. Proc. Nat. Acad. Sci. USA **105**(13), 5012–5015 (2008)
71. Zenil H (to appear), Quantifying Natural and Artificial Intelligence in Robots and Natural Systems with an Algorithmic Behavioural Test. In Bonsignorio FP, del Pobil AP, Messina E, Hallam J (eds.), Metrics of sensory motor integration in robots and animals, Springer
72. Zenil, H., Delahaye, J.-P.: On the algorithmic nature of the world. In: Dodig-Crnkovic, G., Burgin, M. (eds.) Information and Computation. World Scientific Publishing Company (2010)
73. Zenil, H., Hernandez-Quiroz, F.: On the possible computational power of the human mind. In: Gershenson, C., Aerts, D., Edmonds, B. (eds.) Worldviews, Science and US, Philosophy and Complexity. World Scientific (2007)
74. Zenil, H., Marshall, J.A.R., Tégner, J.: Approximations of algorithmic and structural complexity validate cognitive-behavioural experimental results (submitted, preprint available at http://arxiv.org/abs/1509.06338)
75. Zenil, H., Villarreal-Zapata, E.: Asymptotic behaviour and ratios of complexity in cellular automata rule spaces. Int. J. Bifurcat. Chaos **13**(9) (2013)
76. Zenil, H.: Algorithmic Complexity of Animal Behaviour: From Communication to Cognition. In: Theory and Practice of Natural Computing Second International Conference Proceedings, TPNC 2013. Cáceres, Spain, 3–5 Dec (2013)
77. Zenil, H.: Algorithmicity and programmability in natural computing with the game of life as an in silico case study. J. Exp. Theor. Artif. Intell. **27**, 109–121 (2015)
78. Zenil, H.: Compression-based Investigation of the dynamical properties of cellular automata and other systems. Complex Syst. **19**(1), 1–28 (2010)
79. Zenil, H., Gershenson, C., Marshall, J.A.R., Rosenblueth, D.: Life as thermodynamic evidence of algorithmic structure in natural environments. Entropy **14**(11), 2173–2191 (2012)
80. Zenil, H., Soler-Toscano, F., Dingle, K., Louis, A.A.: Correlation of automorphism group size and topological properties with program-size complexity evaluations of graphs and complex networks. Phys. A Stat. Mech. Appl. **404**, 341–358 (2014)

81. Zenil, H.: What is nature-like computation? Behav. Approach Notion Programmability Philos. Technol. **27**(3), 399–421 (2014)
82. Zenil, H., Soler-Toscano, F., Delahaye, J.-P., Gauvrit, N.: Two-dimensional kolmogorov complexity and validation of the coding theorem method by compressibility. PeerJ Comput. Sci. **1**, e23 (2015)
83. Zenil, H., Marshall, J.A.R.: Some aspects of computation essential to evolution and life. Ubiquity (ACM) **2013**, 1–16 (2013)

Using Computational Models of Object Recognition to Investigate Representational Change Through Development

Dean Petters, John Hummel, Martin Jüttner, Elley Wakui
and Jules Davidoff

Abstract Empirical research on mental representation is challenging because internal representations are not available to direct observation. This chapter will show how empirical results from developmental studies, and insights from computational modelling of those results, can be combined with existing research on adults. So together all these research perspectives can provide convergent evidence for how visual representations mediate object recognition. Recent experimental studies have shown that development towards adult performance levels in configural processing in object recognition is delayed through middle childhood. Whilst part-changes to animal and artefact stimuli are processed with similar to adult levels of accuracy from 7 years of age, relative size changes to stimuli result in a significant decrease in relative performance for participants aged between 7 and 10. Two sets of computational experiments were run using the JIM3 artificial neural network with adult and 'immature' versions to simulate these results. One set progressively decreased the number of neurons involved in the representation of view-independent metric relations within multi-geon objects. A second set of computational experiments involved decreasing the number of neurons that represent view-dependent (non-relational) object attributes in JIM3's surface map. The simulation results which show the best qualitative match to empirical data occurred when artificial neurons representing metric-precision relations were entirely eliminated. These results therefore provide

D. Petters (✉)
Birmingham City University, Birmingham, UK
e-mail: dean.petters@bcu.ac.uk

J. Hummel
University of Illinios at Urbana-Champaign, Champaign, USA

M. Jüttner
University of Aston, Birmingham, UK

E. Wakui
University of East London, London, UK

J. Davidoff
Goldsmiths College, London, UK

© Springer International Publishing AG 2017
G. Dodig-Crnkovic and R. Giovagnoli (eds.), *Representation and Reality in Humans,*
Other Living Organisms and Intelligent Machines, Studies in Applied Philosophy,
Epistemology and Rational Ethics 28, DOI 10.1007/978-3-319-43784-2_8

further evidence for the late development of relational processing in object recognition and suggest that children in middle childhood may recognise objects without forming structural description representations.

1 Introduction

Only a fraction of a second passes from when a person sees a familiar visual object to when they can then name it. Despite it being relatively quick, the process of visual object recognition is complex, with multiple sub-processes, some occurring in parallel. Multiple forms of representation are invoked in object recognition, from the point of initially perceiving an object to finally being able to provide the name for it. A very rough framework to start understanding the complexities of recognition is to consider how recognition processes get started and how they complete. First, a person perceives an object is present in sense data. Then this perceptual pattern is compared with a representation of some kind for that object type in long-term memory. When a match is found between these two representations of the object currently present in perception, recognition has occurred. However, this brief sketch is an oversimplification because perceptual processes do not complete before memory retrieval starts. Rather, perception and memory retrieval occur together and influence each other. In addition this rough distinction between perceptual and memory processes leaves open many further questions. For example, can the perceptual component of the overall recognition process be broken down to more basic sub-processes?; how are visual features in sense data selected and bound [44, 48]?; how are part-relations bound [28]?; and, what is the role of attention in object recognition [34]? Questions also arise from considering how percepts are matched to memories: does object recognition rely on one, two or many mediating representations, in perception, and in long-term memory [21]?; and what backwards projections or 'top-down' influences from memory retrieval to perception exist [19]?

The main question that this chapter is concerned with arises from the finding that object recognition performance changes through adolescence. So this chapter explores possible developmental trajectories for how object recognition representations change during this period in development. Specifically, this chapter will use computational modelling to attempt to explain empirical evidence for differences in the visual representations used for object recognition in middle childhood (ages 7–10) and adulthood. In attempting to answer this specific question, some general issues in recognition processes and representations for recognition will need to be examined. So before the developmentally focussed question can be answered we will consider what representations are used when adults recognise objects. Visual object recognition has been far more intensively studied in adults than children, and current theories propose that adults use a variety of representations when recognising objects. These include compositional representations which describe objects as three-dimensional (3D) structures in terms of the interrelationship of their parts

[4, 29, 36], and image-based representations which capture two-dimensional (2D) object views [18, 37].

1.1 Compositionality in Visual Object Representations

Humans are highly accomplished at recognising objects visually. Familiar objects can be recognised in novel viewpoints, and new unseen members of familiar categories are also often recognised with speed and accuracy. Structural description theories explain this impressive performance by proposing that recognition of objects occurs through intermediate object representations that are compositional in nature and are abstracted from sensory data. Formal logics and natural languages demonstrate compositionality because the meaning of linguistic or logical expressions with multiple parts is determined not just by the meaning of those parts but the way they are put together. In addition to language, compositionality is found in a diverse range of other entities in the world including visual objects. So in a recognition task an object's identity can be determined by identifying relations between its component parts, not just the nature of those parts viewed in isolation [4]. In our interactions with objects the perception of compositionality can be manifested across multiple modalities [32]. We can perceive visual compositionality in scenes and objects and thus form structural descriptions. Structural description recognition processes that are mediated by compositional representations are also termed analytic processes because they specify the relations among an object's parts explicitly and independently ([43], p. 257).

The 'recognition by components' (RBC) theory of object recognition is distinguished from other structural description theories of object recognition because it postulates that geons (geometric components derived from readily detectable properties of edges) are the fundamental unit of representation for objects [4]. Geons can therefore be compared with phonemes in spoken language. In both systems, a small number of representational primitives can code for a very large number of component representations (words or visual objects, respectively). In the original RBC theory 36 geons are proposed as components for all objects, compared with the 55 phonemes required to represent virtually all words in human speech [4]. A key similarity of these systems is that how the primitives are combined matters. One way in which phonemes and geons differ is that phonemes form words by linkage in serial chains where the order matters. However, visual objects can be formed of multiple geons with several different types of relations, such as larger-smaller, and above-below or beside. For example, consider a typical coffee mug. The spatial relation between the handle and the body of a coffee mug might be explicitly described as a small curved cylinder side attached to a vertical straight cylinder. This structural description would match a whole range of slightly different coffee mugs. So what structural descriptions allow is a generalisation across metric variations that are still within categorical divisions. This property of structural descriptions is related to the issue of view-invariance to rotation discussed in the next section.

Artificial neural networks can represent visual compositionality and hence model natural cognition [21, 47]. Visual compositionality is also of interest in machine representations because it can facilitate artificial systems extracting verbal descriptions of scenes or objects. Active research questions include the comparative benefits of mechanisms for neural instantiation of visual combinatorial representations [12, 47], and how generalised shape information develops [13].

1.2 Non-compositional Image-Based Object Representations

Visual object recognition that is accomplished through the use of image-based representations provides a complementary capability to recognition relying on compositional representations [21]. This is because abstracting images into parts and relations between parts takes time and is potentially error prone. Avoiding these drawbacks, and just using the actual image in sensory data to match against a similar type of unabstracted view-based representation for objects in long-term memory can therefore afford a faster and potentially less error-prone route to recognition. Image-based recognition will also be advantageous when an object does not possess distinct separable components which can be described compositionally, such as is the case with a cloud or piece of wire. Image based representational theories propose that the particular 2D object views received in sensory data are encoded in memory. This is an opposing account to structural description theories because no abstract representations are held in memory. View-based representations are also described as 'holistic' because, although they may include visual features or fragments, these are not represented independently but rather at fixed locations within the 2D object overall shape. In terms of compositionality, such image-based representations do not get decomposed into recombinable components. So view-based (holistic) coding is analogue rather than compositional because object parts that are big in the image have to be represented as a big proportion of the view-based representation, if one part is above another in the image it has to be positioned above in the representation, not just described as such. In view-based representations object parts are represented in terms of their topological positions in a 2D space described by the outline of the object [39, 41]. When perceptual images are matched to view-based representations in memory this occurs "all of a piece" [43]. This means that an object is recognised according to its overall shape and features within the image but not in terms of the interrelationship of isolatable parts.

To recognise an object such as a horse, a view-based representation of a horse in memory would be matched "*in its entirety*" against the object's perceptual image to determine the degree of fit [39, 41]. As Thoma, Hummel and Davidoff note, the "*process is directly analogous to laying a template for* [a] *horse over the image of the* [] *horse and counting the points of overlap*" ([43], p. 259).

1.3 Differing Predictions for Performance Outcomes from Rotation in Depth

When a known object is presented from an unfamiliar view the 2D view-based representation in memory may not fully match up to the currently sensed image because rotated objects can present different 2D images from the familiar views that are already in memory. Image-based theories propose that known objects seen in unfamiliar views can still be matched and recognised by bringing these perceptual images into line with stored images in memory. This is postulated to occur through processes of normalisation, which can include alignment, mental rotation and view interpolation [8, 30, 31, 37, 38, 45, 46]. An important characteristic of all these normalisation theories is that they predict a linear cost in recognition performance due to rotation. Structural description theories do not make the same prediction of linear cost to rotation. Figure 1 helps illustrates the difference between predictions on rotation costs to recognition performance for view-based and structural description mechanisms. A recognition task based on this object would involve an experiment with a learning phase and a test phase. In the learning phase, a participant would

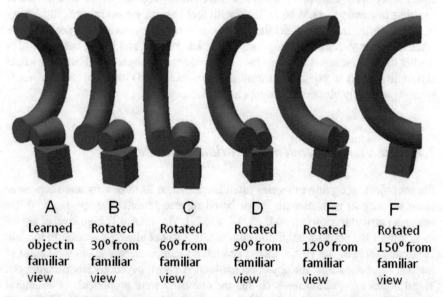

A	B	C	D	E	F
Learned object in familiar view	Rotated 30° from familiar view	Rotated 60° from familiar view	Rotated 90° from familiar view	Rotated 120° from familiar view	Rotated 150° from familiar view

Fig. 1 Figure showing why structural description and image-based theories make different predictions about performance in response to rotation in depth. After the object has been learned in the view shown by stimuli *A*, this and other novel views (*B–F*) can be tested for how quickly they are recognised as the same as the familiar learned object view. View-based theories predict that recognition latencies will increase proportionately with rotation magnitude, from stimuli *B* through to stimuli *F*. Structural description theories predict no added recognition cost from stimuli *B* through to stimuli *D*, as all these stimuli are described with the same structural description, and it is the abstract structural description that is matched to memory

become familiar with object A, perhaps by learning a name or label for this object, but would never see the object in the rotated views B–F. Then in the test phase of the experiment, the participant would have to show recognition of objects, perhaps by assigning them as already learned but rotated objects or novel objects that have never been seen in any view. In the images presented in Fig. 1, a view-based process with a normalisation mechanism with linear cost to rotation would predict stimuli B would take longer to recognise than A, and C longer than B, D longer than C, through to F, but with the same added cost to recognition latencies as the magnitude of rotation increased from presentations of B to a presentation of the F view of the stimuli. However, Fig. 1 shows why changes in viewpoint will often make no difference to the structural description that results when perceiving the object from different rotated views. Despite the fact that the rotations in depth for objects B–F in Fig. 1 result in significant changes to the observable 2D image, object views for stimuli B, C and D generate exactly the same structural description as the original learned familiar object A, with the same relations between the same component parts [large curved cylinder side-attached small truncated cone] and [small truncated cone above small cube]. So for objects B–D recognising the same structural description as A would predict the same recognition latencies. Object E has the truncated cone mostly occluded, and in object F it is completely occluded. So a view-invariant response to rotation would be unlikely with these object views. In summary, structural description models predict that rotation will not affect recognition performance when the same structural descriptions will reliably result, and image-based theories predict that, when novel rotated views are perceived, recognition is not immediate and some process is required to transform the perceived 2D image to check matches against previously viewed 2D exemplars in memory.

1.4   The View Dependence/Invariance Debate

Recent object recognition theories often invoke some form of structured representation activity in parallel with image-based representations (though details differ between particular approaches) [2, 3, 17, 21, 27]. This relatively high level of agreement within the object recognition community was not always the case. In the late 1980s and most of the 1990s, there occurred a vigorous debate about the nature of internal representations for object recognition amongst psychologists carrying out visual object recognition research. On the one hand were proponents of structural description models, such as the RBC theory of object recognition [5–7]. On the other hand were those who promoted view-based theories which proposed representations much closer to the sensory input [18, 37, 38]. The differing predictions illustrated by Fig. 1 led to an approximately 10 years debate in object recognition research. The view-dependence/invariance debate assumed that object recognition performance that was invariant to object rotation provided evidence for structural description representations, and rotational costs in performance were viewed as evidence for image-based representations in recognition. The debate occurred in these terms because researchers who were actually interested in elucidating the nature of internal representations used object recognition performance for known objects in novel rotated viewpoints as a proxy or surrogate for the nature of the internal representations believed to be involved in object recognition.

As the number of published studies increased, significant evidence was found that supported both positions in the view-dependence debate on visual object recognition. Sometimes recognition occurs with little performance cost from stimulus rotation, supporting the view-invariant predictions of the RBC theory. Other studies report a pronounced cost to rotation, with increasing performance costs as the magnitude of the rotation increased. As Hayward noted in 2003:

> the viewpoint debate appears to have run out of steam. It has ended because, on most major issues, the two sides are in basic agreement. Both agree that a change in viewpoint will normally result in viewpoint costs, albeit small in some cases. Both agree that some visual properties, particularly those related to the structure of an object, will be particularly important for generalizing across viewpoint. Finally, neither can deny the findings of the other, both view-invariant and view-specific patterns of data have been replicated so often that it has been difficult for either to argue that their opponents' results are a special case ([17], p. 425).

So a consensus has arisen that both image-based representations and structured representations with some degree of abstraction from the sensory image are involved in visual object recognition. With dual processes mediated by contrasting representations operating in parallel we can produce rotation performance predictions that have a different, more complex pattern. We can expect the fastest recognition to occur when an image is perceived in a familiar view, as both structural description and view-based mechanisms will operate optimally. Next fastest will be images that are rotated but which are still parsed to give the same structural description as the familiar view. After this, rotations that result in occlusions of geons that are clearly seen in the familiar view, or when previously unseen geons become viewable, result in the slowest recognition. The next section will present further convergent evidence for the operation of dual recognition processes. These dual processes (with dual representations) act as complementary solutions to object recognition because the dual processes draw upon the contrasting strengths of image-based and compositional representations.

2 Accumulation of Further Evidence for Multiple Representations Mediating Adult Object Recognition

2.1 Fractionating the Visual Object Recognition System into Independent Components

At the close of the view-dependence/invariance debate, the mixed results from studies of object rotation suggested that both view-invariant and view-dependent mechanisms are likely to be operating. Stankiewicz [33], and Foster and Gilson [15] responded to this state of affairs with research aiming to observe independent dimensions of the object recognition system using experimental manipulations. These experiments thus enable us to see how structural and view information is combined by object recognition processes [17]. Foster and Gilson used 'paper clip' stimuli that did not possess geons, and that varied in their number of parts (structural information) and in view-specific properties such as length, degree of curvature and angle of joints between parts. They then assessed how each of these properties affected discrimination performance. Experiments showed object structure information and image-based information are both independent and are combined additively. Stankiewicz conducted experiments which showed that 3D properties, such as primary axis curvature and aspect ratio, are estimated independently of 2D object image properties. He also showed that 3D shape is estimated independently of object viewpoint. As Stankiewicz notes [33], the results of fractionating the visual object recognition system into independent components provide strong evidence for a dual process model of object recognition, with a view-invariant component that forms structural descriptions and hence compositional representations, and a view-dependent component that forms representations much closer to the unabstracted sensory image.

2.2 Neuropsychology Evidence for Dual Representations in Object Recognition

Evidence for dual processes (and hence dual representations) occurring in visual object recognition is also provided by neuropsychological case studies. Dual process theories of visual object recognition postulate that each process acts upon a different kind of representation. Analytic processes involve perception of object, parts before the formation of compositional representations. Holistic processes do not involve such a decomposition. Davidoff and Warrington found patients who could not recognise individual object parts but who could name whole objects [10, 11]. However, the fact that the intact naming ability only occurred in familiar object views suggests that these patients' abilities resulted from holistic recognition whilst analytic recognition no longer functioned.

2.3 Priming Evidence for Dual Representations in Object Recognition

Perhaps the most comprehensive and persuasive evidence for the dual nature of representations for visual object recognition comes from behavioural studies which measure how priming can improve recognition performance. Priming involves the

repeated viewing of an object, and priming effects help infer the nature of representations because of various changes that can be made in the manner in which the
priming object can be depicted [40]. Priming benefits to recognition performance
are measured as the difference in latencies between repeated and unrepeated (and so
unprimed) probe images [39, 41]. The key finding from priming studies is that performance can be improved by two kinds of priming: (1) by primes that are presented
so that they are attended to and (2) with primes that are presented so that they are not
attended to [43]. Attended primes are believed to activate both structural description
and view-based representations. Unattended primes are presented outside the focus
of attention for a brief enough exposure that attention cannot switch to them. Priming
images in these cases only increase recognition when they exactly match the object
being recognised. However, attended primes can be altered with various modifications that leave the structurally described elements intact and still provide performance improvements. These modifications include being mirror-reflected [23], split
and recombined [43], rotated in the picture plane [42] and rotated in depth [41].

Splitting images and then recombining them is an interesting modification for a
prime because split and recombined images give highly similar structural descriptions if the image is not split where it would naturally be parsed into its geon components. Thoma, Davidoff and Hummel [43] use images where splitting in the middle
resulted in all part shapes being recoverable in the recombined image. So split images
disrupt view-specific matching, and hence result in no performance increase when
unattended. However, they do give a performance increase when attended, as analytic recognition is based on discrete, isolatable parts which are still present in the
split and recombined stimuli [40].

Section 1.4 explained that recognition after rotating objects in depth did not provide a clear-cut distinction between view-invariant and view-dependent processes.
This is because both structural description theories and view-based theories provide
explanations for how humans recognise rotated stimuli. Structural description theories posit representations that do a lot of the 'heavy lifting' in recognition tasks.
So such relatively rich representations only need to be acted upon with relatively
simple processes to produce flexible recognition behaviour. If image-based representations are acted upon with similarly simple processes, they cannot be expected to
perform as flexibly as systems using structural descriptions. But the key issue is that
image-based representations may be operated upon with complex and sophisticated
processes which compensate for the lack of flexibility in the representation. So an
object recognition researcher may be left using view-dependence/invariance to distinguish two kinds of representation that predict similar outcomes because structural
description theories postulate simple operations on "smart" representations, whereas
view-based theories postulate "smart" operations on simple representations ([20], p.
160). However, using priming images rotated in depth works better at distinguishing
underlying representations because it works by priming representations that mediate view-invariant mechanisms and these representations are not involved in view-
dependent mechanisms. So as with priming probes formed from split images, prime
probes that have been rotated in depth still possess the same parts and part-relations
as the unrotated images, and hence give a performance benefit to mechanisms that are

mediated by structural description representations. Rotating objects in depth results in significant changes to the observable 2D image, and this image-based representation has no carry-over effect on the recognition task and so does not give a priming benefit to performance.

Priming studies not only show that the structure of part-relations in priming images is only captured in attended conditions, but also that performance improvements from priming from same-view pairs is equivalent in attended and non-attended conditions (around 50 ms improvement in recognition performance when a same-view image is previously presented attended or unattended). That the priming benefit of a view-based representation is equal with and without attention suggests that the process that provides this benefit occurs independently and that the recognition system has at least two independent components [39, 41]. In summary, the results of many priming studies show that attended prime images reliably primed exactly the same view as well as many modified images that kept the prime images, structural description elements intact, whereas unattended images only primed themselves in exactly the same view [39].

2.4 Brain Imaging Evidence for Dual Representations in Object Recognition

Whilst behavioural studies using priming techniques have made great progress in consolidating knowledge of the representational distinctions in object recognition processes [39]. Thoma and Henson [40] have also extended the behavioural findings from priming experiments by conducting the first brain imaging studies to provide neural evidence for dual processes mediated by contrasting compositional and view-based representations. Their results implicate a ventral stream in attention requiring processing mediated by compositional representations and a dorsal stream implicated in view-based recognition which does not require attention. They adapted priming paradigms which used split and recombined priming stimuli. As Thoma and Henson [40] note:

> The current findings support hybrid models of visual object recognition that include both analytic and holistic object pathways, with the analytic pathway dependent on visual attention. Regions in the left ventral visual stream only showed repetition suppression (RS) from primes in more anterior fusiform regions, and the amount of this RS correlated with the amount of behavioural priming, consistent with an analytic pathway. Regions in the dorsal stream on the other hand, specifically the intraparietal sulcus, showed repetition enhancement (RE) only for intact primes, regardless of attention and the amount of RE correlated with the amount of behavioural priming from uncued, intact primes, consistent with a holistic pathway. ([40], p. 524)

2.5   Summary of Representational Properties of Dual Recognition Systems

Dual process theories have become quite well established through convergent evidence from behavioural, neuropsychological and imaging studies. In these theories, it is hypothesised that the representations mediating recognition differ as a function of whether an object is attended or ignored [35] and multiple visual representations are activated in response to attended objects. These include: (1) compositional structural descriptions with explicit relations that require attention as visual features must be bound into parts and parts then need to be bound to relations; and (2) non-compositional image-based representations of specific views that are activated in response to attended and non-attended objects (that are outside the focus of attention). In addition to being distinguished by whether they require attention, representations also differ in how much they are abstracted from the sensory image. Compositional representations need to abstract away from many details of the image present in sensory data, whilst image-based representations do not need to significantly abstract from sense data. Figure 2 presents some distinctions between compositional and view-based recognition processes.

However, the great majority of this convergent evidence for the operation of dual processes (and dual representations) in visual object recognition has come from studies using adults as participants, leaving open the possibility of different developmental trajectories for image-based and compositional representations. The next section presents relevant empirical evidence from recent studies with adults and children.

Fig. 2 Table summarising the contrasting properties of the representations mediating structural description and view-based recognition processes

Structural description representations	View-based representations
Abstracted away from particular examples in sense data	Based upon view - specific details of an exemplar image with no abstraction
Compositionally formulated representations in perception and memory	Image-based/analogue representations in perception and memory
Analytic (decomposition into parts and recombination with relations between parts)	Holistic (no decomposition into isolatable parts and no explicit relations between parts)
3D representations	2D representations
Response to object rotation is viewpoint invariance performance	Response to rotation is viewpoint dependence performance
Ventral stream implicated	Dorsal stream implicated

3 The Development of Configural Processing in Object Recognition: Recent Empirical Results

A number of behavioural studies suggest there is a retarded developmental trajectory for object recognition, with object recognition skills continuing to significantly improve during adolescence [9, 24, 32]. Recently, Jüttner et al. [26] examined developmental trends associated with identification of correct pictures when presented alongside incorrect distracters (in a 3 alternative-forced-choice (AFC) task). Two distracter types were compared: part-changed stimuli, where one part of the stimuli was substituted for an incorrect part (Fig. 3); and a change to the overall proportions of the object (the configural change condition, Fig. 4).

Fig. 3 Showing an animal version of a part-change stimuli used in human studies. Selecting the 'real' cow image is a non-configural task as only one object-part needs to be checked at a time

Fig. 4 Showing an animal version of a relative size change stimuli used in human studies. Selecting the 'real' fly image is a configural task as recognition results from checking the relative sizes of two (or more) parts

In both part-change and configural (relative size) change conditions, the task is to choose the 'correct' image. So in Fig. 3 the bottom 'cow' is the only image with a cow's head. In Fig. 4 the middle 'fly' is the only one with eyes that are the correct size in proportion to its body. In addition to stimuli derived from a set of naturalistic animal images, experiments were undertaken with stimuli from naturalistic images of defined-base, rigid artefacts (see [26], p. 163 for examples). Responses to defined-base, rigid artefact stimuli (Figs. 5 and 6) showed the same pattern of results as the animal stimuli.

The part-change and configural change sets of experimental stimuli were calibrated to be equally difficult for adults, with an 0.8 mean accuracy set for both conditions. After calibration with adults on upright stimuli, adult performance was recorded on inverted (upside down) versions of the stimuli. Then the same stimuli set was used to assess recognition performance in school children aged between 7

Fig. 5 Showing an artefact version of a part-change stimuli used in human studies. Selecting the 'real' bicycle image is a non-configural task as only one object-part needs to be checked at a time

and 16 years in upright and inverted conditions. Overall, 32 participants were used in each of six age ranges (7–8, 9–10, 11–12, 13–14, 15–16 and adult).

The full description of method and results for these experiments is detailed in [26]. Performance in terms of accuracy, and latency preceding a correct response, show a similar pattern of results to each other, with no evidence of a speed/accuracy trade-off. The key empirical results for younger children (7–10-years-old) are that, whilst part-change performance is marginally lower than adult levels, relative size change performance is significantly lower. For older children (11–16-years-old), part-change performance has reached the adult level whilst relative size change performance is still not fully consolidated [26]. Figure 7 shows mean and standard errors of the recognition accuracy, with results combined across animals and artefacts, as the stimulus type (animal/artefact) did not significantly affect recognition accuracy or latency nor interact with any other experimental variable. Developmental studies have used the differential performance of recognition using configural processing and part-based processing as a surrogate for differing access/use of image-based and structural description representations. This is because, whilst part-based recognition

Fig. 6 Showing an artefact version of a relative size change stimuli used in human studies. Selecting the 'real' motorbike image is a configural task as recognition results from checking the relative sizes of two (or more) parts

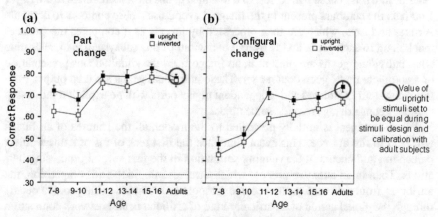

Fig. 7 Results of experiments where participants of different ages were tested with part and configural changed stimuli

only requires focussing on a single object part, configural processing requires attending to at least two parts at the same time.

To evaluate further this possible dual process explanation for these results, this paper now presents simulation results gained by developmentally regressing JIM3 [21], a prominent dual process model that simulates visual object recognition. JIM3 was created as an implementation of RBC theory as a computer simulation, and its use in computational modelling of human vision demonstrates how machine vision algorithms can investigate representations for object recognition.

4 JIM3: A Dual Process Model of Object Recognition

4.1 Introduction to JIM3

JIM3 is an eight-layer artificial neural network model of visual object recognition [21–23]. It takes as input a representation of contours from a single object's image. The output is a representation of an object's identity. Figure 8 (adapted from [21]) shows JIM3's eight layers and the two places where changes were made to developmentally regress the architecture.

4.2 Layers 1–3: From Feature Maps to Independent Geons

The first three layers comprise feature maps and are concerned with grouping local features into sets. These sets correspond to which geons the features arise from. Layer 1 outputs the contours present in the image. Layer 2 uses these contours to compute vertices and axes, which are then processed by layer 3 as it computes the surfaces that belong to each geon. So the overall behaviour of this subsystem is to determine what individual geons are present in an image from the simultaneous presentation of a complete multi-geon contour set. These individual geons are then output from this subsystem as isolated and independent object parts with no explicit relationship to other geons arising from the same object.

When an object is initially presented to the model, all the features of an image will tend to fire at once. This event simulates the first tens of ms of natural object perception and occurs in the running simulation in the first several processing iterations. Then in an attentive process which involves inhibition and competition, the attributes from different geons become temporally separated. This process occurs through the global action of a particular kind of artificial neural network connection termed by [22] as fast enabling links (FELs).

The first three layers of JIM3 act together to output each component geon at a different point in time. If this did not happen and attributes of separate geons fired synchronously, then their attributes would get super-imposed. The three conditions

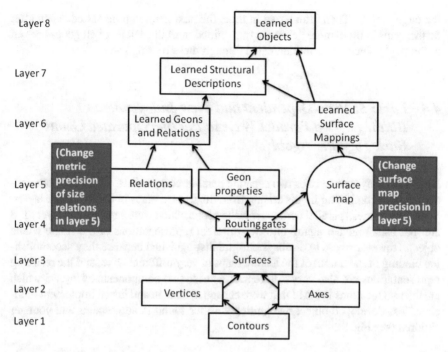

Fig. 8 Diagram of JIM3 showing the two locations in the architecture where changes were made to capture this architecture's performance for an earlier developmental stage

which cause FELs to treat units as from the same geon are: local coarse coding of image contours; cotermination in a intra-geon vertex; and, distant collinearity through lone terminations. The simultaneously firing features become organised so that only the attributes for a single geon fire at one time by an iterative process of competition and inhibition.

4.3 Layer 4: Routing Gates (Passing Each Independent Geon Forward Separated in Time from the Other Geons)

The fourth layer is a set of routing gates that splits the output from the first three layers and sends this output to two separate subsystems in layer 5. The information carried by these routing gates is of attribute sets for individual geons. After an initial period of phase locking, the information about individual geons are sent as temporally separated signals. That is, attributes for individual geons are transmitted together and separated in time from the transmission of attributes describing the other geons present in the target object. This means that at any particular time the output from the routing gates is just an attribute set for one individual geon from the target object. Then after a gap in time, the next geon is transmitted. Then after further gaps in time, more geons are transmitted until the details of all geons present in the target object are communicated through these routing gates.

4.4   Layer 5: View-Dependent and View-Independent Bindings: Two Parallel Ways to Put the Separated Geons Back Together Again

JIM3's fifth layer comprises two separate parallel components. These are both concerned with combining inputs arising from the feature maps in the first three layers. So in both of these parallel components the geons which were separated in layers 1–3 are 'put back together again' into two different representations of the single whole object. However, these two subsystems are distinguished because they accomplishing binding of the output of the feature maps in very different ways, and the resulting representations are also very different. It is these two components of layer 5 which are the two locations in JIM3 that were chosen to change and hence implement models of less developed object recognition abilities found in adolescents and younger children (see Fig. 8).

4.5   The View-Independent Subsystem

A view-independent subsystem called the independent geon array (IGA) acts to form representations of explicit relations between geons, thus dynamically (but slowly) forming a view-independent structural description of the object. It accomplishes binding of the geons which result from the first three layers by identifying how individual geons relate to each other in terms of relative size and relative position within the overall object they originated from. So this attention-requiring component of layer 5 is a serial mechanism rather than the global parallel and distributed processing mechanism that operates in the view-dependent surface map.

This subsystem achieves several important outcomes not achieved by the faster view-dependent system. First, the attribute-relation structure is formed explicitly. Since relations among geons are made explicit they enable humans to appreciate relational similarities between objects independently of whether similar object parts stand in corresponding relations. So we can appreciate two objects are similar if they have a large geon above a small geon, whatever the non-accidental properties of any of the geons. Second, relations are dynamically bound to the geons they describe. So this provides the potential for recognising complex multi-geon objects with a variety of interrelationships between the geons; to do this with static binding mechanisms such as templates might involve an impractically large set of templates ([23], p. 204). Figure 9 presents three different objects all described by the same set of attributes:

(a)   (b)   (c)

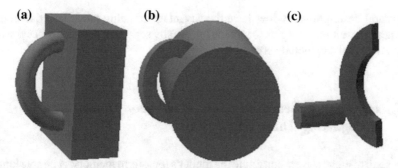

Fig. 9 Figure showing how very different objects can be formed from combination of the same attributes but with different relations between the attributes. Object **a** has a large geon with straight sides and straight cross-section beside a smaller geon which is curved along axis with a round cross-section. Object **b** has a large geon with straight sides and a round cross-section beside a smaller geon which is curved along its axis and with a straight cross-section. Object **c** has a large geon which is curved along its axis with a straight cross section beside a smaller geon with straight sides and round cross section

a large part and a small part, one part curved and one straight edged and one part with a straight cross-section and one part with a curved cross-section. What distinguishes these two objects is how particular attributes are dynamically linked to each other. So in the attributes for object (a), the large geon attribute is linked to straight sides and straight cross-section, whilst the small geon attribute is related to curved edge and curved cross-section. Thirdly, forming relations which are invariant with geon identity and viewpoint allows the formation of a structural description that will remain the same under translation, scale and left–right reflection and is relatively insensitive to rotation in depth [22].

4.6 The View-Dependent Subsystem

The soonest to complete is the surface map representation in the other subsystem in layer 5. This accomplishes a view-dependent static binding of geons by coding where each geon is fixed at a specific position in a holistic surface map. This 2D representation captures the interrelation of geons as they were perceived in one particular view. The mapping from the output of the feature maps in the first subsystem preserves the topological relations and metric properties of the geon attributes but discards their absolute sizes and location in the image. This means that the target image representation in the holistic surface map is invariant with translation and scale. However, because the topological relations and metric properties that are preserved in the holistic surface map come from only one particular view of the object, this representation is sensitive to rotation in depth and the picture plan and left–right reflection ([21], p. 498). Although this second subsystem in layer 5 does not form structural descriptions, it does have the advantage of being much faster, as it does not need to wait for its inputs to include temporally separated geons, a process which takes time and can include errors.

4.7 Layers 6–8: Learning About Multi-geon Objects and Recognising Them When Learnt

The sixth to eighth layers constitute the model's long-term memory. A simple kind of unsupervised Hebbian learning is used to encode the patterns of activation generated in layer 5. Each unit in layer 6 learns to respond to geon shape attributes and relations. Units in layer 7 sum input from layer 6 to reconstruct patterns representing geons and relations into complete structural descriptions of whole objects. These layer 7 units then activate object identity units in layer 8.

5 Simulation Results for Experiments Using Animal and Artefact Stimuli

5.1 Procedure for Simulation Experiments

To simulate the results from the animals and artefacts experiments of [26] we developmentally regressed JIM3 by changing two properties of the model. Figure 8 shows that the locations where the two parameters were changed were both in layer 5 of JIM3. The parameters chosen to make less mature, 'child' versions of JIM3 were the numbers of 'neurons' involved in processing in these two components. It was assumed that, at earlier levels of development, there might be either less resources given to recognition tasks (or perhaps these resources would be used less effectively) and this would be expected to decrease performance.

First, on the assumption that children have a less metrically-precise holistic representation of object shape than do adults, we reduced the number of locations in the model's surface map from 17 (the centre plus two radii and eight orientations away from the centre) to 9 (the centre plus two radii and four orientations), 5 (the centre plus one radius and four orientations); and 1 (a single central location). So with 17 neurons the model's surface map provides the most precise representation of the target object metric properties and when reduced respectively to nine neurons the model loses the eight neurons that provide the most fine-grained metric precision. Then with each further reduction in surface map neurons, it is again the neurons that provide the most metric precision that are removed. Second, on the assumption that children are generally much less relational than adults in their thinking (an assumption for which there is a great deal of empirical support [13]), we removed relation units from the model's independent geon array (IGA) for the child simulations. As a result of this change, the 'child' version of the model has an implicit representation of an object's inter-part relations in the surface map at an adult level, but less resources given to an explicit representation of those relations.

Before these developmentally regressed versions of JIM3 were used, we decided upon a performance measure which would allow straightforward comparison between the performance of JIM3 and the results reported by Jüttner et al.'s experiments with human participants [26]. We also developed a set of stimuli which was calibrated in a similar manner to the calibration carried out in the empirical studies with humans.

5.1.1 Performance Measure

In the original experiments of [26], human subjects (adults and children of various ages) were tested for their ability to choose the correct picture of an animal or an artefact from a display depicting an un-altered picture of that animal or artefact along with two distracters. There were two main conditions arising from use of two different types of distracter: a variant constructed by changing one part of the original object and another variant created by changing the relative part sizes of the original object (and thus effectively changing the metric relations among the object's parts).

JIM3 is not capable of performing this 'choose the correct object out of three' task (instead, it simply views one object at a time and attempts to find the best match in its long term memory (LTM)). Therefore, we developed a performance measure to estimate how well it would perform the choice task based on how well each object matched the correct (trained) object and each of the distracters activated the trained object's representation in the model's LTM. This measure was based on the model's response time to recognize an object (the number of iterations until an object [trained object or distracter variant] activated the corresponding trained object's representation in LTM to criterion [21]). A second possible measure which might be used when the model could not activate the corresponding trained object representation was the model's accuracy (i.e. the likelihood that an object [trained or distracter] would activate the corresponding trained object's representation in LTM). However, this was not used because the simulations typically 'recognised' both target objects and distracters as the target object, with the only distinction between conditions being how many simulation cycles this took (since the distracter objects were not present in the set of recognition targets present in the learning phase).

The logic of these measures is that, the more closely a distracter matches the representation of a trained object in LTM, the more difficult it would be for the model to correctly reject that distracter in favour of the trained target. Accordingly, our RT-based measure of performance consists of the model's RT to 'correctly' recognise a distracter (either non-accidental-property (NAP) or size change) as an instance of the trained target. So although the model did not correctly reject the distracters (even very long durations eventually resulted in recognition of the learned target), it is the closest performance measure to a 'rejection' of a distracter that the current implementation of JIM3 can support.

A drawback of this performance measure is that, since it compares human performance accuracy with simulation timing, it does not provide a straightforward comparison between different types of task that take different amounts of time to be carried out within the simulation. This applies to the upright and inverted stimuli tasks, with inverted stimuli taking longer to be recognised than upright stimuli. This does not of course mean that inverted stimuli are easier to recognise. So within a manipulation this performance measure does allow for comparisons, but between manipulations we cannot say that longer to recognition in JIM3 infers better discrimination performance.

5.1.2 Calibration of Stimuli for Equal Difficulty with the Adult Version of JIM3

The original behavioural experiments involved a calibration stage where part-change and configural change stimuli sets were formed to be of equivalent difficulty. Following this original design, we ran pilot simulations with JIM3 to equate the discriminability of the NAP and size-change variants of the trained stimuli.

Specifically, we made five novel multi-part objects and trained JIM3 to recognize them, along with the dozen or so objects it was trained to recognize in the simulations reported in [21]. We then made two variants of each trained stimulus. An NAP distracter was made by changing one non-accidental property of one geon in the corresponding trained object; and a size-change distracter was made by changing the size of one geon in the corresponding trained object. During piloting we made several variants of each size-change distracter and chose, for the final simulations, the variant whose discriminability from the corresponding trained object most closely matched that of the NAP distracter. That is, following the original experiment, we explicitly equated the NAP and size-change distracters for their discriminability from the corresponding trained objects to adults. For the adult version we used JIM3 in its original 2001 version [21], with σ (the standard deviation on the Gaussian receptive fields of the memory units in layer 6) set to 0.5 and the metric precision and surface map precision set to maximum (adult) values. In Fig. 10 we can see data points emphasised with dashed circles that the performance measures for the NAP changes averaged at 10.94 simulation cycles and were 10.8 for the relative size configural changed stimuli.

5.2 Results of Simulation of Animals and Artefacts Experiment

Figures 10 and 11 show the results of the two sets of computational experiments with JIM3 developmentally regressed from adult level (3) to three lower levels of development (level '0' being the most regressed), with Fig. 10 presenting results

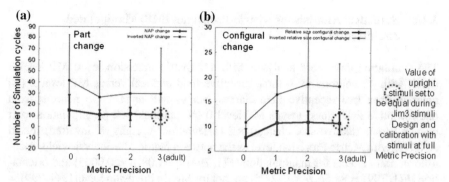

Fig. 10 Showing simulation results of animals and artefacts experiment with Metric Precision at four different levels of 'development'

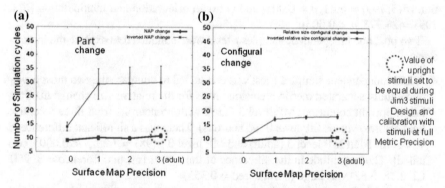

Fig. 11 Showing simulation results of animals and artefacts experiment with the surface map at four different levels of 'development'

with metric relation precision in the IGA decreased and Fig. 11 with surface map precision decreased. Both these graphs show adult results in the upright condition circled with a dashed line—denoting that they were calibrated to be similar in value.

Figure 10 shows that, as metric precision in the IGA is decreased, there is a different pattern of results in the NAP change and configural change conditions. As metric precision tends to zero neurons being used, recognition performance in the configural change condition drops. However, we seem to see a performance increase with the NAP change condition when metric precision is decreased. So these simulated results show the same qualitative pattern found in the empirical results presented in [26].

Figure 11 shows that, as surface map precision is decreased, there is no evidence to suggest a different pattern of results between the NAP change and configural change conditions. This pattern of results is therefore different from the empirical results presented in [26].

5.2.1 Statistical Analysis for Metric-Properties (MP) Manipulated Architecture

The simulation data were analysed with a 4 metric precision level: MP level 3 (adult with 45 neurons from three receptive field classes) versus MP level 2 (30 neurons from two receptive field classes) versus MP level 1 (15 neurons from one receptive field class) versus MP level 0 (no neurons) $\times 2$ (manipulation: part change versus relative size change) $\times 2$ (orientation: upright vs. inverted) mixed ANOVA with Metric Precision level as the between factor. The analysis yielded significant main effects for manipulation $[F(1,799) = 41.08, p < 0.0005]$ and orientation $[F(1,799) = 84.56, p < 0.0005]$ but not for Metric Precision Level $[F(1,799) = 0.571, p = 0.634]$.

Significant interactions were found between metric precision level and manipulation $[F(3,799) = 5.41, p = 0.001]$ and between orientation and manipulation $[F(1, 799) = 24.773, p < 0.005]$.

Two post hoc independent-samples t-tests were conducted to explore the interactions:

- A first independent-samples t-test was conducted to compare the two most developmentally separated metric precision levels for the relative size change manipulation upright condition: MP level 3 (adult with 45 neurons from three receptive field classes) versus MP level 0 (no neurons). There was a significant difference in scores for adult MP level 3 (adult) and MP level 0 $[t(98) = 7.28, p < 0.0005$, two tailed]. The magnitude in the difference of the means (mean difference = 3, 95% CI: 2.18–3.82) was large (eta squared = 0.353).
- A second independent-samples t-test was conducted to compare the two most developmentally separated metric precision levels for the NAP change manipulation upright condition: MP level 3 (adult with 45 neurons from three receptive field classes) versus MP level 0 (no neurons). There was a non-significant difference in scores for adult MP level 3 (adult) (M = 3, and MP level 0 $[t(98) = -1.947, p = 0.054$, two tailed]. The magnitude in the difference of the means (mean difference = 4, 95% CI: −8.072 to 0.77) was large (eta squared = 0.85).

5.2.2 Discussion for MP Manipulated Architecture

The analysis showed that there is a significant difference between relative size changed and NAP changed stimuli, but this main effect may not be a clear match to the required discrepancy between relative size changed and NAP changed conditions specified in Sect. 1. This is because the inverted results may produce much of this main effect difference and the difference between simulations of upright stimuli may not be significant when considered on their own against each other. So a post hoc test, discussed below, provides a finer detailed analysis of the relative size change and NAP change manipulations in the upright condition.

There is also a significant main effect of orientation between upright and inverted stimuli, with the inverted stimuli taking longer to be incorrectly recognised. Our performance measure suggests that, within the same task, taking longer to be recognised is equivalent to a more accurate recognition. But between tasks this relationship does not hold. So since the simulation will actually take longer to recognise inverted stimuli because they are upside down, this main effect does not show that inverted stimuli are easier to discriminate.

There was no main effect for metric precision level. However, this does not mean that there were not differences between the simulated age ranges. A significant interaction was found between metric precision level and manipulation. So as the simulation parameters modelled 'younger' parameters in the relative size change condition, performance decreased and in the NAP change condition performance increased. So these results do match the empirical results, but as noted above, the larger component of this difference between manipulation conditions may have come from the inverted results, as these mean values differ more widely than the upright conditions. The interpretation that the inverted conditions provide most of the difference between manipulation conditions is strengthened by the significant interaction between orientation and manipulation, with inverted relative size changed stimuli having the longest number of simulation cycles to recognition (in the MP = 0 condition over 40 cycles).

The complication in the analysis of considering upright and inverted orientations together was resolved with a post hoc t-test which only looked at upright results to consider whether the 'youngest' MP regressed condition was significantly different from the adult performance level. This gave a very clear result. The relative size change condition showed significantly lower recognition performance for the youngest parameters, whereas the NAP change condition showed no significant difference between youngest and adult parameters, and a large effect size in the opposite direction to the relative size change condition (see Fig. 12). So the metric relation regressed simulations demonstrate a very clear dissociation in performance between the relative size change and NAP change conditions, just as the empirical results with human subjects show.

Metric relation regressed	MP 0	MP 3 (adult)	Δ	Effect size
Part (NAP) change	14.94	10.94	+4*	Large*
Config change	7.8	10.8	−3	Large
Surface map regressed	SMP 0	SMP 3 (adult)	Δ	Effect size
Part (NAP) change	9.2	10.94	−1.74	Moderate
Config change	9.56	10.8	−1.24	Moderate

Fig. 12 Key comparisons from post hoc t-tests. This table presents results from the two computational experiments, one which simulated the developmental regression of metric relation precision (*top* half of table) and the other experiment which regressed surface map precision (* not a significant difference, $p = 0.054$)

5.2.3   Statistical Analysis for Surface-Map (SM) Manipulated Architecture

The simulation cycle data were analysed in a 4 surface map level: SM level 3 (adult with 17 neurons in two further orientations from the center neuron) versus SM level 2 (9 neurons in two further orientations from the centre) versus SM level 1 (5 neurons in one further orientation from the centre) versus SM level 0 (1 neuron with no further orientations from central neuron) $\times 2$ (manipulation: part change vs. relative size change) $\times 2$ (orientation: upright vs. inverted) mixed ANOVA with metric precision level as the between factor. The analysis yielded significant main effects for manipulation $[F(1, 799) = 14.42, p < 0.0005]$ and orientation $[F(1, 799) = 71.81, p < 0.0005]$ and for surface precision level $[F(1, 799) = 6.4, p < 0.0005]$.

Significant interactions were found between manipulation and orientation $[F(1, 799) = 16.13, p < 0.0005]$ and between surface map precision and orientation $[F(1, 799) = 24.773, p < 0.001]$. The interaction between surface map precision level and manipulation was not significant.

Two post hoc independent-samples t-tests were conducted to explore the interaction:

- A first independent-samples t-test was conducted to compare the two most developmentally separated surface map precision levels for the relative size change manipulation upright condition: SM level 3 (adult with 17 neurons) versus SM level 0 (1 neuron). There was a significant difference in scores for SM level 3 (adult) and SM level 0 $[t(98) = 2.81, p = 0.006$, two tailed]. The magnitude in the difference of the means (mean difference $= 1.24$, 95% CI: 0.36–2.11) was moderate (eta squared $= 0.074$).
- A second independent-samples t-test was conducted to compare the two most developmentally separated surface map precision levels for the NAP change manipulation upright condition: SM level 3 (adult with 17 neurons) versus SM level 0 (1 neuron). There was a significant difference in scores for adult SM level 3 (adult) (M =3) and SM level 0 $[t(98) = 2.31, p = 0.023$, two tailed]. The magnitude in the difference of the means (mean difference $= 1.74$, 95% CI: 0.24–3.23) was moderate (eta squared $= 0.047$).

5.2.4   Discussion for SM Manipulated Architecture

The analysis showed that there is a significant difference between relative size changed and NAP changed stimuli; but as in the MP changed architecture, the SM changed inverted results may produce much of this main effect difference. So a post hoc test, reported below, provided a test of this point.

As with the MP regressed architecture, there is also a significant main effect of orientation between upright and inverted stimuli in the SM regressed experiments. The same explanation applies here as above: the inverted condition involves a different task so we cannot conclude inversion increases recognition performance.

There was a main effect for surface map precision level. As the simulation modelled 'younger' versions, performance levels decreased. Again, as with the MP regressed architecture, the SM-changed simulations show an interaction between surface map precision and orientation, with the longest number of simulation cycles recorded in the inverted condition. There is not a significant interaction between surface map precision and manipulation.

The post hoc t-test results highlight that the surface map regressed results do not match the empirical results reported in [26]. This test looked at whether the upright results for the 'youngest' SM regressed condition were significantly different from the adult performance level. Both manipulation conditions were significantly lower performing in the youngest SM condition than the adult condition, with a similar effect size. This pattern of results is clearly different from the empirical results reported by [26].

Figure 12 highlights the results of the post hoc t-tests for both the MP regressed and SM regressed architectures.

6 Conclusions

This paper shows that recent empirical results presented by Jüttner et al. [26] can be explained in terms of dual process models of object recognition. Simulations with the JIM3 artificial neural network suggest that a non-attentive process develops early in humans and allows part-based recognition at adult levels by children in the 7–10 age range. According to this dual process explanation, the observed developmental delay in the relative size change stimuli results from the later development of attention-requiring processes that support perception of relations between object parts and the production of structural descriptions in object perception and recognition.

Removing neurons from the non-attentive surface map in JIM3 did not cause a significant difference to appear in JIM3's performance on the part (NAP) change and configural (relative size) change conditions. However, a notable and surprising result was that it took reducing the neurons all the way to zero in the attention-requiring IGA to bring about a significant difference between these experimental conditions in the other set of computational experiments with JIM3. The psychological inferences that can be taken from this finding are discussed in more detail below. However, just viewing this result from the perspective of processing with machine representations provides a key lesson for artificial systems engineering. This is that the dual processes in JIM3 interact together in producing behaviour so that deficiencies in attention-requiring processes were masked by non-attentive processes. This highlights a more general challenge in empirical research on the structures used to represent reality: how should experimentalists untangle the interacting effects linked to multiple representation types?

The purpose of running simulations with varying precision levels for metric relations in the IGA and the holistic surface map was to see if either of these simulations captured the pattern of results shown in empirical observation of humans. What the human results from [26] showed was that performance for younger participants on configurally changed stimuli decreased compared with adult levels whereas performance on NAP changed stimuli stayed the same. A successful simulation should therefore show equal performance between stimuli distracter types for 'adult' parameters and show a lower performance on relative size change stimuli than part-change stimuli for developmentally younger simulation parameters. As can be seen comparing Figs. 7 and 10, the simulations where metric-relation precision level changes in the IGA were decreased provide a good qualitative fit to the pattern observed with Jüttner et al.'s artefact and animal stimuli [26]. Since the human participants performed a different task than did the model, it is impossible to provide a precise quantitative fit between the empirical and simulation data.

The limitations in this particular modelling exercise using JIM3 are of four types. Firstly, the task that the simulation carried out was probably more simple than various strategies likely used by the human experimental participants to eliminate distracters. In the JIM3 experiments time to recognition is always taken for stimuli presented on their own. The 'choose one from three task' gives more potential for using complex memory retrieval strategies than simply measuring time to recognition for a single object. In addition, which strategies might be used in either task is likely to change through development independently of the changes to resources given over to metric relation or surface map precision. Developing proficiency in metacognition and increasing cognitive resources have been presented as competing explanations in memory development [14]. The simulations reported here present development just in terms of an increase in the numbers of neurons used for recognition in the IGA and the holistic surface map. So this explains changes in performance over development just in terms of differences in cognitive resources. We can also imagine an analogous theory of development from 'increasing metacognition' when attempting to explain developmental trajectories in object recognition.

Secondly, the images that JIM3 learns and then recognises are simpler than the naturalistic 2D images used by [26]. The naturalistic images possess difference in texture and colour which the stimuli used by JIM3 do not possess.

Thirdly the modelling exploration has been set up as a two-horse race, to decide which of these changes to JIM3 provides the best fit for the pattern of empirical results for adults and children described by [26]. Each of these regressions was 'clean' in the sense that only one parameter at a time was regressed. In a real infant we might expect both MP and SM precision to decrease as well as there being a number of other changes that involve lower recognition performance for younger participants. For example, on the assumption that children have less stable and/or precise memories for objects than do adults, we might change σ on the Gaussian receptive fields in layer 6 of JIM3 from 0.5 (the value in the adult simulations) to 1.0. This increase would have the effect of making any given unit in layer 6 more tolerant of deviations from its preferred pattern (corresponding to the centre of the distribution). Possible future computational experiments with JIM3 might therefore involve co-varying the two existing changes with each other and with changes in σ. However, preliminary experiments have shown that decreasing σ on its own does not cause relative size stimuli to be processed less effectively, with mean simulation runs actually higher for relative size stimuli at a value of σ that gives minimal recognition performance.

Lastly, both the empirical results and the modelling research do not rule out the impact that differing life experiences and consequent encoding differences in memories might have on the performance of JIM3 after layer 5.

These four limitations of: (1) task and strategy simplicity, (2) 'clean' changes to parameters, (3) image simplicity and (4) learning experience in the simulation being equal between regressed and adult architectures might all be expected to increase recognition performance in JIM3 compared with human performance. So it may be as a result of a combination of these factors that it took decreasing the metric precision neurons to zero to get a large drop in performance. Alternatively, the finding that only the 'MP=0' condition provides a large decrement in performance may suggest that children of age 7–9 years really do have a much lower than previously expected ability to make metric judgements in visual object recognition. That this is not apparent in day-to-day life or in other kinds of object recognition experiment may be because this lower ability will only be apparent when children view objects in such a way that their highly performing 2D systems cannot quickly produce recognition. Otherwise partial orderings rather than absolute metric judgements may suffice. So one suggestion for future work is to adapt JIM3 so that it can support more complex tasks and more complex strategies, with image simplicity matched, with many parameters being systematically changed during simulations, and with learning regimes matched to those that the adult participants experienced. Some of these suggestions have already been carried out; for example, experiments have been conducted which control for differing previous experience with novel objects (see [26] experiment 3 and [25]). The finding that JIM3 needs to have no metric relation precision to qualitatively match 7–9-years-old human performance might also suggest new empirical studies where participants learn novel objects but are then presented with very different views of these objects so that the view-dependent system would not be expected to maintain high performance levels.

In addition to just thinking about the four limitations noted above for how the computer simulation matches the task used in the human experiments, we can also consider that the human experiments are a limited approach in capturing the complexities of object recognition in more ecologically valid contexts. For instance, the human recognition task modelled in this paper involves a participant sitting passively whilst being presented with images, which do not move and cannot be acted upon or manipulated. This is partly done to conserve clear experimental control between the experimental conditions. However, it does have the downside of limiting possible mechanisms of active perception, such as the development of sensorimotor contingencies, and so limiting the role of active perception mechanisms which may not rely on explicit representations. Future work may involve more active experimental tasks, and modelling these observations with robots rather than disembodied simulations.

There are also a number of deeper issues linked to the core features of JIM3. For example, in JIM3, both the view-dependent and view-independent routes through the architecture use geons as a fundamental representational unit. However, it is not a settled issue what the basic level in structural descriptions in visual object recognition

are. For example, children from 3 to 4 made less use than adults of the shape boundaries that distinguish different types of geons [1]. So to model children's performance
we might want to relax the requirement that geons are a fundamental representational
unit at earlier stages in development. In addition, it is also worth noting that JIM3
possesses surfaces in layer 3 of the architecture, but these surfaces are only used
in the assignment of geons before layer 4, rather than primitives for the spatial relationships recorded in the view-independent component of layer 5. However, surfaces
have been proposed as representational primitives within spatial relationships [27].

Secondly, in JIM3 there is limited opportunity for processing in later layers to
influence earlier processing in an on-line dynamic fashion. For example, top-down
effects of memory on processing before layer 5 through backward projections do not
occur in JIM3. We might imagine that attention emerges moment to moment as an
internal representation of an object emerges, a dynamic process not captured within
JIM3. Instead, in JIM3, attention is 'on full' as the object starts to be represented.

Lastly, JIM3 is a dual process model where each process is supported by different hardware, in the form of separate neural networks in layer 5. Other dual process
theories have a similar arrangement. For example, object perception and action are
proposed to occur in two separate dorsal and ventral streams [16, 40]. Alternatively,
the idea of dual processes can be de-linked from the idea of dual 'systems'. It may
be different processing occurs at different times on a common substrate. So 'dual
process–one system' could be a design schema for a new object recognition system
where compositional and non-compositional processes are separated in time but not
space. Alternatively, as Thoma and Henson's imaging results suggest, there may be
two streams: dorsal and ventral, where the dorsal stream is involved in solely view-
based recognition and the ventral stream involves some view-based as well as compositional recognition. Evidence for this complex arrangement is presented by Thoma
and Henson, who noted that in their imaging studies: "*the ventral stream regions
also showed greater RS from intact than split primes, which would not be expected
if these regions utilised purely structural representations*" ([40], p. 524). This finding makes intuitive sense if we think of attention building up over the briefest of
moments in time rather than starting 'on full'. So before attention can link object
parts dynamically with relations, this system will already be decomposing parts and
these isolated and independent parts may trigger backward activation from memory traces before fully compositional memories are matched to fully compositional
perceptual representations.

So, in summary, a version of JIM3 with regressed metric relation precision in the
IGA has been shown to provide a better match to empirical results than a regressed
holistic surface map version. An interesting finding is that even small numbers of
neurons present in the IGA can provide similar level of recognition performance to
an 'adult' JIM3 with its full complement of neurons. Though the lessons for human
psychology from this are still to be worked out, this work does provide an example
for research in machine representation of the benefits of dual representation systems.
Future work has also been suggested that: (1) would involve adapting JIM3 to more
closely match the types of task and stimuli and learning pattern used in empirical
studies of object recognition development; (2) that would involve empirical testing

of younger adolescents with stimuli that have been rotated so that the view-dependent mechanisms do not provide an effective route to recognition; and (3) would involve developing alternatives to JIM3 that support surfaces as a representational primitive, provide more backward projections to provide top-down effects of existing knowledge, and development of dual process–single system models where differences in processing exist across time but not across resources.

Philosophers have long theorised about compositionality and its benefits. This research illustrates the challenges in investigating how object representations develop. These include that, in natural systems, there is no transparent access to internal representations; performance on simple behavioural tasks, such as measuring view-dependence/invariance to object recognition of rotated images, can act as a poor surrogate for internal representations; multiple representational forms can interact to produce complex behavioural patterns; and the existing implemented computational models do not always neatly fit completely with emerging empirical paradigms. However, using a variety of investigative methods, including priming experiments, neuropsychological studies, brain imaging and computational modelling can provide convergent evidence and an elaborated view of how neural systems can support representational diversity in humans, other animals and machines.

References

1. Abecassis, M., Sera, M.D., Yonas, A., Schwade, J.: What's in a shape? children represent shape variability differently than adults when naming objects. J. Exp. Child Psychol. **78**, 213–239 (2001)
2. Barenholtz, E., Tarr, M.J.: Reconsidering the role of structure in vision. In: Markman, A., Ross, B. (eds.) The Psychology of Learning and Motivation, vol. 47. Elsevier, Amsterdam (2006)
3. Barenholtz, E., Tarr, M.J.: Visual judgment of similarity across shape transformations: evidence for a compositional model of articulated objects. Acta Psychol. **128**, 331–338 (2008)
4. Biederman, I.: Recognition by components: a theory of human image understanding. Psychol. Rev. **94**, 115–147 (1987)
5. Biederman, I., Bar, M.: One shot viewpoint invariance in matching novel objects. Vis. Res. **39**, 2885–2889 (1999)
6. Biederman, I., Gerhardstein, P.C.: Recognising depth rotated objects: evidence and conditions for three-dimensional viewpoint invariance. J. Exp. Psychol. Hum. Percept. Perform. **19**, 1162–1182 (1993)
7. Biederman, I., Gerhardstein, P.C.: Viewpoint dependent mechanisms in visual object recognition: reply to Tarr and Bulthoff. J. Exp. Psychol. Hum. Percept. Perform. **21**, 1506–1514 (1995)
8. Bulthoff, H.H., Edelman, S.: Psychophysical support for a two-dimensional view interpolation theory of object recognition. Proc. Natl. Acad. Sci. U.S.A. **89**, 60–64 (1992)
9. Davidoff, J., Roberson, D.: A theory of the discovery and predication of relational concepts. J. Exp. Child Psychol. **85**, 217–234 (2002)
10. Davidoff, J., Warrington, E.K.: The bare bones of object recognition: implications from a case of object recognition impairment. Neuropsychologia **26**, 279–292 (1999)
11. Davidoff, J., Warrington, E.K.: A particular difficulty in discriminating between mirror images. Neuropsychologia **39**, 1022–1036 (2001)
12. Doumas, L., Holyoak, K., Hummel, J.: The problems of using associations to carry binding information. Behav. Brain Sci. **29**, 74–75 (2006)

13. Doumas, L., Hummel, J., Sandhofer, C.: A theory of the discovery and predication of relational concepts. Psychol. Rev. **115**, 1–43 (2008)
14. Flavell, J.H.: First discussant's comment: what is memory development the development of? Hum. Dev. **14**, 272–278 (1971)
15. Foster, D., Gilson, S.: Recognizing novel three-dimensional objects by summing signals from parts and views. Proc. R. Soc. Lond. B **269**, 1939–1947 (2002)
16. Goodale, M.A., Milner, A.D.: Separate visual pathways for perception and action. Trends Neurosci. **15**(1), 20–25 (1992)
17. Hayward, W.: After the viewpoint debate: where next in object recognition. Trends Cogn. Sci. **7**, 425–427 (2003)
18. Hayward, W., Tarr, M.: Testing conditions for view point invariance in object recognition. J. Exp. Psychol. Hum. Percept. Perform. **23**, 1511–1521 (1997)
19. Heinke, D., Humphreys, G.W.: Attention, spatial representation and visual neglect: simulating emergent attention and spatial memory in the selective attention for identification model (SAIM). Psychol. Rev. **110**, 29–87 (2003)
20. Hummel, J.: Where view-based theories of human object recognition break down: the role of structure in human shape perception. In: Dietrich, E., Markman, A. (eds.) Cognitive Dynamics: Conceptual Change in Humans and Machines, pp. 157–185. Lawrence Erlbaum (2000)
21. Hummel, J.: Complementary solutions to the binding problem in vision: implications for shape perception and object recognition. Vis. Cogn. **8**, 489–517 (2001)
22. Hummel, J., Biederman, I.: Dynamic binding in a neural network for shape recognition. Psychol. Rev. **99**, 480–517 (1992)
23. Hummel, J., Stankiewicz, B.J.B.J.: Categorical relations in shape perception. Spat. Vis. **10**, 201–236 (1996)
24. Juttner, M., Muller, A., Rentschler, I.: A developmental dissociation of view-dependent and view-invariant object recognition in adolescence. Behav. Brain Res. **175**, 420–424 (2006)
25. Juttner, M., Petters, D., Wakui, E., Davidoff, J.: Late development of metric part-relational processing in object recognition. J. Exp. Psychol. Hum. Percept. Perform. **40**, 1718–1734 (2014)
26. Juttner, M., Wakui, E., Petters, D., Kaur, S., Davidoff, J.: Developmental trajectories for part-based and configural object recognition in adolescence. Dev. Psychol. **49**(1), 161–176 (2013)
27. Leek, E.C., Reppa, I., Arguin, M.: The structure of three-dimensional object representations in human vision: evidence from whole-part matching. J. Exp. Psychol. Hum. Percept. Perform. **31**, 668–684 (2005)
28. Logan, G.D.: Spatial attention and the apprehension of spatial relations. J. Exp. Psychol. Hum. Percept. Perform. **20**, 1015–1036 (1994)
29. Marr, D., Nishihara, H.K.: Representation and recognition of the spatial organization of three-dimensional shapes. Proc. R. Soc. Lond. B **200**, 269–294 (1978)
30. Olshausen, B., Anderson, C., Van Essen, D.: A neurobiological model of visual attention and invariant pattern recognition based on dynamic routing of information. J. Neurosci. **57**, 4700–4719 (1993)
31. Poggio, T., Edelman, S.: A network that learns to recognize three-dimensional objects. Nature **343**, 263–266 (1990)
32. Rentschler, I., Juttner, M., Osman, E., Mller, A., Caelli, T.: Development of configural 3D object recognition. Behav. Brain Res. **149**, 107–111 (2004)
33. Stankiewicz, B.: Empirical evidence for independent dimensions in the visual representation of three-dimensional shape. J. Exp. Psychol. Hum. Percept. Perform. **28**, 913–932 (2002)
34. Stankiewicz, B.J., Hummel, J.: Automatic priming for translation- and scale-invariant representations of object shape. Vis. Cogn. **6**, 719–739 (2002)
35. Stankiewicz, B.J., Hummel, J., Cooper, J.E.: The role of attention in priming for left-right reflections of object images: evidence for a dual representation s of object shape. J. Exp. Psychol. Hum. Percept. Perform. **24**, 732–744 (1998)
36. Sutherland, N.S.: Outlines of a theory of visual pattern recognition in animals and man. Proc. R. Soc. Lond. B **171**, 95–103 (1968)

37. Tarr, M., Bulthoff, H.: Is human object recognition better described by geon-structure-descriptions or by multiple views? J. Exp. Psychol. Hum. Percept. Perform. **21**, 1494–1505 (1995)
38. Tarr, M., Pinker, S.: Mental rotation and orientation dependence in shape recognition. Cogn. Psychol. **21**, 233–282 (1989)
39. Thoma, V., Davidoff, J.: Object recognition: attention and dual routes. In: Osaka, I., Rentschler, I., Biederman, I. (eds.) Object Recognition, Attention and Action
40. Thoma, V., Henson, R.N.: Object representations in ventral and dorsal visual streams: fMRI repetition effects depend on attention and part-whole configuration. Neuroimage **57**, 513–525 (2011)
41. Thomai, V., Davidoff, J.: Priming of depth-rotated objects depends on attention and part changes. Exp. Psychol. **53**, 31–47 (2006)
42. Thomai, V., Davidoff, J., Hummel, J.: Priming of plane-rotated objects depends on attention and view familiarity. Vis. Cogn. **15**, 179–210 (2007)
43. Thomai, V., Hummel, J., Davidoff, J.: Evidence for holistic representations of ignored images and analytic representations of attended images. J. Exp. Psychol. Hum. Percept. Perform. **30**, 257–267 (2004)
44. Triesman, A.M., Gelade, G.: A feature-integration theory of attention. Cogn. Psychol. **12**, 97–136 (1980)
45. Ullman, S.: Object representations in ventral and dorsal visual streams: fMRI repetition effects depend on attention and part-whole configuration. Cognition **32**, 193–254 (1989)
46. Ullman, S.: Three-dimensional object recognition based on the combination of views. Cognition **67**, 21–44 (1998)
47. van der Velde, F., de Kamps, M.: Neural blackboard architectures of combinatorial structures in cognition. Behav. Brain Sci. **29**, 37–108 (2006)
48. Wolfe, J.M., Cave, K.R., Franzel, S.L.: Guided search: an alternative to the feature integration model for visual search. J. Exp. Psychol. Hum. Percept. Perform. **15**, 419–433 (1989)

Part III
Natural Sciences Perspectives

The Quantum Field Theory (QFT) Dual Paradigm in Fundamental Physics and the Semantic Information Content and Measure in Cognitive Sciences

Gianfranco Basti

Abstract In this paper we explore the possibility of giving a justification of the "semantic information" content and measure, in the framework of the recent coalgebraic approach to quantum systems and quantum computation, extended to QFT systems. In QFT, indeed, any quantum system has to be considered as an "open" system, because it is always interacting with the background fluctuations of the quantum vacuum. Namely, the Hamiltonian in QFT always includes the quantum system and its inseparable thermal bath, formally "entangled" like an algebra with its coalgebra, according to the principle of the "doubling" of the degrees of freedom (DDF) between them. This is the core of the representation theory of cognitive neuroscience based on QFT. Moreover, in QFT, the probabilities of the quantum states follow a Wigner distribution, based on the notion and measure of quasiprobability, where regions integrated under given expectation values do not represent mutually exclusive states. This means that a computing agent, either natural or artificial, in QFT, against the quantum Turing machine paradigm, is able to change dynamically the representation space of its computations. This depends on the possibility of interpreting QFT system computations within the framework of category theory logic and its principle of duality between opposed categories, such as the algebra and coalgebra categories of QFT. This allows us to justify and not only to suppose, like in the "theory of strong semantic information" of L. Floridi, the definition of modal "local truth" and the notion of semantic information as a measure of it, despite both measures being defined on quasiprobability distributions.

G. Basti (✉)
Faculty of Philosophy, Pontifical Lateran University, Vatican City, Italy
e-mail: basti@pul.it

© Springer International Publishing AG 2017   177
G. Dodig-Crnkovic and R. Giovagnoli (eds.), *Representation and Reality in Humans,*
Other Living Organisms and Intelligent Machines, Studies in Applied Philosophy,
Epistemology and Rational Ethics 28, DOI 10.1007/978-3-319-43784-2_9

1 Introduction: A Paradigm Shift

Perhaps the best synthesis of the current paradigm shift in fundamental physics is the positive answer that it seems necessary to give to the following question: "Is physics legislated by cosmogony?". Such a question is the title of a visionary paper written in 1975 by J.A. Wheeler and C.M. Patton and published in the first volume of a successful series of the Oxford University about quantum gravity [1].

Such a revolution, suggesting a dynamic justification of the physical laws, fundamentally amounts to the so-called *information-theoretic approach* in quantum physics as the natural science counterpart of a *dual ontology* taking information and energy as two fundamental magnitudes in basic physics and cosmology. This approach started from Richard Feynman's influential speculation that a quantum computer could simulate any physical system [2]. This is the meaning of the famous "it from bit" principle posited by R. Feynman's teacher, Wheeler [3, p. 75]. The cornerstones of this reinterpretation are, moreover, D. Deutsch's demonstration of the universality of the quantum universal Turing Machine (QTM) [4], and C. Rovelli's overall development of a *relational* quantum mechanics (QM) [5]. An updated survey of such an informational approach to fundamental physics is provided in the recent collective book, edited by H. Zenil, and with contributions, among others, from R. Penrose, C. Hewitt, G. J. Chaitin, F. A. Doria, E. Fredkin, M. Hutter, S. Wolfram, S. Lloyd, besides D. Deutsch himself [6].

There are, however, several theoretical versions of the information-theoretic approach to quantum physics. It is not important to discuss all of them here (for an updated list in QM, see, for instance [7]), even though all can be reduced to essentially two:

1. The first one is the classical "infinitistic" approach to the *mathematical physics* of information in QM. Typical of this approach is the notion of the *unitary evolution* of the *wave function*, with the connected, supposed *infinite* amount of information it "contains" being "made available" in different spatiotemporal cells via the mechanism of the "decoherence" of the wave function. Finally, essential for this approach is the necessity of supposing *an external observer* ("information for whom?" [7]) for the foundation of the notion and of the measure of information. This is ultimately Shannon's purely syntactic measure and notion of information in QM [5]. Among the most prominent representatives of such an approach, we can quote the German physicist Zeh [8, 9] and the Swedish physicist Tegmark [10].

2. The second approach, the emergent one today, is related to a "finitary"[1] approach to the *physical mathematics* of information, taken as a fundamental physical magnitude together with energy. It is related to quantum field theory (QFT), because of the possibility it gives of spanning the microphysical, macrophysical, and even the cosmological realms, within only one quantum theoretical framework, differently from QM [17].

In this chapter we discuss the relevance of this second approach for the theory of *semantic information* in both biological and cognitive sciences.

2 From QM to QFT in Fundamental Physics

The notion of the quantum vacuum is fundamental in QFT. This notion is the only possible explanation, at the fundamental microscopic level, of the *third principle of thermodynamics* ("The entropy of a system approaches a constant value as the temperature approaches zero"). Indeed, the Nobel Laureate Walter Nernst first discovered that, for a given mole of matter (namely an ensemble of an Avogadro number of atoms or molecules) for temperatures close to absolute zero, T_0, the variation of the entropy ΔS would become infinite (through division by 0).

Nernst demonstrated that, to avoid this catastrophe, we have to suppose that the molar heat capacity C is not constant at all, but vanishes, in the limit $T \to 0$, to make ΔS finite, as it has to be. This means, however, that near absolute zero, there is a mismatch between the variation of the bodies content of energy and the supply of energy from the outside. We can only avoid such a paradox by supposing that such a mysterious inner supplier of energy is the vacuum. This implies that absolute zero is unreachable. In other terms, there is an unavoidable fluctuation of the elementary constituents of matter. The ontological conclusion for fundamental physics is that we can no longer conceive physical bodies as isolated.

> The vacuum becomes a bridge that connects all objects among them. No isolated body can exist, and the fundamental physical actor is no longer the atom, but the field, namely the atom space distributions variable with time. Atoms become the "quanta" of this matter field, in the same way as the photons are the quanta of the electromagnetic field [18, p. 1876].

---

[1]For the notion of "finitary" computation, as distinguished from "infinitistic" (second-order computation) and "finististic" (Turing-like computation), see [11]. This notion depends on the category theory (CT) interpretation of logic and computation [12], as far as based on Aczel's non-well founded (NWF) set theory [13], justifying a *coalgebraic semantics* in quantum computing [14], as far as based on the CT principle of the *dual equivalence* between a Boolean initial algebra and a final coalgebra [15, 16]. The key notion of the *doubling of the degrees of freedom* between a q-deformed Hopf algebra and a q-deformed Hopf coalgebra, as representing each quantum system in quantum field theory, perfectly satisfies such a logic, as we see below.

For this discovery, eliminating once and forever the notion of "inert isolated bodies" of Newtonian mechanics, Walter Nernst, a chemist, is one of the founders of modern quantum physics.

Therefore, the theoretical, core difference between QM and thermal QFT can be essentially reduced to the criticism of the classical interpretation of QFT as a "second quantization" of QM. In QFT, indeed, the classical Stone von Neumann theorem [19] does not hold. This theorem states that, for system with a *finite* number of degrees of freedom, which is always the case in QM, the representations of the canonical commutation relations (CCRs)[2] are all *unitarily equivalent to each other*, so as.to justify the exclusive use of Shannon information in QM.

On the contrary, in QFT systems, the number of degrees of freedom is not finite, so that infinitely many unitarily inequivalent representations of the canonical commutation (bosons) and anticommutation (fermions) relations exist. Indeed, through the principle of *spontaneous symmetry breaking* (SSB) in the vacuum ground state, infinitely (not denumerable) many quantum vacuum conditions, compatible with the ground state, exist there. Moreover, this not holds only in the relativistic (microscopic) domain, but also applies to nonrelativistic many-body systems in condensed matter physics, i.e. in the macroscopic domain, and even on the cosmological scale [17, pp. 18. 53–96].

Indeed, starting from the discovery, during the 1960s, of dynamically generated long-range correlations mediated by *Nambu–Goldstone bosons* (NGBs) [20, 21], and hence their role in the local gauge theory through the Higgs field, the discovery of these collective modes deeply changed fundamental physics. Above all, it appears as an effective, alternative method to the classically Newtonian paradigm of perturbation theory, and hence to its postulate of the asymptotic condition.

In this sense, "QFT can be recognized as an *intrinsically thermal* quantum theory" [17, p. ix]. Of course, because of the intrinsic character of the thermal bath, the whole QFT system can recover the classical Hamiltonian character, because of the necessity of still satisfying the energy balance condition of each QFT (sub) system with its thermal bath ($\Delta E = 0$), mathematically formalized by the "algebra doubling" between a q-deformed Hopf algebra and its "dual" (see note 2) q-deformed Hopf coalgebra, where q is a thermal parameter [22].

Therefore, in QFT an uncertainty relation holds, similar to the one of Heisenberg, relating the uncertainty on the number of field quanta to that of the field phase, namely

$$\Delta n \Delta \varphi \geq \varphi(\hbar),$$

---

[2]It is useful to recall here that the *canonical variables* (e.g. position and momentum) of a quantum particle do not commute among themselves, like in classical mechanics, because of Heisenberg's uncertainty principle. The fundamental discovery of D. Hilbert consists in demonstrating that each canonical variable of a quantum particle commutes with the Fourier transform of the other (such a relationship constitutes a CCR), allowing a geometrical representation of all the states of a quantum system in terms of a commuting variety, i.e. the relative "Hilbert space".

where n is the number of quanta of the force field, and φ is the field phase. If $\Delta n = 0$, φ is undefined so that it makes sense to neglect the waveform aspect in favor of the individual, particle-like behavior. On the contrary, if $\Delta\varphi = 0$, n is undefined because an extremely high number of quanta are oscillating together according to a well-defined phase, i.e. within a given phase coherence domain. In this way, it would be nonsensical to describe the phenomenon in terms of individual particle behavior, since the collective modes of the force field prevail.

In QFT there is a duality between *two dynamic entities*: the fundamental force field and the associated quantum particles that are simply the quanta of the associated field that is different for different types of particles. In this a way, quantum entanglement does not imply any odd relationship between particles like in QM, but is simply an expression of the unitary character of a force field. To sum up, according to this more coherent view, the Schrödinger wave function of QM appears to be only a statistical coverage of the finest structure of the dynamic nature of reality.

3 QFT of Dissipative Structures in Biological Systems

3.1 Order and Vacuum Symmetry Breakdowns

It is well known that a domain of successful application of QFT is the study of the microphysics of condensed matter, that is in systems displaying at the macroscopic level a high degree of coherence related to an *order parameter*. The "order parameter", which is the macroscopic variable characterizing the new emerging level of matter organization, is related to the *matter density distribution*. In fact, in a crystal, the atoms (or molecules) are "ordered" in well-defined positions, according to a *periodicity law* individuating the crystal lattice.

Other examples of such ordered systems in the condensed matter realm include magnets, lasers, superconductors, etc. In all these systems, the emerging properties related to the respective order parameters are neither the properties of the elementary constituents, nor their "summation", but new properties depending on *the modes in which they are organized*, and hence on *the dynamics controlling their interactions*. In this way, for each new macroscopic structure, e.g. crystal, magnet or laser, there corresponds a new "function" the "crystal function", the "magnet function", etc.

Moreover, all these emerging structures and functions are controlled by *dynamic parameters*, that in engineering terminology, we can define as *control parameters*. Changing one of them, the elements can be subject to different dynamics with different collective properties, and hence exhibit different collective behaviors and functions. Generally, the temperature is the most important of them. For instance, crystals beyond a given critical temperature—that is different for different materials—lose their crystal-like ordering, and the elements acquire as a whole the macroscopic structure-functions of an amorphous solid or, for higher temperatures, they lose any static structure, acquiring the behavior-function of a gas.

So, any process of *dynamic ordering*, and of *information gain*, is related with a process of *symmetry breakdown*. In the magnet case, the "broken symmetry" is the rotational symmetry of the magnetic dipole of the electrons, and the "magnetization" consists in the correlation among all (most) electrons, so that they all "choose", among all the directions, that of the magnetization vector.

To sum up, any dynamic ordering among many objects implies an "order relation", i.e. a *correlation* among them. What, in QFT, at the *mesoscopic/macroscopic* level is denoted as *correlation waves* among molecular structures and their chemical interactions, at the *microscopic* level any correlation, and more generally any interaction, are as many coherent oscillation modes of force fields, mediated by *quantum correlation particles*. They are called "Goldstone bosons" or "Nambu–Goldstone bosons (NGBs)" [20, 21, 23], with mass—even though always very small (if the symmetry is not perfect in finite spaces)—or *without mass at all* (if symmetry is perfect, in the abstract infinite space). The lower the inertia (mass) of the correlation quantum, the greater the distance over which it can propagate, and hence the distance over which the correlation (and the ordering relation) constitutes itself.

However, an important caveat is necessary regarding the different role of Goldstone bosons as quantum correlation particles, and the bosons of the different energy fields of quantum physics (Quantum Electro-Dynamics (QED), and Quantum Chromo-Dynamics (QCD)). These latter are the so-called *gauge bosons*: the photons γ of the electromagnetic field, the gluons g of the strong field, the bosons $W^{\pm}$ and the boson Z of the electroweak field, and the scalar Higgs boson H^0 of the Higgs field, common to all these interactions.

The gauge bosons are properly mediators of *energy exchanges* among the interacting elements they correlate, because they are effectively quanta of the energy field they mediate (e.g. the photon is the quantum of the electromagnetic field). Therefore, the energy quanta are bosons that can change the *energy state* of the system. For instance, in QED of atomic structures, they are able to change the fundamental state (minimum energy) into one of the excited states of the electronic "cloud" around the nucleus.

On the contrary, NGB correlating quanta are not mediators of interactions among elements of the system. They determine only the *modes of interaction* among them. Hence, any symmetry breakdown in the QFT of condensed matter of chemical and biological systems has one only gauge boson mediator of the underlying energy exchanges, the photon, since they are all electromagnetic phenomena. Therefore, the phenomena involved here, from which the emergence of *macroscopic* coherent states derives, implies the generation, effectively the *condensation*, of correlation quanta with negligible mass, in principle null: the NGB, indeed. This is the basis of the fundamental "Goldstone theorem" [24, 25]. NGBs acquire different names for the different modes of interaction, and hence of the coherent states of matter they determine: *phonons* in crystals, *magnons* in magnets, *polarons* in biological matter, etc. Indeed, what characterizes the coherent domains in living matter is the phase coherence of the *electric dipoles* of the organic molecules and of the water, in which only the biomolecules are *active*. Therefore, although the correlation quanta are real particles, observable with the same techniques (diffusion, scattering, etc.), not only in QFT of condensed matter, but also in QED and in QCD like the other quantum particles, wherever we have to deal with broken symmetries [21], nevertheless they do not exist *outside* the system they are correlating. For instance, without a crystal structure (e.g. when heating a diamond over 3545 °C), we still have the component atoms, but no longer phonons. In this regard, the correlation quanta differ from energy quanta, like photons. Because the gauge bosons are *energy* quanta, they cannot be "created and annihilated" without residuals.

So, in any quantum process of particle "creation/annihilation" in quantum physics, what is conserved is the energy/matter, mediated by the energy quanta (gauge bosons), not their "form", mediated by the NGB correlation quanta. Also in this regard, a dual ontology (matter/form) is fundamental for avoiding confusion and misinterpretation in quantum physics.

Moreover, because the mass of the correlation quanta is in any case negligible (or even null), *their condensation does not imply a change of the energy state of the system*. This is the fundamental property for understanding how, not only the stability of a crystal structure, but also the relative stability of the structures/functions of living matter, at different levels of self-organization (cytoskeleton, cell, tissue, organ, etc.), can depend on such basic *dynamic* principles. In fact, all this means that, if the symmetric state is a fundamental state (a minimum of the energy function corresponding to a *quantum vacuum* in QFT of dissipative systems), also the ordered state, after symmetry breakdown and the instauration of the ordered state, remains *a state of minimum energy*, thus being *stable* in time. In kinematic terms, it is a *stable attractor* of the dynamics.

3.2 Doubling of Degrees of Freedom (DDF) in QFT and in Neuroscience

We said that the relevant quantum variables in biological systems are the *electrical dipole vibrational modes* in the water and organic molecules, constituting the oscillatory "dynamic matrix" in which also neurons, glia cells, and the other mesoscopic units of brain dynamics are immersed. The condensation of massless NGB (polarons)—controlling the electrical dipole coherent oscillation modes, and corresponding, at the mesoscopic level, to the long-range correlation waves observed in brain dynamics—depends on the triggering action of an external stimulus for symmetry breakdown of the quantum vacuum of the corresponding brain state. In such a case, the "memory state" corresponds to a coherent state for the basic quantum variables, whose mesoscopic order parameter displays itself as the amplitude and phase modulation of the carrier signal.

In the classical Umezawa model of brain dynamics [26], however, the system suffered from an "intrinsic limit of memory capacity". Namely, each new stimulus produces an associated polaron condensation, cancelling the preceding one, for a sort of "overprinting". *This limit does not occur in dissipative QFT where the many-body model predicts the coexistence of physically distinct patterns, amplitude modulated and phase modulated.* That is, by considering the brain as it is, namely an "open", "dissipative" system continuously interacting with its environment, there does not exist only one ground (quantum vacuum) state, like in the thermal field theory of Umezawa, where the system is studied at equilibrium; On the contrary, in principle, there exist infinitely many ground states (quantum vacuums), thus giving the system a potentially infinite capacity of memory. To sum up, the solution to the overprinting problem relies on three facts [27]:

1. In a dissipative (nonequilibrium) quantum system, there are (in principle) infinitely many quantum vacuum's (ground or zero-energy) states, on each of which a whole set of nonzero energy states (or "state space" or "representation states") can be built.

2. Each input triggers one possible irreversible time evolution of the system, by inducing a "symmetry breakdown" in one quantum vacuum, i.e. by inducing in it an ordered state, a coherent behavior, effectively "freezing" some possible degrees of freedom of the behaviors of the constituting elements (e.g. by "constraining" them to oscillate on a given frequency). At the same time, the input "labels" dynamically the induced coherent state, as an "unitary non-equivalent state" of the system dynamics. In fact, such a coherent state persists in time as a ground state (polarons are not energetic bosons, but Nambu-Goldstone bosons) thus constituting a specific "long-term" memory state for such a specific coupling between the brain dynamics and its environment. On the other hand, a brain that is no longer dynamically coupled with its environment is either in a pathological state (schizophrenia) or simply dead.

3. At this point, the DDF principle emerges as both a physical and mathematical necessity of such a brain model: physical, because a dissipative system, even though in nonequilibrium, must anyway satisfy the *energy balance*; mathematical, because the zero energy balance requires a "doubling of the system degrees of freedom". The *doubled* degrees of freedom, say $\tilde{A}$ (the tilde quanta, where the nontilde quanta A denote the brain degrees of freedom), thus represent the environment to which the brain state is coupled. The environment (state) is thus represented as the "time-reversed *double*" of the brain (state) on which it is impinging. The environment is hence "modeled on the brain", but according to the finite set of degrees of freedom *the environment itself elicited* in the brain.

What is relevant for our aims is that, each set of degrees of freedom A and for its "entangled doubled" $\tilde{A}$, there is a relater *unique number* $\mathcal{N}$, i.e. $\mathcal{N}_A, \mathcal{N}_{\tilde{A}}$, which in modul, $|\mathcal{N}|$, *univocally, identifies* i.e. *dynamically labels*, a given *phase coherence domain*, i.e. a quantum system state entangled with its thermal bath state, in our case, *a brain state matching its environment state*. This depends on the fact that, generally, in the QFT mathematical formalism, the number $\mathcal{N}$ is a numeric value expressing the NGB condensate value on which a phase coherence domain *directly depends*. In an appropriate *set-theoretic interpretation*, because for each "phase coherence domain" x, $|\mathcal{N}|$ effectively *identifies univocally* such a domain, it corresponds to an "identity function Id_x" that, in a "finitary" coalgebraic logical calculus, corresponds to the *predicate satisfied by such a domain because it identifies it univocally*. In other words, Vitiello's reference to the predicate "magnet function" or "crystal function" we quoted at the beginning of Sect. 3.1 are not metaphors, but are expressions of a fundamental formal tool—the "co-membership notion"—of the coalgebraic predicate calculus (see Sect. 5.2). Regarding the DDF applied to the quantum foundation of cognitive neuroscience, we have illustrated elsewhere its logical relevance, for an original solution of the reference problem (see [28, 29]).

There exists a huge amount of experimental evidence in brain dynamics of such phenomena, collected by W. Freeman and his collaborators. This evidence found, during the last ten years, its proper mathematical modeling in the dissipative QFT approach of Vitiello and his collaborators, justifying the publication during recent years of several joint papers on these topics (see, for a synthesis, [30, 31]).

To sum up [32], Freeman and his group used several advanced brain imaging techniques such as multielectrode electroencephalography (EEG), electrocorticograms (ECoG), and magnetoencephalography (MEG) study what neurophysiologist to generally consider the *background activity* of the brain, often filtering it out as "noise" with respect to the synaptic activity of neurons they are exclusively interested in. By studying these data with computational tools of signal analysis with which physicists, differently from neurophysiologists, are acquainted, they discovered the massive presence of patterns of AM/FM phase-locked oscillations. They are intermittently present in resting and/or awake subjects, as well as in the same subject actively engaged in cognitive tasks requiring interaction with the environment. In this way, we can describe them as features of the background activity of brains, modulated in amplitude and/or in frequency by the "active engagement" of a brain with its surroundings. These "wave packets" extend over coherence domains covering much of the hemisphere in rabbits and cats [33–36], and regions of linear size of about 19 cm in human cortex [37], with near-zero phasedispersion [38]. Synchronized oscillations of large-scale neuron arrays in the β and γ ranges are observed by MEG imaging in resting and motor-task-related states of the human brain [39].

4 Semantic Information in Living and Cognitive Systems

4.1 QFT Systems and the Notion of Negentropy

Generally, the notion of information in biological systems is a synonym of the *negentropy* notion, according to E. Schrödinger's early use of this term. Applied, however, to QFT foundations of dissipative structures in biological systems, the notion of negentropy is not only associated with the *free energy*, as Schrödinger himself suggested [40], but also with the notion of *organization*, as the use of this term by A. Szent-György first suggested [41]. The notion of negentropy is thus related with the constitution of *coherent domains* at different space–time scales, as the application of QFT to the study of dissipative structures demonstrates, since the pioneering work by Frölich [42, 43].

In this regard, it is important to emphasize also the key role of the notion of *stored energy* that such a multilevel spatiotemporal *organization* in coherent domains and subdomains implies (i.e. the notion of quantum vacuum "foliation" in QFT), as distinct from the notion of *free energy* of classical thermodynamics [44]. Namely, as we know from the discussion above, the constitution of coherent domains allows chemical reactions to occur at *different timescales*, with a consequent energy release, thus becoming immediately available exactly *where/when it is necessary*. For instance, resonant energy transfer among molecules typically occurs in 10^{-14} s, whereas the molecular vibrations themselves die down, or thermalize, in a time between 10^{-9} and 10^1 s. Hence, this is a 100% highly efficient and highly specific process, being determined by the frequency of the vibration itself, given that resonating molecules can attract one another. Hence, the notion of "stored energy" is meaningful at every level of the complex spatiotemporal structure of a living body, from a single molecule to the whole organism.

This completes the classical thermodynamic picture of Szilard [45] and Brillouin [46], according to which the "Maxwell demon" for getting information to compensate the entropic decay of the living body must consume free energy from the environment. This means an increase of the global entropy according to the Second Law. However, this has to be completed in QFT with the evidence coming from the Third Law discussed in this paper.

This occurs at the maximum level in the biological realm in human brain dynamics. To illustrate this point as DDF applied in neuroscience, Freeman and his collaborators spoke about "dark energy" for the extreme reservoir of energy hidden in human brain dynamics. The human brain indeed has 2% of the human body mass, but dissipates 20–25% of the body resting energy. This depends on the extreme density of cells in the cortices ($10^5/mm^3$), with an average of 10^4 connections [47].

To conclude this discussion, we showed that the "dual paradigm" related to the QFT interpretation of the "information-theoretic" approach to quantum physics does not depend on the distinction between "energy" and "information", like in the QM interpretation, where the "information" notion and measure—differently from the "energy" ones—are "observer-related", and therefore, logically, only "syntactic". In the QFT interpretation, where "information" is a physical magnitude, i.e. a thermodynamic *negentropy*, the duality concerns the two components of the negentropy notion and measure. These are, respectively, the *energetic* component (quantum "gauge bosons") and the *ordering* component (quantum "Nambu–Goldstone bosons") of a phase coherence domain, including the two entangled quantum states of the system and of its environment.

On the other hand, precisely because "ordering" is also a fundamental semantic notion in set-theoretic logic, the "semantic information" notion and measure strictly depend on the *logical* and *mathematical* notion of "duality". This duality in category theory logic concerns two opposed categories, specifically, in theoretical computer science (TCS), the notion of the "dual equivalence" between an algebra and its coalgebra, on which the notion of "local truth" and "finitary computation", on the one hand, as well as the notion and measure of *semantic information*, on the other, strictly depend.

These two notions of "duality", physical and logical, are however strictly interconnected in QFT, because both depend on the notion of *NGB condensates*, as constituting, respectively, the "ordering" component of the negentropy in information physics, and the sufficient condition for interpreting QFT systems as computing systems. A short introduction to all these notions will be the object of the rest of this paper, in the framework of the recent "coalgebraic approach" to quantum computing in TCS.

4.2 Syntactic Versus Semantic Information in Quantum Physics

4.2.1 Shannon's Syntactic Theory of Information in QM and in Mathematical Communication Theory

The Shannon nature of the notion and measurement of information that can be associated with decoherence in QM, overall in the relational and hence computational interpretations of QM illustrated above, has been emphasized [5]. In fact, in both cases, the "information" can be associated with the uncertainty H removal, in the sense that, the "more probable" or "less uncertain" an event/symbol is, the less informative (or, psychologically, less "surprising") its occurrence is. Mathematically, in the mathematical theory of communication (MTC), the information H associated with the ith symbol x among N (=alphabet), can be defined as

$$H = \sum_{i=1}^{N} p(x_i) I(x_i) = - \sum_{i=1}^{N} p(x_i) \log p(x_i),$$

where $p(x_i)$ is the relative probability of the ith symbol x with respect to the N possible ones, and I is the information content associated with the symbol occurrence, that is, the inverse of its relative probability (the less probable it is, the more informative its occurrence is). The information amount H thus has the dimensions of a statistical *entropy*, being very close to the thermodynamic entropy S of statistical mechanics:

$$S = - k_B \sum_{i} p(x_i) \log p(x_i),$$

where x_i are the possible microscopic configurations of the individual atoms and molecules of the system (microstates) which could give rise to the observed macroscopic state (macrostate) of the system, and k_B is the Boltzmann constant. Based on the correspondence principle, S is equivalent in the classical limit, i.e.

whenever the classical notion of probability applies, to the QM definition of entropy
by John von Neumann:

$$S = -k_B \mathrm{Tr}(\rho \log \rho),$$

where ρ is a density matrix and Tr is the trace operator of the matrix. Indeed, it was
von Neumann himself who suggested to Claude Shannon to denote as "entropy" the
statistical measure of information H he discovered. The informativeness associated
with (the occurrence of) a symbol in the MTC (or with an event in statistical
classical and quantum mechanics) is only "syntactic" and not "semantic" [48, p. 3].
Effectively, the symbol (event) occurs as *uninterpreted* (context independent) and
wellformed (determined), according to the *rules* of a *fixed* alphabet or code (i.e.
according to the *unchanged laws* of physics).

Anyway, starting from the pioneering works of Mackay [49] and of Carnap and
Bar-Hillel [50], in almost any work dealing with the notion of information in
biological and cognitive systems, the vindication of its *semantic/pragmatic* character is a leit motiv. Particularly, because information concerns here self-organizing
and complex processes, in them the "evolution of coding", and the notion of "local
(contingent) truth" (semantics), in the sense of *adequacy* for an optimal fitting with
the environment (pragmatics), are essential [51–53]. More specifically, in QFT
differently from QM, the pragmatic information content is significant, defined as the
ratio of the rate of energy dissipation (power) to the rate of decrease in entropy
(negentropy) [53] a measure generally considered in literature as the proper
information measure of self-organizing systems. Evidently, in the DDF formalism
of QFT, in the relationship between a quantum system and its thermal bath (environment), and specifically, in neuroscience, the relationship between the brain and
its contextual environment, the notion and measure of pragmatic information, as
described in [53], play an essential role [47].

What is to be emphasized here, above all, is that the Wigner function (WF) in QFT,
from which the probabilities of the physical states are calculated, is deeply different
from the Schrödinger wave function of QM, not only because the former, differently
from the latter, is defined on the phase space of the system; What is much more
fundamental is that the WF uses the notion of *quasiprobability* [54], and not the notion
of probability of the classical Kolmogorov axiomatic theory of probability [55].

Indeed, the notion of quasiprobability allows regions integrated under given
expectation values to not represent *mutually exclusive states*, thus violating one of
the fundamental axioms of Kolmogorov's theory i.e. the separation of variables in
such distributions is not fixed, but, as is the rule in the case of phase transitions, can
evolve dynamically (see the QFT interpretation of the "quantum uncertainty principle" at the end of Sect. 2). From the computability theory standpoint, this means
that a physical system in QFT, against the TM and QTM paradigms, is able to
change dynamically the "basic symbols" of its computations, since new collective
behaviors can emerge from individual ones, or vice versa. In this way, this justifies
the definition of the information associated with a WF as a "semantic information
content".

The semantic information in QFT computations hence satisfies, from the logical standpoint, the notion of *contingent*, or better, *local truth*, thus escaping from the Carnap and Bar-Hillel paradoxes (CBPs) [50]. To introduce this notion, it might be pedagogically useful to discuss briefly the "theory of strong semantic information" (TSSI) developed by L. Floridi, essentially because it shares with QFT the same notion of quasiprobability. In the QFT usage of the quasiprobability notion there is no necessity of violating also the other axiom of Kolmogorov's axiomatic theory of probability, i.e. the axiom excluding the "negative probabilities". On the contrary, Floridi uses the notion of negative probabilities [56], so that the reference to his theory has only a "pedagogical" value in the present context.

4.2.2 Floridi's Semantic Information Theory

Following the critical reconstruction of both theories (CSI and TSSI) by Sequoiah-Grayson [57], the CSI approach is based on Carnap's theory of intensional modal logic [58]. In this theory, given n individuals and m monadic predicates, we have 2^{nm} possible worlds and 2^m Q-predicators, intended as individuations of possible types of objects, given a conjunction of primitive predicates either unnegated or negated. A full sentence of a Q-predicator is a Q-sentence, hence a possible world is a conjunction of n Q-sentences, as each Q-sentence describes a possible existing individual. The *intension* of a given sentence is taken to be the set of possible worlds that make it true, i.e. are included by the sentence. This is in relation with the notion of *semantic information* in CSI, here referred to as the *content* of a declarative sentence s and denoted by "Cont(s)". In this way, the CBP consists in the evidence that, because an always true sentence is true for all possible worlds, i.e. it does not exclude any world, it is empty of any semantic content (effectively, it is a tautology) the maximum semantic content is for the always false (i.e. contradictory) sentence, because it excludes any possible world.

In Carnap & Bar Hillel terms, "a self-contradictory sentence asserts too much: it is too informative for being true" [50, p. 229]. Effectively, it is well known also to common sense that tautologies have no information content. What is paradoxical for common sense is that contradictions have the maximum information content. For logicians, however, who know the famous pseudo-Scotus law, according to which anything can be derived from contradictions (i.e., the so-called "explosion principle"), this conclusion is not surprising, once we have defined the information content of a sentence s, Cont(s), as the set of all sentences (possible worlds) belonging to the same universe W of the theory excluded by s.

Of course, the limit of CSI consists in its abstraction, namely in the *logical* notion of truth, and the a priori probability that it supposes. Surprisingly, but not contradictorily, it is just this supposition of *a logical notion of truth* (=true in all possible contexts, or "worlds" in modal logic terms) that makes it impossible to use truth as a necessary condition for meaningfulness in CSI.

What makes the TSSI of Floridi and followers interesting is that it offers a theory and measures of the semantic information for *contingent* and not *necessary* propositions namely for propositions that are not *logically* true, i.e. true for all possible worlds, in contrast to both tautologies (i.e. logical laws) and/or general ontology propositions which are true for whichever "being as being". Namely, both the propositions of all empirical sciences and the propositions of specific ontologies are *true* for objects *actually* existing (or that *existed*, or that *will exist*) only in *some* possible worlds—in the limit *one*: the actual, "present" world. In other terms, the scientific and ontological theories are "models" (i.e. theories true only for a limited domain of objects), precisely because both have semantic content, differently from tautologies. I developed elsewhere [59] a formal ontology of the QFT paradigm in natural sciences, in which this notion of truth is logically and ontologically justified, as an alternative to Carnap's logical atomism, i.e. alternative to the formal ontology of the Newtonian paradigm in natural sciences, on which both CSI and BCP depend.

Hence, it is highly significant to develop a *theory* and a *measure* of information content such as TSSI, compatible with what S. Sequoiah-Grayson defines as the *contingency requirement of informativeness* (CRI), supposed in TSSI. Unfortunately, a requirement such as CRI cannot be *supposed*, but only *justified*, as G. Dodig-Crnkovic indirectly emphasizes in her criticism of TSSI [60], and this is the limit of TSSI. In fact, the CRI states [57]: «A declarative sentence *s* is informative *iff s* individuates at least some but not all w_i from W (where $w_i \in W$)». Sequoiah-Grayson recognizes that CRI in TSSI is an idealization. However, he continues,

> Despite this idealization, CRI remains a convincing modal intuition. For a declarative sentence *s* to be informative, in some useful sense of the term, it must stake out a claim as to which world, out of the entire modal space, is in fact the actual world.

This requirement is explicitly and formally satisfied in the formal ontology of the "natural realism" as an alternative to the "logical atomism" of CSI [59, 61]. Effectively the main reason, Floridi states, leading him to defend the TSSI is that only such a theory having truthfulness as *necessary condition* for meaningfulness can be useful in an epistemic logic. In it, indeed, the entire problem consists in the justification of the passage from belief as "opinion" to belief as "knowledge", intended as a *true* belief.

That a CRI is operating in TSSI is evident from the "factual" character of the semantic information content in it, and of its probabilistic measure. Starting from the principle that semantic information σ has to be measured in terms of distance of σ from w, we have effectively four possibilities. Using the same example of Floridi [56, p. 55ff.], let us suppose that there are exactly three people in the room: this is the situation denoted in terms of the actual world w. The four possibilities for σ as to w are:

(T) There are *or* there are not people in the room;
(V) There are some people in the room;
(P) There are three people in the room;
(F) There are *and* there are not people in the room.

By defining θ as the distance between σ and w, we have: θ (T) = 1; θ (V) = 0.25 (for the sake of simplicity); θ (P) = 0; θ (F) = −1. From these relations it is possible to define the *degree of informativeness* i of σ, that is:

$$i(\sigma) = 1 - \theta(\sigma)^2.$$

The graph generated by the equation above (Fig. 1a) shows this as θ ranges from the necessary false (F) (=contradiction) to the necessary true (T) (=tautology), both showing the maximum distance from the contingent true (P).

To calculate the quantity of semantic information contained in σ relative to $\iota(\sigma)$, we need to calculate the area delimited by the equation above, that is, the definite integral of the function $\iota(\sigma)$ on the interval [0, 1]. On the contrary, the amount of vacuous information, which we denote as β, is also a function of θ. More precisely, it is a function of the distance of θ from w, i.e.

$$\int_0^\theta \iota(\sigma)\mathrm{d}x = \beta.$$

It is evident that, in the case of (P), $\beta = 0$. From α and β, it is possible to calculate the amount of semantic information carried by σ, i.e. γ, as the difference between the maximum information that can be carried in principle by σ and the vacuous information carried effectively by σ, that is, in bit:

$$\gamma(\sigma) = \log(\alpha - \beta).$$

Of course in the case of (P):

$$\gamma(P) = \log(\alpha).$$

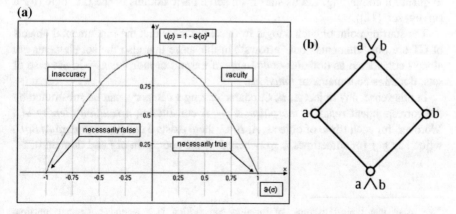

(a)

(b)

Fig. 1 **a** Degree of informativeness. From [56, p. 56], **b** Boolean lattice in equation logic

That confirms CRI in TSSI, that is, the contingently true proposition, namely denoting the actual situation w and/or expressing the true knowledge of w, carries the maximum semantic information about w.

5 Coalgebraic Semantics of Quantum Systems

5.1 Category Theory Logic and Coalgebraic Semantics

To satisfy Dodig-Crnkovic's criticism about the necessity of a *formal justification* of the notion of "local (contingent) truth" theory in logic and computability theory, let us start from the extension of the Boolean lattice (matrix) of Fig. 1b from the propositional calculus (Boolean equation logic) to the *monadic predicate calculus*, that is, where the proposition is $b = \neg a$. In such a case, the *meet* of the lattice $(a \land b)$ would correspond to the *always false* proposition $(a \land \neg a)$, and the *join* $(a \lor b)$ would correspond to the *always true* proposition $(a \lor \neg a)$ of the quasiprobability distribution of Fig. 1a, while the maximum of this distribution corresponds to the assertion of $|a|$ (and not of $a \mid b$, as in the lattice in figure) as "locally true". To make this representation computationally effective , it is necessary that we are allowed to associate this maximum to a measure of the *maximum of entropy* expressing the "matching" (convergence till equivalence) of the results of two "concurrent computations" of a system and of its environment, as the result of the "physical work" of the phase space dynamic reconfiguration (phase transition), consuming all the available "free energy", generated by the original "mismatch" between them.

What is highly significant for our aims is that in a way completely independent from quantum physicists—at least till the very last years (see Sect. 5.2 below)—logicians and computer scientists developed in the context of CT logic a coalgebraic approach to Boolean algebra semantics that only recently started to be applied also to quantum computing. Let us start from some basic notions of the CT logic (for a survey, see [12]).

The starting point of such a logic from set theory is that the fundamental objects of CT are not "elements" but "arrows", in the sense that also the set elements are always considered as domains-codomains of *arrows* or *morphisms*—in the case of sets, domains-codomains of *functions*.

In this sense, any object A, B, C, characterizing a category, can be substituted by the correspondent *reflexive morphism* $A \rightarrow A$ constituting *a relation identity* Id_A. Morover, for each triple of objects, A, B, C, there exists a *composition map* $A f B g C$, written as $g \circ f$ (or sometimes $f; g$), where B is the codomain of f and domain of g.[3]

---

[3]We recall that typical example of function composition is a recursive, iterated function: $x_{n+1} = f(x_n)$.

Therefore, a *category* is any structure in logic or mathematics with structure-preserving morphisms, e.g. in set-theoretic semantics, all the models of a given formal system, because sharing the same structure constitutes a category. In this way, some fundamental mathematical and logical structures are also categories: **Set** (sets and functions), **Grp** (groups and homomorphisms), **Top** (topological spaces and continuous functions), **Pos** (partially ordered sets and monotone functions), **Vect** (vector spaces defined on numerical fields and linear functions), etc.

Another fundamental notion in CT is the notion of *functor*, F, that is, an operation mapping objects and arrows of a category **C** into another **D**, $F: \mathbf{C} \to \mathbf{D}$, so as to preserve compositions and identities. In this way, between the two categories, there exists a *homomorphism up to isomorphism*. Generally, a functor F is *covariant*, that is, it preserves arrows, directions and composition orders (e.g. in the QM attempt of interpreting thermodynamics within kinematics [62]); i.e. if $f: A \to B$, then $FA \to FB$; if $f \circ g$, then $F(f \circ g) = Ff \circ Fg$; if id_A, then $Fid_A = id_{FA}$. However, two categories can be equally homomorphic up to isomorphism if the functor G connecting them is *contravariant*, i.e. *reversing* all the arrows, directions and the composition orders, i.e. $G: \mathbf{C} \to \mathbf{D}^{\mathrm{op}}$:

if $f: A \to B$, then $GB \to GA$; if $f \circ g$, then $G(g \circ f) = Gg \circ Gf$; but if id_A, then $Gid_A = id_{GA}$.

Through the notion of contravariant functor, we can introduce the notion of *category duality*. Namely, given a category **C** and an *endofunctor* $E: \mathbf{C} \to \mathbf{C}$, the contravariant application of E links a category to its opposite, i.e. $E^{\mathrm{op}}: \mathbf{C} \to \mathbf{C}^{\mathrm{op}}$. In this way it is possible to demonstrate the *dual equivalence* between them, in symbols: $\mathbf{C} \rightleftharpoons \mathbf{C}^{\mathrm{op}}$. In CT semantics, this means that, given a statement α defined on **C**, α is true *iff* the statement α^{op} defined on $\mathbf{C}^{\mathrm{op}}$ is also true. In other terms, truth is invariant for such an exchange operation over the statements, that is, they are *dually equivalent*. In symbols: $\alpha \rightleftharpoons \alpha^{\mathrm{op}}$, as distinguished from the ordinary equivalence of the logical tautology: $\alpha \leftrightarrow \beta$, defined within the very same category.

A particular category, indeed, that is interesting for our aims is the category of algebras, **Alg**. They constitute a category because any algebra $\mathcal{A}$ can be defined as *a structure on sets* characterized by an endofunctor projecting all the possible combinations (Cartesian *products*) of the subsets of the carrier set, on which the algebra is defined, onto the set itself, that is, $\mathcal{A} \otimes \mathcal{A} \to \mathcal{A}$. The other category interesting for us is the category of coalgebras **Coalg**. Generally, a coalgebra can be defined as a structure on sets, whose endofunctor projects from the carrier set onto the *coproducts* of this same set, i.e. $\mathcal{A} \to \mathcal{A} \otimes \mathcal{A}$. Despite appearances, an algebra and its coalgebra *are not dual.* This is the case, for instance, of a fundamental category of algebras in physics, that is, the category of *Hopf algebras*, **Halg**, generally used in dynamic system theory both in classical and in quantum mechanics, as we know. Each *HAlg* is essentially a *bi-algebra* because it includes two types of operations on/from the carrier set, where—because they are used to represent energetically closed systems —products (algebra, e.g., for calculating the energy of a single particle in a quantum state) and coproducts (coalgebra, e.g., for calculating the total energy of two particles in the same quantum state) can be defined on the same basis, and therefore *commute* among themselves. That is, there exists a complete *symmetry* between a *HAlg* and its *HCoalg*, so that they are *equivalent* and not *dually equivalent*. In this sense, any Hopf algebra is said to be *self-dual*, that is, isomorphic with itself. To make, on the contrary, a Hopf algebra dually equivalent with its coalgebra, as we know from thermal QFT, we have to introduce a q-deformation, where q is a thermal parameter. In this case, both coproducts and then products do not commute among themselves. In fact, in the case of coproducts used for calculating a quantum state total energy, they are associated to a system state energy and to its thermal-bath state energy, so that they cannot commute among themselves.

More generally, indeed, it is possible to define a dual equivalence between two categories of algebras and coalgebras by a contravariant application of the same functor. We might give two examples of this notion, the first in mathematics and computability theory concerning Boolean algebras, and the second in computational physics concerning QFT.

5.2 Coalgebraic Semantics of a Boolean Logic for a Contravariant Functor

The first example, concerning Boolean algebras, depends essentially on the fundamental representation theorem for Boolean algebras demonstrated in 1936 by the American mathematician M. Stone, five years after having demonstrated with John von Neumann the fundamental theorem of QM we quoted in Sect. 2. Indeed, the Stone theorem associates each Boolean algebra B to its Stone space $S(B)$ [63]. Therefore, the simplest version of the Stone representation theorem states that every Boolean algebra B is *isomorphic* to the algebra of partially *ordered by inclusion* closed-open (clopen) subsets of its Stone space $S(B)$, effectively an *ultrafilter*[4] of the power set of a *given set (interval) of real numbers* defined on $S(B)$.

Because each monotone function between a Boolean algebra A and a Boolean algebra B corresponds to a continuous function from $S(B)$ to $S(A)$ in the opposite direction so to make them *dual*, we can state that each endofunctor Ω in the category of the coalgebras on Stone spaces, **SCoalg**, *induces* a contravariant functor in the category of the Boolean algebras, **BAlg** [64, 65]. In CT terms, the theorem— effectively demonstrated by Abramsky in 1988—states the *dual equivalence* between them for the contravariant application of the "Vietoris functor" $\mathcal{V}$, i.e. **SCoalg**$(\mathcal{V}) \rightleftharpoons BAlg(\mathcal{V}^*)$. Let us deepen this fundamental point, by summarizing the essential steps leading to this result.

It is difficult to exaggerate the fundamental importance of the Stone theorem that, according to computer scientists, inaugurated the "Stone era" in computer science. In

---

[4]We recall here that by an "ultrafilter" we mean the maximal partially ordered set defined on the power set of a given set ordered by inclusion, and excluding the empty set.

Particular, this theorem demonstrated definitively that Boolean logic semantics requires only *a first-order semantics* because it requires only *partially ordered* sets and not *totally ordered sets*. This result is particularly relevant for the foundations of computability theory. Indeed, the demonstration of the fundamental Lövenheim–Skolem theorem (1921) blocked the research program of E. Schröder of the so-called algebra of logic in the foundations of mathematics and of calculus [64], because it demonstrated that algebraic sets are not able to deal with *non-denumerable sets*, e.g. with the *totality* of real numbers. For this reason, and the subsequent fundamental demonstrations of Tarski's theory of truth as correspondence (1929) [65], and of Gödel's incompleteness theorems (1931) [66], the set-theoretic semantics migrated to higher-order logic, in order to grant the *total ordering* of sets, by some foundation axiom, e.g. the *axiom of regularity* in Zermelo-Fraenkel (ZF) set-theory. In this way, no *infinite chain of inclusions* among sets is allowed in *standard* set theory, so as to separate the semantic "set ordering" from the complete "set enumerability".[5]

Therefore, the further step for making the Stone theorem computationally effective for a Boolean first-order semantics, avoiding the limits of the Turing-like computation scheme, where a UTM is effectively a second-order TM as to an infinity of first-order TMs, and then strictly dependent on Gödel and Tarski theorems, is the definition of *nonstandard* set theories without foundation axioms. In this way, we allow infinite chains of set inclusions, according to the original intuition of the Italian mathematician E. De Giorgi [67, 68]. The most effective among the non-standard set theories is Aczel's set theories of *non-well-founded (NWF) sets* based on the *anti-foundation axiom* (AFA) [13]. The AFA, indeed, allowing set *self-inclusions* and therefore infinite chains of set inclusions, makes it also possible to define the powerful notion of set *co-induction* by *co-recursion,* dual to the algebraic notion of *induction* by *recursion,* both as formal methods of set definition and proof [15, 68, 69] (see Appendix A.1).

In this sense, the key role of the AFA is threefold:

1. Above all, it grants the *compositionality* of the set inclusion relations by prohibiting that the ordinary transitivity rule (TR), $\langle \forall u, v, w((uRv \wedge vRw) \rightarrow uRw) \rangle$, where R is the inclusion relation and u, v, w are sets—holds in set inclusions, because TR supposes the set total ordering. In this way, because only the "weaker" transitivity of the Euclidean rule (ER) $\langle \forall u, v, w((uRv \wedge uRw) \rightarrow vRw) \rangle$ between inclusions is allowed here, this means that the representation of sets ordered by inclusion as *oriented graphs,* in which the nodes are sets and the edges are inclusions with only one root (in our case the set u), *always* satisfies an "ascendant–descendant relationship" without "jumps" (each descendant always

---

[5]Two corollaries of the Lövenheim–Skolem theorem, demonstrated by Skolem himself in 1925, are significant for our aims, i.e. (1) that only *complete* theories are *categorical*, and (2) that the *cardinality* of an algebraic set depends intrinsically by the algebra defined on it. Think, for instance of the principle of *induction by recursion* for Boolean algebras, allowing a Boolean algebra to *construct* the sets on which its semantics is justified, blocking however Boolean computability on *finite* sets. It is evident that Zermelo's strategy of migrating to second-order set-theoretic semantics grants categoricity to mathematics on an *infinitistic* basis.

has its own ascendant i.e. they form a *tree*). This is the core of the "compositionality" of the *inclusion operator* of a coalgebra defined on NWF sets, i.e. the basis of the so-called tree-unfolding of NWF sets, starting from an "ultimate root" similar to the *universal set V*—which is allowed here, because of the possibility of set self-inclusion,[6] i.e. the disjunction of all sets forming the universe of the theory, like the "join" of a Boolean lattice. All this is the basis for extending the duality between the category **Stone** of Stone spaces, and the category **BAlg**, to the dual equivalence between the category of the coalgebras **Coalg** and the category of the algebras **Alg**, for an induced contravariant functor Ω^*, i.e. **Coalg(Ω)** $\rightleftharpoons$ **Alg(Ω^*)** [16, p. 417ff].[7]

2. Secondly, the AFA and the "final coalgebra theorem" justify *the coalgebraic interpretation of modal logic* in the framework of *first-order logic* (see the fundamental van Benthem theorem in this regard [70]) because the principle of set unfolding for partially ordered sets within an unbounded chain of set inclusions gives us an algebraically "natural" interpretation of the modal *possibility operator* "$\diamond$", in the sense that $< \diamond\alpha >$ means that "α is true in *some* possible worlds" [71–74], so as to give a computationally effective (first-order logic, where the predicate calculus is complete) justification to Thomason's early program of "reduction of the second-order logic to the modal logic" [75], made effective by another celebrated theorem, the Goldblatt–Thomason theorem. Because any set tree can be modeled as a *Kripke model*, this theorem defines rigorously which elementary classes of Kripke models are modally definable (for a deep discussion of this theorem, see [76, pp. 33–43]. For an intuitive treatment of these notions, see Sect. A.2 in the Appendix).

3. Thirdly, in the fundamental paper of 1988 [11], Abramsky first suggested that the endofunctor of modal coalgebras on Stone spaces, defined on NWF sets, is the so-called Vietoris functor $\mathcal{V}$.[8] In this way we can extend the duality between coalgebras and algebras for the induction of a contravariant functor Ω^*, to the *dual equivalence* between modal coalgebras on Stone spaces and modal Boolean algebras for the induction of a contravariant functor $\mathcal{V}^*$, i.e., $SCoalg(\mathcal{V})$ $\rightleftharpoons$

---

[6]Recall that set self-inclusion is not allowed for standard sets because of Cantor's theorem. This impossibility is the root of all semantic antinomies in standard set theory, from which the necessity of a second-order set-theoretic semantics ultimately derives.

[7]This depends on the trivial observation that a coalgebra $C = \langle C, \gamma : C \rightarrow \Omega C \rangle$, where γ is a transition function characterizing C, over an endofunctor $\Omega: C \rightarrow C$ can be seen also as an algebra in the opposite category C^{op}, i.e. **Coalg(Ω)** = **(Alg(Ω^{op})^{op}** [16, p. 417].

[8]The fundamental property of $\mathcal{V}$ is that it is the counterpart of the power set functor $\wp$ in the category of the topological spaces (i.e. for continuous functions) such as the Stone space category, **Stone**. This functor maps a set S to its power set $\wp(S)$ and a function $f: S \rightarrow S'$ to the image map $\wp f$ given by $(\wp f)(X) := f[X] (= \{ f(x) \mid x \in X\})$. Applied to Kripke's relational semantics in modal logic, this means that Kripke's *frames* and *models* are nothing but "coalgebras in disguise". Indeed, a *frame* is a set of "possible worlds" (subsets, s) of a given "universe" (set, S) and a binary "accessibility" relation R between worlds, $R \subseteq S \times S$. A Kripke's *model* is thus a frame with an *evaluation function* defined on it. Now R can be represented by the function $R[\bullet]: S \rightarrow \wp(S)$, mapping a point s to the collection $R[s]$ of its successors. In this way, frames in modal logic correspond to coalgebras over the *covariant* power set functor $\wp$. For such a reconstruction see [16, p. 391].

BAlg($\mathcal{V}^*$) [16, p. 393ff]. This depends on the fact that $\mathcal{V}$ is a functor defined on a particular category of topological spaces, the category of the vector spaces **Vect** we introduced in Sect. 5.1. Vector spaces are fundamental in physics; also the Hilbert spaces of the quantum physics mathematical formalism belong to this category, and, overall, the topologies of Stone spaces are the same of the C^*-algebras associated to Hilbert spaces in quantum physics. On the other hand, the morphisms characterizing the vector space category are, indeed, linear functions, so if we apply to modal coalgebras van Benthem's "correspondence theorem" [77] and the consequent "correspondence theory" [70] between the modal logic and the decidable fragments of the first order monadic predicate calculus, associating each axiom of modal calculus with a first-order formula (see Appendix A.2 for some examples), we obtain the following amazing result that Abramsky first suggested [11], and Kupke et al. developed [74]. Namely, we can formally justify the modal coinduction (tree unfolding) of predicate domains so as to justify *the modal operators* of "possible converse membership" or "possible co-membership", $\langle \ni \rangle$, and of "actual co-membership", i.e. $\neg \langle \neg \ni \rangle$, that is, $[\ni]$, where the angular and square parentheses are reminders of the possibility-necessity, "$\Diamond$-$\Box$" operators, respectively [16, p. 392ff].

What, intuitively, all this means for our aims is that, because modal coalgebras admit only a *stratified (indexed) usage* of the necessity operator $\Box$ and of the universal quantifier $\forall$, since a set *actually* exists as far as effectively *unfolded* by a co-inductive procedure, the semantic evaluations in the Boolean logic effectively consist in a *convergence* between an inductive "constructive" procedure, and a co-inductive "unfolding" procedure, namely they effectively consist in the super-position *limit/colimit* between two concurrent inductive/coinductive computations (see Appendix A.1). This is the core of Abramsky's notion of *finitary objects* as "limits of finite ones", definable only on NWF sets, finitary objects that according to him are the most proper objects for mathematical modeling of computations [11].

This is also the core of the related notion of *duality* between an *initial Boolean algebra*, starting from a *least* fixed point, $x = f(x)$, and its *final Stonean coalgebra*, starting from a *greatest* fixed-point (see Appendix A.1), at the basis of the notion of *universal coalgebra* as a "general theory of both computing and dynamic systems" [15]. This theory allows one to justify a *formal semantics of computer programming* as satisfaction of a given program onto the physical states of a computing system, outside the Turing paradigm. Indeed, this approach systematically avoids the necessity of referring to an UTM for formally justifying the *universality* in computations, because of the possibility of referring to algebraic and co-algebraic universality.[9] At the same time, this theory is able to give a *strong formal foundation* to the notion of *natural computation*, as far as we extend such a coalgebraic

---

[9]However, see the fundamental remarks about the limits of *decidability* and *computability* in this first-order modal logic semantic approach in [76], in which it is said, just in the conclusion, that one of the most promising research programs in this field is related to the coalgebraic approach to modal logic semantics.

semantics to quantum systems and quantum computation. This research program was inaugurated by S. Abramsky and his group at Oxford only a few years ago, both in fundamental physics [78], and in QM computing [14], even though it has its most natural implementation in a QFT foundation of both quantum physics and quantum computation [79].[10]

From the standpoint of the *natural ontology* of cognitive neuro dynamics in the framework of a QFT foundation of it (see Sect. 3.2 above), all this, roughly speaking, means that it is *logically* true that the *(sub)class* of horses is a *member* of the *(super)class* of mammalians *iff*, dually, it is *ontically* (dynamically) true that a *co-membership* of the *species* of horses to the *genus* of mammalians occurs, from some step n onward of the universe evolution (="natural unfolding" of a biological evolution tree), I.e.

$$\ominus_{\forall n(n > m)} \left(\underbrace{horse \in mammmalian}_{Algebra(\Omega^*)} \quad \underbrace{\Omega^* \leftarrow \Omega \leftrightarrows}_{Onto - logical iff} \quad \underbrace{horse \ni mammalian}_{Co - Algebra(\Omega)} \right)$$

In other terms, we are faced here with an example of a "functorially induced" homomorphism, from a coalgebraic *natural structure* of *natural kinds* (genera/species) into a *logical structure of predicate domains* (class/subclass), as an example of modal *local truth*, applied to a theory of the ontological natural realism, in the framework of an evolutionary cosmology [59, 82],[11] where it is nonsensical to use not indexed (absolute) modal operators and quantifiers, given that emergence of physical laws depends on the universe evolution. In parenthesis, this also gives a solution to the otherwise unsolved problem, in Kripke's relational semantics, of the denotation of *natural kinds* (the denoted objects of common names, such as "horses" or "mammalians" in our example) and of the connected Kripke's and Putnam's *causal theory of reference* (see on this point my previous discussion about these problems in [28]). Finally, this gives a logical interpretation as *predicate* (e.g. "being horse") of the "doubled number", i.e. $\mathcal{N}_A, \mathcal{N}_{\tilde{A}}$, as identity functions relative to two mirrored (doubled) sets of degrees of freedom, A and $\tilde{A}$, one relative to a logical realm (the Algebra(Ω^*)), the other to its dual natural realm (the Coalgebra(Ω)), the latter satisfying (making true) *naturally*—i.e. *dynamically* in

---

[10]This depends on the fact that *contravariance* in QM algebraic representation theory can have only an *indirect justification*, as Abramsky elegantly explained in his just quoted paper. QM algebraic formalism is, indeed, intrinsically based on von Neumann's *covariant* algebra, so that only Hopf algebras' self-duality are "naturally" (in the algebraic sense of the allowed functorial transforms) justified in it [62, 80, 81].

[11]In this regard, the famous Aristotelian statement synthesizing his "intentional" approach to epistemology—"not the stone is in the mind, but the form of the stone"—has an operational counterpart in the homomorphism algebra coalgebra of QFT neuro dynamics.

this QFT implementation—the former (see above, Sect. 3.2). The co-membership relation in the coalgebraic half has its physical justification in QFT by the general principle of the *"foliation* of the QV" at the ground state, and of the relative Hilbert space into physically inequivalent subspaces", allowing "the building up" via SSB of ever more complex phase coherence domains in the QV, given their stability in time. They do not depend, indeed, on any energetic input (they depend on as many NGB condensates $|\mathcal{N}|$, each correspondent to a SSB of the QV at the ground state), but on as many "entanglements" with stable structures of the environment [83]. This justifies Freeman and Vitiello in suggesting that this is the fundamental mechanism of the formation of the so-called long-term memory traces in brain dynamics [47], i.e. the formation of the "deep beliefs" in our brains by which each of us interprets the world, based on her/his past experience, using the recently diffused AI jargon in artificial neural network computing [84].

Anyway, apart from this "ontological" exemplification, which is useful however to connect the present discussion with the rest of this chapter, all this means extending to a Boolean lattice L of the monadic predicate logic the modal semantics notions of co-induction and/or of "tree unfolding", so as to give a formal justification of the modal notion of "local truth" also in a computational environment.[12] Indeed, because such a co-inductive procedure of predicative domains justification is defined on NWF sets supporting set self-inclusion, i.e. $x \rightarrow \{x\}$, for each of these co-induced domains also the relative Id_x, i.e. the relative predicate φ, is defined, without any necessity to refer to Fregean second-order axioms, such as the comprehension axiom of ZF set theory, i.e. $< \forall x \exists y \ x \in y \equiv \varphi x >$. This justifies the general statement that, in CT coalgebraic semantics, there exists a Tarski-like model theory [12], without, however, the necessity to refer to higher-order languages for justifying the semantic meta language [86], according to Thomason's reduction program.

We can thus conclude this section by affirming that the previous discussion satisfies the first requirement of Dodig-Crnkovic's criticism to a theory of semantic information at the end of Sect. 4.2.2, i.e. the necessity of a *formal justification* of the theory of "local truth", essential for the notion and measure of Floridi's *semantic information* that can be naturally given in the context of a coalgebraic (modal) semantics of predicate logic. Quoting the first concluding remark of V. Goranko and M. Otto's contribution to the *Handbook of Modal Logic* devoted to the Kripke model theory of modal logic [87, p. 323], we can conclude too:

> Modal logic is local. Truth of a formula is evaluated at a current state (possible world); this localization is preserved (and carried) along the edges of the accessibility relations by the restricted, relativized quantification corresponding to the (indexed) modal operators.

---

[12]This result has been recently formally obtained [85]. For an intuitive explanation of this result, see below the two Appendices, Sects. 7.1–7.2.

5.3   Coalgebraic Semantics of Quantum Systems

We now have only one last step to perform: implementing the theory of local truth in a QFT system, that is, demonstrating that the curve of the quasi probability diagram of Fig. 1a as an information measure of the degree of semantic informativeness represents a measure of maximal entropy, expressing the fact that a given *dynamic cognitive system* (e.g. brain dynamics in the QFT interpretation depicted in Sect. 3.2) consumed all the *free-energy* deriving from the mismatch with its thermal bath, for the reorganization "work" (in the thermodynamic sense) of its inner state, so to match the outer state (thermal-bath), and then minimizing the free energy of the whole system (brain + thermal bath). In other terms, we have to interpret the maximal entropy physical measure as a logical measure of *maximal local truth* in the statistical sense. To sum up, we have to interpret consistently a QFT dynamic system as a *computing system*.

In the light of the discussion above it is necessary and sufficient for such an aim to demonstrate that the collections of the "q-deformed Hopf algebras" and the "q-deformed Hopf coalgebras" of the QFT mathematical formalism constitute two *dually equivalent categories* for the contravariant application of the same functor T, that is, the contravariant application of the so-called *Bogoliubov transform*. This is the classical QFT operator of "particle creation-annihilation", where the necessity of such a contravariance depends on the constraint of satisfying anyway the *energy balance principle*, i.e. q-**HCoalg**$(T) \rightleftharpoons q$-**HAlg**(T^*). It is useful to recall here that the q-*deformation* parameter characterizing each pair of q-deformed Hopf algebra coalgebra is physically a *thermal parameter*, so as to constitute the "evolution parameter" of the universe in a QFT interpretation of cosmology, via SSBs of the QV, according to Wheeler's suggestion with which we started our paper that in the new physics paradigm "cosmogony is the legislator of physics".

On the other hand, mathematically, this parameter is related to the "Bogoliubov angle", θ, characterizing each different application of the transform—where, as we know, the angle, with the frequency and the amplitude, are the three main parameters characterizing generally the phase of a given waveform. For a systematic presentation of the QFT mathematical formalism, see [17, pp. 185–235].

The complete justification of a coalgebraic interpretation of this mathematical formalism is given elsewhere [79], because we cannot develop it here. Nevertheless, at least two points of such a justification are important to emphasize, for justifying the interpretation of the maximal entropy in a QFT system as a semantic measure of information, i.e. as a statistical measure of *maximal local truth* in a CT coalgebraic logic for QFT systems.

Firstly, the necessary condition to be satisfied in order that a coalgebra category for some endofunctor Ω, i.e. **Coalg**(Ω), can be interpreted as a *dynamic* and/or *computational* system, is that it satisfies the formal notion of *state transition system* (STS). Generally a STS is an abstract machine characterized as a pair $(S, \rightarrow)$, where S is a set of states, and $((\rightarrow) \subseteq S \times S)$ is a transition binary relation over S. If p, q belong to S, and (p, q) belongs to $(\rightarrow)$, then $(p \rightarrow q)$, i.e. there is a transition over S. To allow a dynamic/computational system to be represented as a STS on a functorial coalgebra for some functor Ω, it is necessary that the functor admits a *final coalgebra* [16, p. 389]; I.e.:

Definition 1: (Definition of final coalgebra for a functor). A functor $\Omega : \mathbf{C} \rightarrow \mathbf{C}$ is said to admit a final coalgebra iff the category **Coalg(Ω)** has a final object, that is, a coalgebra $\mathbb{Z}$ such that from every coalgebra $\mathbb{A}$ in **Coalg(Ω)**, there exists a unique homomorphism, $!\mathbb{A}\colon \mathbb{A} \rightarrow \mathbb{Z}$.

This property has a very intriguing realization—and this is the sufficient condition to satisfy for formalizing a QFT system as a computing system—into the final coalgebra associated with a particular abstract machine, the so-called infinite state black-box machine $\mathbb{M}\langle M, \mu \rangle$ [16, p. 395]. It is characterized by the fact that the machine internal states, $x_0, x_1, ...$, cannot be directly observed, but only some values ("colors", c_n) associated with a state transition μ. I.e. $\mu(x_0) = (c_0, x_1)$, $\mu(x_1) = (c_1, x_2)$, ... In this way, the only "observable" of this dynamics is the infinite sequence of behaviors or *stream beh* $(x_0) = (c_0, c_1, c_2, ...) \in C^\omega$ of value combinations or "words" over the dataset C. The collection C^ω forms a *labeled* STS for the functor $C \times \mathcal{I}$, where $\mathcal{I}$ is the set of all the identity functions (labels), as far as we endow C^ω with a transition structure γ splitting a stream $u = c_0 c_1 c_2, ...$ into its "head" $h(u) = c_0$, and its *tail* $t(u) = c_1 c_2 c_3....$ If we pose $\gamma(u) = (h(u), t(u))$, it is possible to demonstrate that the behavior map $x \mapsto beh(x)$ is the unique homomorphism from $\mathbb{M}$ to this coalgebra $\langle C, \gamma \rangle$, that is, the final coalgebra $\mathbb{Z}$ in the category **Coalg**$(C \times \mathcal{I})$.[13]

The abstract machine $\mathbb{M}$ is used in TCS for modeling the *coalgebraic semantics* of programming relative to infinite datasets, so-called *streams*: think, for instance, of the internet and more generally of all the ever-growing databases ("big data") [15]. The application of $\mathbb{M}$ for characterizing the QFT dynamics as a "computing dynamics" is evident in the light of the discussion above because we are allowed to interpret the thermodynamic functor T (Bogoliubov transform) characterizing the category *q*-**HCoalg**(T) as a functor able to associate the observable c of each "word" (phase coherence domain) of the QFT infinite dataset C, i.e. the infinite CCRs characterizing the QV, with the correspondent I_c, so that $T = (C \times \mathcal{I})$. Indeed, each I_c corresponds in the QFT formalism to the NGB condensate numerical value $|\mathcal{N}|$ univocally identifying each phase coherence domain, i.e. a "word" of the QV "language". In this way, the QV, because it is endowed with the SSB state transition —effectively a phase-transition—structure γ, selecting every time one CCR (*head*) from the others (*tile*), corresponds to the final coalgebra $\mathbb{Z}$ of the category *q*-**HCoalg**(T). Moreover, the dynamics of $\mathbb{M}_{QFT}$ is a *thermo*-dynamics; i.e. its state (phase) transition is "moved" by the II Principle (energy equipartition), in a way

---

[13]In parenthesis, in the machine $\mathbb{M}$ the general coalgebraic principle of the *observational* (or *behavioral*) *equivalence* among states holds in the following way. Indeed, for every two coalgebras (systems) $\mathbb{S}_1, \mathbb{S}_2 \in \mathbf{Coalg}(C \times \mathcal{I})$, $(!c_{\mathbb{S}_1} = !c_{\mathbb{S}_2}) \Rightarrow (!x_{\mathbb{S}_1} = !x_{\mathbb{S}_2})$. All scholars agree that this has an immediate meaning for quantum systems logic and mathematics, as a further justification for a coalgebraic interpretation of quantum systems.

that must satisfy, on the one hand, the "energy arrow contravariance" related to the I Principle, and, on the other one, without consuming all the QV energy "reservoir" as requested by the III Principle.[14] All this implies the necessity of doubling the behavior map, i.e. $x \mapsto beh(x, \tilde{x})$, and all the related objects and structures—i.e. the necessity of "echoing" each word of the QV language—so as to satisfy finally the "dual equivalence" characterizing the QFT categorical formalism, i.e. q-**HCoalg** $(T) \rightleftharpoons q$-**HAlg**(T^*). In logical terms, the functor induction $T \leftarrow T^*$ means that the semantics (coalgebra) induces its own syntax (algebra). This, on the one hand, justifies the computer scientist's interest toward a coalgebraic approach to quantum computation for managing streams, and on the other, demonstrates that the QFT interpretation of this approach is the more promising one. In fact, what we intended using the metaphor of "word echoing" within the model of the $\mathbb{M}_{QFT}$ is effectively the DDF principle determining the *dynamic choice*, observer independent, of the structure (syntax) of the "composed Hilbert space" of a QFT system as based on the *dual equivalence* (semantics) of one pair q-HCoalg$(T) \rightleftharpoons q$-HAlg(T^*) representing the whole dissipative system [79].

   All this is related to the second, final, observation, justifying the interpretation of the maximal entropy in a QFT "doubled" system as a semantic measure of information, i.e. as a statistical measure of *maximal local truth* in the CT coalgebraic logic. In the QFT mathematical formalism, this maximum of the entropy measure is formally attained when the above-illustrated DDF principle (far-from-equilibrium energy balance) between a system (algebra) and its thermal bath (coalgebra) is *dynamically* (=automatically) satisfied. This means that we are allowed to interpret the QFT *qubit* of such a natural computation as an "evaluation function" in the semantic sense. Indeed, in the QFT "composed Hilbert space" including also the thermal bath degrees of freedom, $\tilde{A}$, i.e. $\mathcal{H}_{A,\tilde{A}} = \mathcal{H}_A \otimes \mathcal{H}_{\tilde{A}}$, for calculating the static and dynamic entropy associated with the time evolution generated by the free energy, i.e. $|\phi(t)\rangle, |\psi(t)\rangle$, of the qubit mixed states $|\phi\rangle, |\psi\rangle$, one needs to double the states by introducing the tilde states $|0\rangle$ and $|1\rangle$, relative to the thermal bath, i.e. $|0\rangle \rightarrow |0\rangle \otimes |0\rangle$ and $|1\rangle \rightarrow |1\rangle \otimes |1\rangle$. This means that such a QFT version of a qubit implements effectively the CNOT (controlled NOT) logical gate, which flips the state of the qubit, conditional on a *dynamic* control of an effective input matching [79].

6   Conclusion

I used many times in this paper, also in the title, the expression "new paradigm". Th. Kuhn, who coined the fortunate expression "paradigm shift", said that the "scientific community" has to decree this shift every time it happens in the history

---

[14]A condition elegantly satisfied in the QFT formalism by the *fractal structure* of the systems phase space and, therefore, by the *chaotic character* of the macroscopic trajectories (phase transitions) defined on it, generally [88], and specifically in dissipative brain dynamics [32].

of science. Therefore, if we agree in giving to the Stockholm Royal Academy the honor and the duty of representing the scientific community at its higher levels, it decreed just a few months ago that we are living one of these turns characterizing the history of modern science. In the official conference press release for announcing to the world that the 2015 Nobel Prize in Physics was awarded to the physicists T. Kajita and A. B. McDonald for their observational discovery of the neutrino mass, the Academy stated that "the new observations had clearly showed that the Standard Model cannot be the complete theory of the fundamental constituents of the universe" [89].

In this paper we defended the idea that QFT interpreted as a "thermal field theory" is a candidate for constituting, above all, the proper theory of the "physics beyond the Standard Model" because it is able to give physics a strong formal alternative to the "perturbative methods" and their "asymptotic states" that lie at the basis of the Standard Model "mechanistic" interpretation of the statistical distinction between "fermions" and "bosons", in terms of "particles" and of "force field quanta", respectively. The validity of perturbative methods relies indeed on the possibility of correctly defining asymptotic states for the system, namely states defined in infinitely distant space time regions, so as to make interactions negligible, in the presumption that this representation is not falsifying the nature of the physical system to be represented. In this light, the "paradigm turn" with which today we are faced is therefore not only with respect to the Standard Model physics (QED and QCD, i.e. the so-called standard QFT), but also to QM, the "many-body dynamics" extension of classical mechanics, from Laplace on. Therefore, the presumption of correctly representing a system as isolated lies at the bottom of the same origins of modern physics and modern calculus. This presumption cannot hold, however, in the case of QFT systems, as they are intrinsically "open" to background QV fluctuations, or, more generally, when we have to reckon with system phase transitions. In all these cases, the QFT alternative picture of representing both fermions and bosons as quanta of the relative force field is more suitable. On the other hand, we showed that the paradigm shift we are discussing, because it involves the foundations of modern science, also involves the foundations of mathematics and computability theory, as far as both are related to non standard set theories.

The alternative formalism offered by thermal QFT is, therefore, the doubled algebra representation of a quantum system and of its thermal bath, through the mathematical formalism of the DDF between a q-deformed Hopf algebra and its q-deformed Hopf coalgebra, as illustrated in this chapter. Such formalism has been successfully applied not only in fundamental physics, the physics of neutrino oscillations included [17, pp. 91–95], but also in condensed matter physics, biological matter, and brain dynamics. For this reason, in this chapter we deepened the possibility of justifying in such a formalism also the notion and measure of "semantic information", generally associated with biological and neural information processing. For this aim, we discussed an information-theoretic interpretation of the DDF in QFT systems, in the framework of the coalgebraic approach to quantum computation, recently introduced as an alternative to the information-theoretic interpretation of QM systems as quantum Turing machines. This allowed us to give a formal justification of the notion of "local truth", associated with the measure of semantic information that is therefore interpreted as a measure of maximal entropy, because it minimizes the "free energy" associated with the mismatch with the system environment thermal bath. Practically, this measure expresses the "entanglement" that dynamically occurred, as signaled by the flipping of the associated qubit, between the degrees of freedom of the system and of its thermal bath, within the same representation space including both [79]. This result is ultimately based on the possibility of justifying the dual equivalence between the categories of the q-deformed Hopf algebras and the q-deformed Hopf coalgebras, allowing one to interpret quantum computations of QFT systems in the framework of category theory logic. This initial result opens the way to new promising scenarios in quantum natural and artificial computation to be explored in the near future.

Appendix A

A.1 Induction and Coinduction as Principles of Set Definition and Proof for Boolean Lattices

The collection of clopen subsets of a Stone space, to which a Boolean algebra is isomorphic, according to the Stone theorem is effectively an ultrafilter U (or the maximal filter F) on the power set, $\wp(S)$, of the set S. Namely, it is the *maximal partially ordered set* (maximal poset) within $\wp(S)$ ordered by inclusion, i.e. $(\wp(S),$ $\subseteq)$, with the exclusion of the empty set. Any filter F is *dual* to an *ideal* I, simply obtained in set (order) theory by inverting all the relations in F, that is, $x \leq y$ with $y \leq x$, and by substituting *intersections* with *unions*. From this derives that each ultrafilter U is dual to a greatest ideal that, in Boolean algebra, is also a *prime* ideal, because of the so-called *prime ideal theorem*, effectively a corollary of the Stone theorem, demonstrated by himself. All this, applied to the Stone theorem, means that the collection of partially ordered clopen subsets of the Stone space, to which a Boolean algebra is isomorphic, corresponds to a Boolean logic complete lattice L for a *monadic first-order predicate logic*. From this, the definition of *induction* and *coinduction* as dual principles of set definition and proof is immediate, as soon as we recall that the fixed point of a computation F is given by the equality $x = F$ (x) [68, p. 46]:

Definition 2 (*sets inductively/coinductively defined by F*). For a complete Boolean lattice L whose points are sets, and for an endofunction F, the sets

$$F_{\text{ind}} := \bigcap \{x | F(x) \leq x\}$$
$$F_{\text{coind}} := \bigcup \{x | x \leq F(x)\}$$

are, respectively, the sets *inductively* defined by a *recursive F*, and *coinductively* defined by a *co-recursive F*. They correspond, respectively, to the *meet* of the pre-fixed point and the *join* of the post-fixed points in the lattice *L*, i.e. the least and greatest fixed points, if *F* is monotone, as required from the definition of the category **Pos** (see above, Sect. 5.1).

Definition 3 (*induction and coinduction proof principles*). In the hypothesis of Definition 2, we have:

$$if\ F(x) \leq x\ then\ F_{ind} \leq x \quad \text{(induction as a method of proof)}$$
$$if\ x \leq F(x)\ then\ x \leq F_{coind} \quad \text{(coinduction as a method of proof)}$$

These two definitions are the basis for the *duality* between an *initial algebra* and its *final coalgebra*, as a new paradigm of computability, i.e. Abramsky's *finitary* one, and henceforth for the duality between the *universal algebra* and the *universal coalgebra* [15].

A.2 Extension of the Coinduction Method to the Definition of a Complete Boolean Lattice of Monadic Predicates

The fundamental result of the above-quoted Goldblatt–Thomason theorem and van Benthem theorem is that a set-tree of NWF sets—effectively a set represented as an oriented graph where nodes are sets, and edges are inclusion relations with subsets governed by Euclidean rule—corresponds to the structure of a Kripke *frame* of his relational semantics, characterized by a set of "worlds" and by a two-place accessibility relation *R* between worlds, e.g. the second graph from left below corresponds to the graph of the number 3, with $u = 3$, $v = 2$, $w = 1$. Therefore for understanding intuitively the extension of the coinduction method to the domains of monadic predicates of a Boolean lattice, let us start from (1) the "Euclidean rule (ER)" $\langle \forall u,v,w\ ((uRv \wedge uRw) \rightarrow vRw) \rangle$ (see the second from left graph below), driving all the NWF set inclusions and that is associated by van Benthem's correspondence theorem to the modal axiom **E** (or **5**): $\langle \Diamond \alpha \rightarrow \Box \Diamond \alpha \rangle$, of the modal propositional calculus, and (2) from the "seriality rule (SR)" $\langle \forall u \exists v\ (uRv) \rangle$ (an example of this axiom is given by the fourth or fifth graph below)—that has an immediate physical sense, because it corresponds to whichever energy conservation principle in physics, e.g. the I Principle of Thermodynamics—and that is associated to the modal axiom **D**: $\langle \Box \alpha \rightarrow \Diamond \alpha \rangle$. The straightforward first-order calculus, by which it is possible to formally justify the definition/justification by coinduction (tree unfolding) of an *equivalence class* as the domain of a given monadic predicate, through the application of the two above rules to whichever triple of objects $\langle u, v, w \rangle$, is the following:

For ER, $\langle \forall u,v,w\ ((uRv \wedge uRw) \rightarrow vRw) \rangle$; hence, for seriality, $\langle \forall u,v\ (uRv \rightarrow vRv) \rangle$; finally: $\langle \forall u,v,w\ [((uRv \wedge uRw) \rightarrow (vRw \wedge wRv \wedge vRv \wedge wRw)) \leftrightarrow$

$((v \equiv w) \subset u)]$, I.e. $(v \equiv w)$ constitutes an equivalence subclass of u, say **Y**, because a "generated" transitive[15]-symmetric-reflexive relation holds among its elements, which are therefore also "descendants" of their common "ascendant", u. More intuitively, using Kripke's relational semantics graphs for modal logics, where ‹u, v, w› are also "possible worlds" (models) of a given universe W, and where R is the two-place "accessibility relation" between worlds, the above calculus reads:

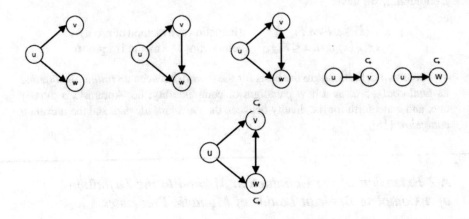

The final graph constitutes a Kripke-like representation of the **KD45** modal system, also defined in literature as "secondary **S5**", since the equivalence relationship among all the possible worlds characterizing **S5** here holds only for a subset of them, in our example, the subset of worlds $\{w, v\}$.

References

1. Patton, C.M., Wheeler, J.A.: Is physics legislated by cosmogony? In: Isham, C.J., Penrose, R., Sciama, D.W. (eds.) Quantum Gravity, pp. 538–605. Clarendon Press, Oxford (1975)
2. Feynman, R.: Simulating physics with computers. Int. J. Theor. Phys. **21**, 467–488 (1982)
3. Wheeler, J.A.: Information, physics, quantum: the search for links. In: Zurek, W.H. (eds.) Complexity, Entropy, and the Physics of Information. Addison-Wesley, Redwood City (1990)
4. Deutsch, D.: Quantum theory, the church-turing principle and the universal quantum computer. Proc. R. Soc. Lond. A **400**, 97–117 (1985)
5. Rovelli, C.: Relational quantum mechanics. Int. J. Theor. Phys. **35**, 1637–1678 (1996)
6. Zenil, H. (ed.): A computable universe. Understanding and exploring nature as computation. Foreword by Sir Roger Penrose. World Scientific Publishing, Singapore (2013)
7. Fields, C.: If physics is an information science, what is an observer? Information **3**(1), 92–123 (2012)

---

[15]Remember that the transitive rule in the NWF set theory does not hold only for the inclusion operation, i.e. for the superset/subset ordering relation.

8. Zeh, H.D.: Wave function: 'it' or 'bit'? In: Barrow, J.D., Davies, P.C.W., Harper Jr., C.L. (eds.) Science and Ultimate Reality, pp. 103–120. Cambridge University Press, Cambridge (2004)

9. Zeh, H.D.: Quantum discreteness is an illusion. Found. Phys. **40**, 1476–1493 (2010)

10. Tegmark, M.: How unitary cosmology generalizes thermodynamics and solves the inflationary entropy problem (2011). http://arxiv.org/pdf/1108.3080.pdf. [Accessed 16 March 2012]

11. Blackburn, P., van Benthem, J.: Modal logic: a semantic perspective. In: Blackburn, P., van Benthem, F.J.F., Wolter, F. (eds.) Handbook of Modal Logic, pp. 2–85. Elsevier, Amsterdam (2007)

12. Abramsky, S., Tzevelekos, N.: Introduction to categories and categorical logic. In: Coecke, B. (ed.) New Structures for Physics. Lecture Notes in Physics, vol. 813, pp. 3–94. Springer, Berlin-New York (2011)

13. Forti, M., Honsell, F.: Set theory with free construction principles. Scuola Normale Superiore —Pisa Classe de Scienza. Serie IV, vol. 10, pp. 493–522 (1983)

14. Abramsky, S.: Coalgebras, Chu spaces, and representations of physical systems. J. Phil. Log. **42**(3), 551–574 (2013)

15. Aczel, P.: Non-well-founded sets. In: CLSI Lecture Notes, vol. 14 (1988)

16. Aczel, P., Mendler, N.P.: A final coalgebra theorem. In: Category Theory and Computer Science. London, UK (1989)

17. Blasone, M., Jizba, P., Vitiello, G.: Quantum field theory and its macroscopic manifestations. Boson condensation, ordered patterns and topological defects. Imperial College Press, London (2011)

18. Del Giudice, E., Pulselli, R., Tiezzi, E.: Thermodynamics of irreversible processes and quantum field theory: an interplay for understanding of ecosystem dynamics. Ecol. Model. **220**, 1874–1879 (2009)

19. Von Neumann, J.: Mathematical Foundations of Quantum Mechanics. Princeton University Press, Princeton (1955)

20. Goldstone, J.: Field theories with superconductor solutions. Nuovo Cimento **19**, 154–164 (1961)

21. Goldstone, J., Salam, A., Weinberg, S.: Broken symmetries. Phys. Rev. **127**, 965–970 (1962)

22. Vitiello, G.: Links. Relating different physical systems through the common QFT algebraic structure. Lect. Notes Phys. **718**, 165–205 (2007)

23. Nambu, Y.: Quasiparticles and gauge invariance in the theory of superconductivity. Phys. Rev. **117**, 648–663 (1960)

24. Itzykson, C., Zuber, J.: Quantum Field Theory. McGraw-Hill, New York (1980)

25. Umezawa, H.: Advanced field theory: micro, macro and thermal concepts. American Institute of Physics, New York (1993)

26. Umezawa, H.: Development in concepts in quantum field theory in half century. Math. Jpn. **41**, 109–124 (1995)

27. Vitiello, G.: The dissipative brain. In: Globus, G.G., Pribram, K.H., Vitiello, G. (eds.) Brain and Being—At the Boundary Between Science, Philosophy, Language and Arts, pp. 317–330. John Benjamins, Amsterdam (2004)

28. Basti, G.: Intelligence and reference. Formal ontology of the natural computation. In: Dodig-Crnkovic, G., Giovagnoli, R. (eds.) Computing Nature. Turing Centenary Perspective, pp. 139–159. Springer, Berlin-Heidelberg (2013)

29. Basti, G.: A change of paradigm in cognitive neurosciences comment on: "Dissipation of 'dark energy' by cortex in knowledge retrieval" by Capolupo, Freeman and Vitiello. Phys. Life Rev. **5**(10), 97–98 (2013)

30. Freeman, W.J., Vitiello, G.: Nonlinear brain dynamics as macroscopic manifestation of underlying many-body field dynamics. Phys. Life Rev. **3**(2), 93–118 (2006)

31. Freeman, W.J., Vitiello, G.: Dissipation and spontaneous symmetry breaking in brain dynamics. J. Phys. A: Math. Theor. **41**(30), 304042 (2008)

32. Vitiello, G.: Coherent states, fractals and brain waves. New Math. Nat. Comput. **5**(1), 245–264 (2009)
33. Freeman, W.J.: Origin, structure, and role of background EEG activity. Part 1. Analytic amplitude. Clin. Neurophysiol. **115**, 2077–2088 (2004)
34. Freeman, W.J.: Origin, structure, and role of background EEG activity. Part 2. Analytic phase. Clin. Neurophysiol. **115**, 2089–2107 (2004)
35. Freeman, W.J.: Origin, structure, and role of background EEG activity. Part 3. Neural frame classification. Clin. Neurophysiol. **116**, 111–1129 (2005)
36. Freeman, W.J.: Origin, structure, and role of background EEG activity. Part 4. Neural frame simulation. Clin. Neurophysiol. **117**, 572–589 (2006)
37. Freeman, W.J., Burke, B.C., Holmes, M.D., Vanhatalo, S.: Spatial spectra of scalp EEG and EMG from awake humans. Clin. Neurophysiol. **114**, 1055–1060 (2003)
38. Freeman, W.J., Ga'al, G., Jornten, R.: A neurobiological theory of meaning in perception. Part 3. Multiple cortical areas synchronize without loss of local autonomy. Intern. J. Bifurc. Chaos **13**, 2845–2856 (2003)
39. Freeman, W.J., Rogers, L.J.: A neurobiological theory of meaning in perception. Part 5. Multicortical patterns of phase modulation in gamma EEG. Int. J. Bifurc. Chaos **13**, 2867–2887 (2003)
40. Schrödinger, E.: What is Life. Cambridge University Press, Cambridge (1944)
41. Szent-György, A.: An Introduction to Sub-molecular Biology. New York (1960)
42. Frölich, H.: Long range coherence and energy storage in biological systems. Int. J. Quantum Chem. **2**, 641ff. (1968)
43. Frölich, H. (ed.): Biological Coherence and Response to External Stimuli. Springer, Berlin (1988)
44. Ho, M.W.: What is (Schrödinger's) negentropy. Modern Trends in BioThermoKinetics, vol. 3, pp. 50–61 (1994)
45. Szilard, L.: On the decrease of entropy content in a thermodynamical system by the intervention of intelligent beings. Behav. Sci. **9**(4), 301–310 (1964)
46. Brillouin, L.: Science and Information Theory. Academic Press, New York (1962)
47. Capolupo, A., Freeman, W.J., Vitiello, G.: Dissipation of dark energy by cortex in knowledge retrieval. Phys. Life Rev. **10**(1), 85–94 (2013)
48. Shannon, C.E.: The mathematical theory of communication. University of Illinois Press, Urbana (1949)
49. MacKay, D.M.: Information, Mechanism, and Meaning. MIT Press (1969)
50. Carnap, R., Bar-Hillel, Y.: An outline of a theory of semantic information. In: Language and Information: Selected Essays on Their Theory and Application, pp. 221–274. Addison-Wesley, Reading (1964)
51. Barbieri, M.: Semantic Codes. An Introduction to Semantic Biology. Cambridge University Press, Cambridge (2003)
52. Deacon, T.W.: Incomplete Nature. How Mind Emerged from Matter. Norton, New York (2011)
53. Atmanspacher, H., Scheingraber, H.: Pragmatic information and dynamical instabilities in multimode continuous-wave dye laser. Can. J. Phys. **68**, 728–737 (1990)
54. Cahill, K.E., Glauber, R.J.: Density operators and quasiprobability distributions. Phys. Rev. **177**(5), 1882–1902 (1969)
55. Kolmogorov, A.N.: Foundations of the Theory of Probability. Second English edition. Chelsea Publishing, New York (1956)
56. Floridi, L.: Semantic conceptions of information. In: Zalta, E.N. (ed.) Stanford Encyclopedia of Philosophy. Spring 2011 Ed, pp. 1–70 (2011)
57. Sequoiah-Grayson, S.: The metaphilosophy of information. Mind. Mach. **17**(3), 331–344 (2007)
58. Carnap, R.: Meaning and Necessity: A Study in Semantics and Modal Logic. Chicago University Press, Chicago (1956)

59. Basti, G.: From formal logic to formal ontology. The new dual paradigm in natural sciences. In: Proceedings of 1st CLE Colloquium for Philosophy and History of Formal Sciences, Campinas, 21–23 March 2013, Campinas-Rome (2016)
60. Dodig-Crnkovic, G.: System modelling and information semantics. In: Bubenkojr, J. et al. (ed.) Proceedings of the Fifth Conference for the Promotion of Research in IT. New Universities and University Colleges in Sweden, Lund, Studentlitteratur (2005)
61. Basti, G.: The Formal Ontology of the Natural Realism. Sociedade Brasileira de Historia de Matematica, Campinas (2014)
62. Connes, A., Rovelli, C.: Von Neumann algebra automorphisms and time-thermodynamics relation in general covariant quantum theories. Class. Quantum Gravity **11**, 2899–2918 (1994)
63. Stone, M.H.: The theory of representation for Boolean algebras. Trans. Am. Math. Soc. **40**, 37–111 (1936)
64. Abramsky, S.: A Cook's tour of the finitary non-well-founded sets (original lecture: 1988). In: Artemov, S., Barringer, H., d'Avila, A., Lamb, L.C., Woods, J. (eds.) Essays in Honor of Dov Gabbay, vol. I, pp. 1–18. Imperial College Publication, London (2005)
65. Venema, Y.: Algebras and co-algebras. In: Blackburn, P., van Benthem, F.J.F., Wolter, F. (eds.) Handbook of Modal Logic, pp. 331–426. Elsevier, Amsterdam (2007)
66. Schröder, E.: Vorlesungen über die Algebra der Logik. (Exakte Logik), vol. 1. B. G. Teubner, Leipzig (1890)
67. Tarski, A.: The concept of truth in formalized languages. In: Corcoran, J. (ed.) Logic, Semantics, Metamathematics, 2, pp. 152–278. Indianapolis, Hackett (1983)
68. Gödel, K.: Über formal unentscheidbare Sätze der Principia Mathematica und verwandter Systeme, I. Monatshefte für Mathematik und Physik, vol. 38, pp. 173–98 (1931)
69. Sangiorgi, D.: Origins of bisimulation and coinduction. In: Sangiorgi, D., Rutten, J. (eds.) Advanced Topics in Bisimulation and Coinduction, pp. 1–37. Cambridge University Press, Cambridge (2012)
70. Rutten, J.J.M.: Universal coalgebra: a theory of systems. Theoret. Comput. Sci. **249**(1), 3–80 (2000)
71. Van Benthem, J.: Correspondence theory. In: Gabbay, D., Guenthner, F. (eds.) Handbook of Philosophical Logic: Vol 2, pp. 167–247. Dordrecht, Reidel (1984)
72. Moss, L.S.: Coalgebraic logic. Ann. Pure Appl. Logic **96**, 277–317 (1999)
73. Moss, L.S.: Erratum to "Coalgebraic logic". [Ann. Pure Appl. Logic 96 (1999) 277-317]. Ann. Pure Appl. Logic **99**, 241–259 (1999)
74. Kurz, A.: Specifying coalgebras with modal logic. Theoret. Comput. Sci. **210**, 119–138 (2001)
75. Kupke, C., Kurz, A., Venema, Y.: Stone coalgebras. Theor. Comput. Sci. **327**, 109–134 (2004)
76. Thomason, S.K.: Reduction of second-order logic to modal logic. Math. Logic Q. **21**(1), 107–114 (1975)
77. Van Benthem, J.: Modal correspondence theory. Ph.D. thesis, Dept. of Mathematics. Amsterdam University Press, Amsterdam (1976)
78. Coecke, B. (ed.) New Structures for Physics. Lecture Notes in Physics, vol. 813. Springer, Berlin-New York (2011)
79. Basti, G., Capolupo, A., Vitiello, G.: Quantum field theory and coalgebraic logic in theoretical computer science. Progress in Biophysics and Molecular Biology (preprint) (2017). https://doi.org/10.1016/j.pbiomolbio.2017.04.006
80. Heller, M.: Algebraic self duality as the 'ultimate explanation'. Fund. Sci. **9**, 369–385 (2004)
81. Vitiello, G.: Fractals, coherent states and self-similarity induced noncommutative geometry. Phys. Lett. A **376**, 2527 (2012)
82. Basti, G.: L'idea di scienza di Maritain fra passato e futuro. Aquinas **58**(1–2), 117–165 (2015)
83. Sakellariadou, M., Stabile, A., Vitiello, G.: Noncommutative spectral geometry, algebra doubling, and the seed of quantization. Phys. Rev. D **84**, 045026 (2011)

84. Hinton, G.E., Osindero, S., Teh, Y.-W.: A fast learning algorithm for deep belief nets. Neural Comput. **18**(7), 1527–1554 (2006)
85. Majid, S.: Quantum groups primer. Cambridge University Press, Cambridge (2002)
86. Schröder, L., Pattinson, D.: Coalgebraic correspondence theory. In: Onge, L. (ed.) Foundations of Software and Computational Structures. Lecture Notes in Computer Science, vol. 6014, pp. 328–342. Springer, Berlin-New York (2010)
87. Goranko, V., Otto, M.: Model theory of modal logic. In: Blackburn, P., van Benthem, F.J.F., Wolter, F. (eds.) Handbook of Modal Logic, pp. 252–331. Elsevier, Amsterdam (2007)
88. Bruno, A., Capolupo, A., Kak, S., Raimondo, G., Vitiello, G.: Geometric phase and gauge theory structure in quantum computing. J. Phys. Conf. Ser. **306**, 012065 (2011)
89. Subrata, G., et al.: Design and construction of a brain-like computer: a new class of frequency-fractal computing using wireless communication in a supramolecular organic, inorganic system. Information **5**, 28–100 (2014)
90. The 2015 Nobel Prize in Physics—Press Release. Nobelprize.org. Nobel Media AB 2014. Web. 13 Dec 2015. http://www.nobelprize.org/nobel_prizes/physics/laureates/2015/press. html, Nobelprize.org, 6 Oct 2015. [Online]. [Accessed 13 Dec 2015]

Reality Construction in Cognitive Agents Through Processes of Info-computation

Gordana Dodig-Crnkovic and Rickard von Haugwitz

Abstract *What is reality for an agent? What is minimal cognition? How does the morphology of a cognitive agent affect cognition?* These are still open questions among scientists and philosophers. In this chapter we propose the idea of info-computational nature as a framework for answering those questions. Within the info-computational framework, information is defined as a structure (for an agent), and computation as the dynamics of information (information processing). To an agent, nature therefore appears as an informational structure with computational dynamics. Both information and computation in this context have broader meaning than in everyday use, and both are necessarily grounded in physical implementation. Evolution of increasingly complex living agents is understood as a process of morphological (physical, embodied) computation driven by agents' interactions with the environment. It is a process much more complex than random variation; instead the mechanisms of change are morphological computational processes of self-organisation (and re-organisation). Reality for an agent emerges as a result of interactions with the environment together with internal information processing. Following Maturana and Varela, we take cognition to be the process of living of an organism, and thus it appears on different levels of complexity, from cellular via organismic to social. The simpler the agent, the simpler its "reality" defined by the network of networks of info-computational processes, which constitute its cognition. The debated topic of consciousness takes its natural place in this framework, as a process of information integration that we suggest naturally evolved in organisms with a nervous system. Computing nature/pancomputationalism is sometimes confused with panpsychism or claimed to necessarily imply panpsychism, which we show is not the case. Even though we focus on natural systems in this chapter, the info-computational approach is general and can be used to model both biological and artifactual cognitive agents.

G. Dodig-Crnkovic (✉) · R. von Haugwitz
Chalmers University of Technology and University of Gothenburg,
Gothenburg, Sweden
e-mail: dodig@chalmers.se

R. von Haugwitz
e-mail: rickardv@chalmers.se

© Springer International Publishing AG 2017
G. Dodig-Crnkovic and R. Giovagnoli (eds.), *Representation and Reality in Humans,
Other Living Organisms and Intelligent Machines*, Studies in Applied Philosophy,
Epistemology and Rational Ethics 28, DOI 10.1007/978-3-319-43784-2_10

1 Introduction

For a cognitive agent living in an environment, in order to produce efficient behaviour, it needs to interact with the environment and make sense of those interactions. Interactions are based on sensors that enable perception and actuators that enable action. What an agent perceives of the environment will vary depending on its sensory apparatus, and the context in which it needs to interpret and act upon incoming information. This is true for biological and artificial agents alike. Any cognitive agent is therefore faced with the problem of constructing, internally, its own version of reality. This does not itself imply the need for representation in the classical sense, as information may be encoded in agent's physical bodily structures directly affecting its behaviour. It does, however, imply a form of information processing as the essential property of cognition: information about the environment is processed so as to produce behaviour that aids the survival of the agent.

In this chapter, we will argue that information processing in general, and in the context of cognition in particular, is best understood as computation. This leads to a wide notion of computation known as info-computation, in which all natural processes may be viewed and reasoned about as computational processes, or natural computing. This view of computation is broader than physical computation presented in [1] claiming that *"Physical computing is the use of a physical system to predict the outcome of an abstract evolution"*. On this view there is a human using a physical system, and intrinsic physical processes going on in natural systems are not automatically qualified as computations, which is the case in the info-computational framework of computing nature. The taxonomy of [2, 3] shows the relationships between different models of computation.

From our standpoint of computing nature, we attempt to formulate the process of reality construction in a cognitive agent as a network of networks of hierarchically more complex info-computational processes. This understanding of computation is essentially different from the classical view of early computationalism [4], which focussed on cognition understood as human thinking. Our notion of cognition follows Maturana and Varela's view that *cognition can be found in all living organisms*, as *"Living systems are cognitive systems, and living as a process is a process of cognition. This statement is valid for all organisms, with or without a nervous system"* [5]. So cognition = life, argued even by [6], and thus follows evolution of life/cognition from abiogenesis via complexification through interactions. Life/cognition in our approach is characterised by information processes in networks of agents—starting from non-living parts of a cell to cells, organs, organisms and eco-systems—on different levels of organisation. Interesting to notice is that this idea can be traced back to Aristotle's concept of *soul* that is exactly what makes the difference between animate and inanimate. "The soul of an animate organism, in this framework, is nothing other than its system of active abilities to perform the vital functions that organisms of its kind naturally perform, so that when an organism engages in the relevant activities (e.g. nutrition, movement or thought) it does so in virtue of the system of abilities that is its soul" [7].

"The relation between soul and body, on Aristotle's view, is also an instance of the more general relation between form and matter: thus an ensouled, living body is a particular kind of in-formed matter" (ibid). This view connects smoothly to the idea of information as a structure that defines the form (relations).

There is a widespread misconception that all computational theories of mind necessarily imply purely syntactical manipulation of symbols that objectively represent the (external) environment. Of course, no such thing is actually possible in a real agent: no symbolic representations of unknown entities could exist a priori, but would have to be constructed, and refined, during the lifetime of the agent through interaction with the environment. As the environment can never be observed directly, completely or even reliably, any representations, symbolic or not, would necessarily be dependent on the subject who constructs it. In other words, although the environment must be assumed to exist as a source of observations, it is the subjective *relationship* to the environment, as well as to the received relationships of the community of practice of the social group that agents typically acquire via language, that the cognitive agent has access to. The aforementioned problem of constructing symbolic representations from non-symbolic sensory data is known as the *symbol-grounding problem*. The symbol-grounding problem is addressed by the theory of *enactivism* [5, 8], which states that all of cognition is a process of representation-free sensorimotor perception and response, with no intermediate symbol manipulation. Dynamical, analogue systems like neural networks are often classified under this category.

Identifying computation with the Turing machine model, there does indeed seem to be a dichotomy between computational and enactivist theories of cognition, and we would argue that enactivism is correct in its basic assumptions, see [9] p. 153. However, modern flavours of computationalism have expanded to include all forms of information processing as computation. Since information to a general agent need not come in the form of discrete symbols, there is no reason to assume that the processing of that information to produce behaviour necessitates the introduction of discrete symbols (although it is certainly useful when dealing with logic and mathematics, the original applications of computational models). In this view, neural networks and other dynamical systems are computational because they process information by changing their internal states and outputting previously inaccessible information. It follows that other forms of information processing, such as non-deterministic models, are also taken as computational, and the dichotomy between computational and enactivist theories of mind breaks down. A detailed overview of computational theories of cognition is given by Fresco [10] and comparison between different computational models, including cognitive computing, in [2].

The chapter is structured as follows: we start by defining cognition as a process of living. The next step is to ask what constitutes minimum cognition within the framework of computing nature/natural computationalism. Many still today believe that the only truly cognitive beings are humans. Adopting Maturana and Varela's view, we argue that it is much more explanatory fruitful to see cognition as developing in degrees during evolution. The subsequent section elaborates on the info-computational structure of reality for a cognitive agent. Living beings are parts of eco-systems, and they appear in different networks that interact and sustain each other, thus one section is dedicated to computation (information processing) in networks of agents. Organisms develop and evolve through morphological/morphogenetic processes that can be modelled as info-computation; so one section elaborates such processes. After making this connection between growth of form, evolution and development in living (cognitive) systems, we return to the main theme of our study: generation of reality for an agent: from raw data to information and cognition. In the subsequent section we make the connection of computational models of natural mind to its physical substrate, all the way down to quantum. To meet the criticism that pancomputationalism by necessity implies panpsychism such as furthered by [11], we address the distinction between pancomputationalism/computing nature and panpsychism. We finish with the conclusion and some remarks on future work.

2 Inanimate Versus Living/Cognitive Agents

When discussing cognition as a bioinformatic process, we use the notion of *agent*, i.e. a *system able to act on its own behalf* [12]. In this sense agency can be ascribed even to such simple non-living objects as elementary particles, which act upon each other and change their states by exchanging force carriers (elementary particles). It is important to keep in mind that this definition of agency is more general than the majority of other definitions, such as the notion of agency presented in [13], as we include both non-living and living agents. Cruse makes a distinction between reactive systems and cognitive systems, and thus excludes not only inanimate agents but also the simplest organisms without a nervous system [14]. Froese and Ziemke, in addition to *adaptivity*, emphasise organismic *constitutive autonomy* as necessary leading principles of enactive cognition [15]. It is worth mentioning that there is a gradual transition between inorganic systems and reactive living systems. The main difference between inanimate (material) agency and living agency is that a living agent acts in interplay with its environment *so as to persist*. (As a side note: Apoptosis, the process of programmed cell death, is an exception to this general tendency of life to survive. Apoptosis can be a part of the development of an organism, such as formation of fingers, where cells "in between" recede. Apoptosis can also be found in messenger bacteria losing their lives for the good of the colony, or it may be the consequence of damage or disease of a cell. In the big picture, it is a tautology that living beings are actively engaged in their own survival).

Inanimate agency is found in agents whose existence does not depend on their *active engagement*. An atom's interactions with the world do not actively support its continued existence. On the other hand, for living agents, it is characteristic that they have the possibility of choice of anticipated future outcomes, and they in general tend to make choices so as to survive. They *adapt* to the dynamical environment and *learn*. Living beings are open thermodynamic systems that actively "work" on their own survival. This is what Maturana and Varela termed "autopoiesis." Agency in the case of living agents has been explored in [16, 17]. We have no reason to believe that inanimate objects possess any anticipation. However, even the simplest living organisms such as bacteria exhibit ability to anticipate, as shown in the work of Ben Jacob et al. [18, 19].

This sharp differentiation between living and non-living agency might remind someone of vitalism, the belief that there is a fundamental difference between non-living and living entities, which vitalists ascribed to unexplained "élan vital" possessed by life. However, this is not the case. Living agency originates as an emergent capacity of specific autopoetic, operationally closed [20] structures that have circular self* properties (self-organisation, self-reflection, self-healing etc.) These properties are based on common physics and chemistry, and there is nothing mysterious about them. Just the opposite, with the info-computational framework we hope to contribute to understanding of morphogenetic processes that lead from inorganic to organic matter—abiogenesis, thus the transition from inanimate to living agency.

This chapter addresses the question of reality for different classes of cognitive/living agents, suggesting that study of cognition in nature is helping us construct inanimate cognitive agents. The world as it appears to a cognising agent depends on the type of interaction through which the agent acquires information [12]. Agents communicate by exchanging messages (information) that help them coordinate their actions [21] based on the (partial) information they possess about the world originating in their own inputs and information processing which they share as a part of distributed, social cognition. Increasing levels of cognition developed in living organisms during evolution, starting from basic behaviour such as found in bacteria and later on in insects, to increasingly complex behaviour in higher organisms such as mammals.

It is interesting to notice, that even though relatively simple organisms such as insects have a nervous system and brain, they lack the limbic system that in humans controls emotional response to stimuli, suggesting that simple organisms do not process physical stimuli emotionally. Bacteria communicate through exchange of molecules, by which they can display complex behaviour through quorum sensing [22], which activates gene expression in a bacterium based on the strength of a signal from the rest of the swarm. The same mechanisms are still in use further up the evolutionary ladder, with ants and honey bees capable of quorum sensing through pheromone exchange [23]. In the body of a multicellular organism, neurotransmitters and hormones exchanged between individual cells play a crucial role in the organism's function, thus revealing the principle that the later steps of evolution build upon the preceding ones. Interestingly, emotion appears on a higher level of evolution, mediated by neurotransmitters. It has been shown to have a modulating effect on learning [24], an important ability of adaptive agents in a dynamic environment. Part of the explanation as to why emotion appears gradually along the evolutionary ladder, from simple arousal and primal fear to complex social emotions, may thus be that it serves to enhance the adaptability and learning capacities of an organism in an increasingly complex and unpredictable environment—for as the capability of the organism to perceive and interact with its environment expands and becomes more complex, so does the reality for that organism, in a circular process (ibid).

The framework for the discussion in this chapter is *computing nature* in the form of *info-computationalism* [25]. It takes *reality* to be *information* for an agent with a *dynamics* of information understood as *computation*. Information (for an agent) is taken to be a *structure,* while computation is its *dynamics*. Information is observer relative, and so is computation, [12, 26, 27]. There is no information without physical implementation [28–30] and this approach is in line with the view of embodied embedded cognition [31, 32].

While speaking of general computational systems, we refer by *agent* to an entity capable of *acting* upon the environment through its output (which must take the form of physical interaction), as is usual in agent-based models. In case of a *cognising agent,* we refer to such an agent, which is taking actions according to some intrinsic goal(s). It should therefore not be interpreted as implying that a molecule in a molecular network, while acting in the sense of agent in the agent-based model, makes a *choice* between different actions (that would implying panpsychism, which is a non-scientific and non-naturalist doctrine). This can be compared with the actor model of concurrent computation [33] where the "actors" are the universal computational primitives. Responding to a message it receives, an actor can send messages, create new actors, and determine how to respond to the next message. A more detailed distinction between general computational *agents*, *actors and cognising agents* and their thorough characterisation deserves elaboration and remains to be done in forthcoming work, although a treatment of the subject can be found in [34].

The related question is what can we learn from nature in building artifactual cognitive agents with different degrees of cognition? Is synthetic cognition possible at all, given that cognition in living organisms is a deeply biologically rooted process? Recent advances in natural language processing present examples of developments towards machines capable of "understanding natural language", "translating" and "speaking" so that humans can understand. However mechanisms behind these capacities in machines are very different from biological ones. Along with reasoning, language is considered a high-level cognitive activity of which only humans are capable. Would it be possible to implement other cognitive functions in a machine, including consciousness, which many believe is the ultimate and exclusivly human characteristic? Can AI "jump over" evolutionary steps in the development of cognition and decouple it from metabolism and reproduction? For non-living cognitive agents such as intelligent robots, cognitive computing is being developed with the hope of implementing cognition on a basis different from the biological one.

3 Minimal Cognition

For naturalism, *nature is the only reality*; in short: there are no miracles [35]. It describes nature through its structures, processes and relationships using a scientific approach. Naturalism studies the evolution of the entire natural world, including the life and development of humanity as a part of nature. Social and cultural phenomena are studied through their physical manifestations. If cognition is a result of natural *physico-chemical* and *biological* processes, the important question is: *what is minimal cognition?* In other words: *what are the minimum requirements that we should place on a system to call it cognitive?* We mentioned already that for Maturana and Varela minimal cognition corresponds to minimal life as an emergent property of autopoetic molecular processes. Similar approaches can be found in [16, 36], who developed understanding of minimal life as "a distributed, emergent property based on an organised network of reactions and/or processes" [37]. This network of chemical processes can be modelled as distributed natural computation.

Recently, empirical studies have revealed an unexpected richness of cognitive behaviours (perception, information processing, memory, decision making) in organisms as simple as bacteria swarms and colonies [22, 38, 39]. Even though single bacteria are small, sense only their immediate environment, and live too briefly to be able to memorise a significant amount of data, they are still living organisms and thus possess a degree of cognition in the sense of Maturana and Varela. Already bacterial colonies, swarms and films formed by self-organisation of bacteria exhibit a surprising complexity of behaviours that can undoubtedly be characterised as social cognition. It is a rather new insight that bacteria possess complex cognitive capacities based on information communication that resembles (chemical) language.

Apart from bacteria and similar organisms without a nervous system (such as e.g. slime mould, multinucleate or multicellular Amoebozoa, which recently have been used to compute shortest paths) [40], even plants are typically thought of as living systems without cognitive capacities. However, plants too have been found to possess memory (in their bodily structures that change as a result of past events), the ability to learn (plasticity, ability to adapt through morphodynamics), and the capacity to anticipate and direct their behaviour accordingly. Plants are argued to possess rudimentary forms of knowledge, according to [41–43]. A special field of study is dedicated to *plant signalling*, which involves information communication by chemical or electrical signals within a plant or between plants.

Plant signalling is a special case of *biocommunication* that has been studied by Witzany [21] as signal transduction processes among cells, tissues, organs and organisms in prokaryotes (bacteria), protoctists (eukaryotic microorganisms), fungi, plants, animals and human beings. Within the info-computational framework, communication is a special case of computation, as shown in [44], where communication is an interaction/computation going on between a system and its environment, via sign/signal/message/information exchange. Witzany's work on biocommunication is of great interest for understanding how information processing occurs in different kinds of organisms. Biocommunication (present on different levels of a living organism: intra- and intercellular, interorganismic and transorganismic (trans-species)) includes also genetic code with its evolution and development (editing). In this context, viruses are very interesting as they and their parts present the most abundant genetic matter on Earth, and tools for genome formatting in "natural genetic engineering" [21]. In a similar vein, [45] argues that cells, in order to survive, cannot act blindly, but in a cognitive way, which means that *"they do things based on knowledge of what's happening around them and inside of them. Without that knowledge and the systems to use that knowledge they couldn't proliferate and survive as efficiently as they do."*

In this chapter we take cognition to be the totality of processes of self-generation, self-regulation and self-maintenance on different levels of organisation that enables organisms to adapt and survive using information and matter/energy from the environment. The understanding of cognition as it appears in degrees of complexity in living nature modelled as info-computation can also help us better understand the step between inanimate and animate matter from the first autocatalytic chemical reactions to the first autopoietic proto-cells.

Minimal cognition builds upon *minimal biological agency*, which for Kauffman and Clayton requires that: "such a system should be able to reproduce with heritable variation, should perform at least one work cycle, should have boundaries such that it can be individuated naturally, should engage in self-propagating work and constraint construction, and should be able to choose between at least two alternatives" [46].

4   Natural Info-computation in Networks of Agents. Computation All the Way Down to Quantum

The notion of computation in the info-computational framework of computing nature refers to the most general concept of *intrinsic computation*, that is, spontaneous computation processes in nature, which is used as a basis of specific kinds of *designed computation* found in computing machinery [47]. Intrinsic (natural) computation includes quantum computation [47, 48], molecular processes of self-organisation, self-assembly, developmental processes, gene regulation networks, gene assembly, protein–protein interaction networks, biological transport networks, eco-system networks, symbol manipulation and language processing in distributed networks of agents and similar. Computation can be both continuous (such as found in dynamic systems) and discrete (as found in Turing machines). Some info-computational processes in living agents are sub-symbolic signal processing, and some of them are symbolic (like symbol manipulation of languages) [49].

Info-computationalism, which is a variety of computing nature/ pancomputationalism, models physical nature that spontaneously performs different kinds of computations (through information dynamics) at different levels of organisation. Natural computation is intrinsic, and it is specific for a given physical system. Intrinsic computation(s) of a physical system can be used for designed computation [47, 48] such as that found in computational machinery, which is only a small part of the computation that can be found in nature. Sometimes pancomputationalism is wrongly accused of being vacuous. Natural computationalism is not vacuous in the same way as physics is not vacuous when it makes the claim that the entire physical universe is made of matter/energy. When physical entities exist in nature, unobserved, they are part of *Ding an sich, in Kant's philosophy a thing as it is in itself.* How do we know that they exist? We find out through interactions. What are interactions? They are information exchanges. Epistemologically, constraints or boundary conditions are also information for a system. The bottom layer for the computational universe is the bottom layer of its material substrate that is considered empirically justified.

Within the info-computational framework, computation on a given level of organisation of information presents a realisation/actualisation of the laws that govern interactions between constituent parts. On the basic level, computation is manifestation of causation in the physical substrate. In each next layer of organisation, a set of rules governs the system's switch to the new emergent regime. It remains yet to be established how this process exactly goes on in nature, and how emergent properties arise. Research in natural computing is expected to uncover these mechanisms. In the words of Rozenberg and Kari: *"(O)ur task is nothing less than to discover a new, broader, notion of computation, and to understand the world around us in terms of information processing"* [50]. From the research in complex dynamical systems, biology, neuroscience, cognitive science, networks, concurrency and more, new insights essential for the info-computational universe may be expected in the years to come.

Among models of computation, natural computing has a particular place, as "the field of research that investigates both human-designed computing inspired by nature and computing taking place in nature" [51]. It includes, among others, areas of cellular automata and neural computation, evolutionary computation, molecular computation, quantum computation, nature-inspired algorithms and alternative models of computation. An important characteristic of the research in natural computing is that knowledge is generated bi-directionally, through the interaction between computing and natural sciences. While natural sciences are adopting tools, methodologies and ideas of information processing, computing is broadening the notion of computation, recognizing information processing found in nature as computation [50, 51]. Based on that, Denning argues that computing today is a natural science [52]. Computation found in nature is understood as a physical process, where nature computes with physical bodies as objects. Physical laws govern processes of computation, which necessarily appears on many different levels of organisation of physical systems.

With its *layered computational architecture*, natural computation provides a basis for a unified understanding of phenomena of embodied cognition, intelligence and knowledge generation [53, 54].

Info-computation can be modelled as a *process of exchange of information in a network of informational agents.* Agent-based computational models are a variety of the parallel distributed processing (PDP) model [49]. What can be identified as representation here is a distributed network of activities.

> *Unlike the traditional computer-model, the PDP-approach treats mental representations not as discrete symbols but rather as distributed activity in a network. In this case, a particular representation can be identified by its relations to other representations; each of them can be depicted as a vector in a multidimensional space. Another obvious analogy concerns the development of suchlike systems. According to the holistic approach, this development is based on a process of differentiation which occurs with increasing experience in one field. This is just what happens with PDP networks if their training in one area proceeds. While their ability to differentiate is quite low at the beginning it increases in the learning-process.* [55]

Here "*mental representations*" can be replaced with *representations of the world of any kind* that can be found in cognitive agents. It is not self-evident that networked activity representations in a bacterium should be called "mental". It is *distributed activity in its molecular network* and not a nervous system. We can nevertheless ascribe learning abilities to such simple systems as bacteria swarms that interact with the environment and learn from previous experiences, that is, adapt continuously based on molecular interactions as activities in molecular networks. Their memory consists of their bodily structures, changes in genotype and phenotype.

One sort of concurrent natural computation is found on the quantum-mechanical level where computational agents are elementary particles, and messages (information carriers) are exchanged by force carriers, while different types of computation can be found on other levels of organisation, such as the molecular level where molecules, atoms and elementary particles are exchanged as messages. In biology, information processing is going on in intracellular networks, cells, tissues, organs, organisms and eco-systems, with corresponding agents and message types. In social computing the message carriers are complex chunks of information such as signs, words or sentences and the computational nodes (agents) can be organisms or groups/societies [27].

4.1 Natural Info-computation as Morphological/Morphogenetic Computation

Back in 1952, Turing wrote a paper that may be considered as a predecessor of natural computing. It addressed the process of *morphogenesis*, proposing a chemical model as the explanation of the development of biological patterns such as the spots and stripes on animal skin [56]. Turing did not claim that physical systems producing patterns actually performed computation. Nevertheless, from the perspective of computing nature, we can argue that *morphogenesis is a process of morphological computing*. Physical process—though not computational in the traditional sense, presents natural morphological computation. An essential element in this process is the interplay between the informational structure and the computational processes, information self-structuring and information integration, both synchronic and diachronic, unfolding on different time and space scales in physical bodies. Informational structure present a *program* that governs computational process [57], which in turn changes the original informational structure obeying/implementing/realising physical laws.

Morphology is the central idea in understanding the connection between computation (morphological/morphogenetic) and information. That which is observed as material on one level of analysis represents morphology on the lower level, recursively. So water as material presents arrangements of [molecular [atomic [elementary particle]]] structures. Thus, "'Hardware' [is] made of 'software'" [58].

Info-computationalism describes nature as informational structure—a succession of levels of organisation of (natural) information. Morphological/morphogenetic computing on that informational structure leads to new informational structures via processes of self-organisation of information. Evolution itself is a process of morphological computation on a long-term scale. It is also relevant and instructive to study morphogenesis of morphogenesis (meta-morphogenesis) as done by Sloman in [59].

One of the important characteristics of computing in biological systems is that it is approximate, robust and just "good enough" given the finite resources of time, energy and space. That is why we humans have problems multiplying big numbers but can make reasonable quick assessments of the order of magnitude. This way of computation developed evolutionary by learning processes in living agents. Valiant [60] describes this phenomenon by *ecorithms*—learning algorithms that perform *probably approximately correct computation,* PAC. Unlike the present paradigm of computing machinery, the results are not perfect but just good enough for an agent's purpose.

All living beings possess cognition (understood as all processes necessary for an organism to survive, both as an individual and as a part of a social group—social cognition), in different forms and degrees, from bacteria to humans. Cognition is based on agency; it would not exist without agency. The building block of life, the living cell, is a network of networks of processes, and those processes may be understood as computation. It is not any computation, but exactly that biological process itself, understood as information processing. Now one might ask what would be the point in seeing metabolic processes or growth (morphogenesis) as computation? The answer is that we try to connect cell processes to the conceptual apparatus of concurrent computational models and information exchange that have been developed within the field of computation and not within biology—we talk about "executable cell biology" [61]. Info-computational approach gives something substantial that no other approach gives to biology, and that is the *possibility of studying the real-time dynamics of a system.*

Processes of life and thus cognition are *critically time dependent.* Concurrent computational models are the field that can help us understand real-time interactive concurrent networked behaviour in complex systems of biology and its physical structures (morphology). That is the pragmatic reason why it is well justified to use conceptual and practical tools of info-computation in order to study living beings. Of course, in nature there are no labels saying: this process is computation. We can see as computation, conceptualise in terms of computation, model as computation and *call computation any process in the physical world*. Doing so we both expand our understanding of *natural processes* (physical, chemical, biological and cognitive) and enrich our concept of computation.

5 Info-computational Nature of Reality for a Cognitive Agent

Computational naturalism (computing nature, naturalist computationalism, pan-computationalism) is the view that all of nature is a huge network of computational processes, which, according to physical laws, computes (dynamically develops) its own next state from the current one [62]. Among representatives of this approach are Zuse, Fredkin, Wolfram, Chaitin and Lloyd, who proposed different varieties of computational naturalism. According to the idea of computing nature, one can view the time development (dynamics) of physical states in nature as information processing of natural computation. Such processes include self-assembly, self-organisation, developmental processes, gene regulation networks, gene assembly, protein–protein interaction networks, biological transport networks, social computing, evolution and similar processes of morphogenesis (creation of form). The idea of computing nature and the relationships between the two basic concepts of information and computation are explored in more detail in [12, 26, 27].

Talking about computing nature, we can ask: what is the "hardware" for this computation? The surprising answer is: the software on one level of organisation of information is the hardware for the next highest level in the sense of Kampis's self-modifying systems [57]. And on the basic level, the "hardware" is potential information, the structure of the world that one usually describes as matter–energy [63]. As cognising agents we are interacting with nature through information exchange, and we experience the world as information.

This view of nature as informational structure has been developed as *informational structural realism* by Luciano Floridi [64] and Kennet Sayre [65]. It is a framework that takes information as the *fabric of the universe* (for an agent). Even the physicists Zeilinger [66] and Vedral [67] suggest that *information and reality are one* epistemologically. For a cognising agent in the informational universe, the dynamical changes of its structures make it a huge computational network of networks [12]. The substrate, the "hardware", is information that defines data structures on which computation proceeds. It is important to note the difference between the everyday use of the term information as "the communication or reception of knowledge or intelligence" (Webster dictionary) and Shannon's technical definition of information [68]. Shannon's quantitative model of information transmission between sender and receiver through a noisy channel is agnostic about the meaning of information being transmitted, whereas our model of the information intends to apply to different phases of information communication and processing from the point of view of a cognising agent that makes sense of information.

The info-computational framework [54, 69, 70] is a synthesis of informational structural realism and natural computationalism (computing nature, pancomputationalism)—the view that the universe computes its own next state from the previous one [62, 71]. It builds on two basic complementary concepts: information (structure) and computation (the dynamics) as described in [72].

The natural world for a cognising agent exists as *potential information*, corresponding to Kant's *das Ding an sich*. Through interactions, this potential information becomes actual information for an agent, "a difference that makes a difference" [73]. Similarly, Shannon describes the process as the conversion of *latent information* into *manifest information* [74]. Even though Bateson's definition of information is the most widely cited one, there is a more general definition that subsumes Bateson's definition, including the fact that information is *relational in nature*:

> Information expresses the fact that a system is in a certain configuration that is correlated to the configuration of another system. Any physical system may contain information about another physical system. [75]

Combining the Bateson and Hewitt insights, at the basic level, *information is a difference in one physical system that makes a difference in another physical system. Here "physical system" refers to systems as studied in physics. This comes from the fact that there is no information without physical implementation.* [28–30]

6 From Raw Data to Information to Knowledge

Cognition can be seen as a result of processes of morphological computation on informational structures of a cognitive agent in interaction with the physical world, with processes going on at both sub-symbolic and symbolic levels. This morphological computation establishes connections between an agent's body, its nervous (control) system and its environment [1, 32, 76]. Through the embodied interaction with the informational structures of the environment, via sensorimotor coordination, information structures are induced in the sensory data of a cognitive agent, thus establishing perception, categorisation and learning. Those processes result in constant updates of memory and other structures that support behaviour, particularly *anticipation*. *Embodied* and corresponding *induced* informational structures (in the virtual machine of Sloman) are the basis of all cognitive activities, including consciousness and language as a means of maintenance of "reality".

An essential element in this process is the interplay between the informational structures and the computational processes—*information self-structuring* and *information integration*, both synchronic and diachronic, going on in different time and space scales [26].

From the simplest cognising agents such as bacteria to complex biological organisms with a nervous system and brain, the basic informational structures undergo transformations through morphological computation (developmental and evolutionary form generation).

Here an explanation is in order regarding *cognition that is defined in the general way of Maturana and Varela, who take it to be synonymous with life* [5, 77]. All living organisms possess some degree of cognition and for the simplest ones like bacteria cognition, involves metabolism and (our addition) locomotion [12]. This "degree" is not meant as a continuous function but as a qualitative characterisation that cognitive capacities increase from the simplest to the most complex organisms. The process of interaction with the environment causes changes in the informational structures that correspond to the body of an agent and its control mechanisms, which define its future interactions with the world and its inner information processing. Informational structures of an agent become semantic information first in the case of intelligent agents.

Even though we are far from having a consensus on the concept of information, the most general view is that information is a *structure consisting of data that is the difference that makes a difference for an agent*. Floridi [64] has the following definition of a datum: "*In its simplest form, a datum can be reduced to just a lack of uniformity, that is, a binary difference.*" Bateson's "the difference that makes the difference" [73] is a datum in that sense. Information is both the result of observed differences (*differentiation of data*) and the result of synthesis of those data into a common informational structure (*integration of data*), as argued by Schroeder in [78]. In the process of knowledge generation an intelligent agent moves between those two processes—*differentiation* and *integration* of data, see [25]. For *potential* information to become *actual* there must exist *an agent* from whose perspective the relational structure between the world as potential information and its actualisation in an agent is established. Thus (actual) information is a network of data points observed from an agent's perspective.

There is a distinction between the world as it exists autonomously, independently of any agent, Kantian *Ding an sich*, (thing in itself, noumenon), and the world for an agent, things as they appear through interactions (phenomena). Informational realists [64–67] take the reality/world/universe to be information. Dodig-Crnkovic [27] added by analogy *information an sich* representative of the *Ding an sich* as *potential information*. When does this potential information become *actual* information for an agent? The world in itself as (proto) information gets actualised through interactions of agents. Huge parts of the universe are potential information for different kinds of agents—from elementary particles, to molecules, etc. all the way up to humans and societies. Living organisms as complex agents inherit bodily structures as a result of a long evolutionary development of species. Those structures are embodied memory of the evolutionary past. They present the means for agents to interact with the world, get new information that induces memories, learn new patterns of behaviour and construct knowledge. World via Hebbian learning forms an organism's informational structures. As an example we can mention neural networks that "*self-organize stable pattern recognition codes in real-time in response to arbitrary sequences of input patterns*" [79].

If we say that for something to be information there must exist an agent from whose perspective this structure is established, and we argue that the fabric of the world is informational, the question can be asked: *who/what is the agent*? An agent (an entity capable of acting on its own behalf) can be seen as interacting with the points of inhomogeneity (data), establishing the connections between those data and the data that constitute the agent itself (a particle, a system, an organism). There are myriad agents for which information of the world makes differences—from elementary particles to molecules, cells, organisms, societies, etc.—all of them interact and exchange information on different levels of scale in space and time, and this information dynamics is natural computation.

On the fundamental level of quantum-mechanical substrate, information processes represent laws of physics. Physicists are already working on reformulating physics in terms of information. This development can be related to Wheeler's idea "it from bit" [80]. Actually, the idea can be traced back to 1977 when F. W. Kantor published the book *Information Mechanics* [81] that presents the first formal theory built upon the idea that information might be the fundamental phenomenon underlying the foundation of physics. For more details on physics and information research, see the special issue of the journal *Information* dedicated to matter/energy and information [63], and a special issue of the journal *Entropy* addressing natural/unconventional computing [82] that explores the space of natural computation and relationships between the physical (matter/energy), information and computation.

7 Computing Nature is a Naturalist Framework, not Panpsychism

Some computational (information integration) models of consciousness [83–86] can be interpreted as leading to *panpsychism*—"the belief that the physical universe is fundamentally composed of elements each of which is conscious" [11].

On the other hand, pancomputationalism (natural computationalism, computing nature) is *the doctrine that the whole of the universe, every physical system, computes* [87].

However, computation going on in a hurricane is not like human program execution in current computing machinery. A hurricane computes intrinsically its own next state, by implementation of natural laws on many levels of organisation. The info-computational approach starts bottom-up, from basic natural processes understood as computation. It means that computation appears as quantum, chemical, biological etc., see [2]. Only those transformations of informational structure that correspond to intrinsic processes in natural systems qualify as computation. Studying biological systems at different levels of organisation as *layered computational architectures* gives us powerful conceptual and technological tools for studying real-world systems. Even though we can imagine any sort of mapping, those that do not work spontaneously on the "hardware" of the universe do not count as natural (intrinsic) computations. We can simulate virtual worlds, but computation behind this visualisation relies on the implementation on the basic level on a physical substrate (computer hardware) with causal processes in the physical world.

The info-computational view of nature, where every living organism possesses some extent of cognition, does not contradict embodiment, embeddedness, and the interactive, communicative—enactive view of cognition. It provides also insights about possible mechanisms behind for example concepts of Umwelt, that is, the specific way an organism experiences the world, due to Uexküll [88]. The info-computational approach builds on physical computation that represents all those interactive real-world processes. Nevertheless there is still a lack of a common view of what computation is and what it is developing into, so many critics of computational approaches address early forms of computationalism without taking notice of contemporary alternatives. For example Bishop [11] claims that pan-computationalism necessarily implies panpsychism. Bishop's core argument derives from ideas originally proposed by Putnam [89] that "every open system implements every finite state automaton" (FSA) and thus "psychological states of the brain cannot be functional states of a computer". If a computer instantiates genuine phenomenal states, argues Bishop, then this necessarily implies *panpsychism. And as panpsychism is not an option*:

> *against the backdrop of our immense scientific knowledge of the closed physical world, and the corresponding widespread desire to explain everything ultimately in physical terms, panpsychism has come to seem an implausible view ...—then no computers can be conscious.*

So the whole argument is essentially dependent on Putnam's claim that every open system implements every discrete state machine. Even if that claim were true, state machines are definitely not all that is meant by computing, especially not physical computing that we argue is the fundamental principle in nature. As Bishop points out in the paper, his argument does not apply to dynamical systems. If we look into the taxonomy of computation given in [2], we see that this argument does not apply to many other types of computational model, in spite of its general claim that cognitive computation is fallacious, as it presupposes that computation is only a subset of possible computations that implement finite state automata, and that of course cannot cover all cognitive processes.

However, we agree that *panpsychism is not a good scientific hypothesis*. Instead of opening all doors for investigation of the natural basis of consciousness, it declares consciousness permeating the entire universe as the solution. As a theory, panpsychism belongs to mediaeval tradition—that which is to be explained is postulated.

Info-computationalism/computationalism/pancomputationalism/computing nature provides constructive tools for investigation of cognitive agents at different levels of complexity and presents a fruitful approach and coherent theoretical construal that by no means implies or presupposes panpsychism. It is an open framework for learning based on physics, chemistry, biology and cognitive science expressed in terms of info-computation.

The idea of consciousness as information integration in the work of Giulio Tononi [85] and Christoph Koch [86] has recently been widely debated. While we adopt the concept of cognition as a biological process in all living organisms, as argued by Maturana and Varela [5, 90], it is not at all clear that all cognitive processes in different kinds of organisms are accompanied by consciousness. We suggest that cognitive agents with a *nervous system* are the first step in evolution that enabled *consciousness*. A nervous system, on top of metabolism, which all living beings have, provides an internal "model" of an agent's body in relation to its environment [32]. The nervous system models the "self" of an organism that acts in the context of "other". Unlike reactive processes in simplest organisms, where behaviour is directly driven by chemical and physical relationships ("molecular machinery"), the emergence of a nervous system in more complex organisms acts as an integration mechanism for reactive behaviours, establishing an additional *self-reflective layer of cognitive architecture*. "Neural control essentially grows out of the need for maintaining proper internal value systems" [91]. Starting with the distinctive self, enabled by a nervous system, we can talk about consciousness as the capacity of self-reflection and self-awareness.

8 Conclusions and Future Work

The questions: *What is reality for an agent? What is minimal cognition? How does the morphology of a cognitive agent affect cognition?* that we posed at the beginning of this chapter led us to a discussion of info-computational models of cognition that can be applied to both living organisms and machines. Within the info-computational framework applied to living nature, cognition is understood as the process of living itself. Following Maturana and Varela's argument [5], we understand the entire living world as possessing cognition of various degrees of complexity. In that sense, bacteria possess rudimentary cognition expressed in quorum sensing and other collective phenomena based on information communication and information processing. Cognition is a distributed agent-based network of processes realised through communication via different mediators—from chemical molecules to electrical signals. Comparing processes of communication among cells, tissues, organs and organisms in all kinds of living organisms—from prokaryotes (bacteria), protoctists (eukaryotic microorganisms), fungi, plants, animals, to human beings as described in [21], we notice what von Uexküll already found out: differences in cognition and ways of reality construction dependent on the type of cognising agent [88]. The simplest organisms communicate by "chemical language", exchanging molecules, which also include genetic material. The signal itself is information for an agent, and its processing is physical computation. Memory consists in bodily structures of an organism. As organisms develop more complex structures, their communication becomes more diverse. The Brain of a complex organism consists of neurons that are networked communication computational units. The signalling and information processing modes of a brain are much more complex than the exchange of chemical signals in brainless organisms, and consist of more layers of computational architecture than cognitive processes in a bacterial colony.

Maturana and Varela did not think of cognition as computation, as at the time they developed the theory of autopoiesis, computation was conceived as abstract symbol manipulation, and information was considered in its narrow meaning as a way of verbal communication between humans. With our current understanding of information and computation, we claim that cognition is a process of natural/physical/intrinsic/embodied computation [92]. The broader view of computation as a physical process with emergent properties, as found in info-computationalism, is capable of representing processes of life studied in bioinformatics and biocomputation. Reality for an agent is info-computational in nature, and it is established as a result of the interactions of the agent with the environment as the information processes in agents, own intrinsic structures that take the form of anticipation, reasoning etc.

Computing nature consists of physical structures that form levels of organisation, on which computation processes differ. It has been argued that on the lower levels of organisation, finite automata or Turing machines might be adequate models of local processes, while on the level of the whole brain, non-Turing computation is necessary, according to Ehresmann [93] and Ghosh et al. [94].

We also asked a question regarding non-biological cognition: *Is synthetic cognition possible at all, given that cognition in living organisms is a deeply biologically rooted process?* Currently we observe advances in natural language processing, indicating developments of machines capable of "understanding natural language", "translating", "speaking", "object recognition" and "learning". The mechanisms behind those capacities in machines are very different from corresponding mechanisms in living organisms. We asked: *Would it be possible to implement other cognitive functions in a machine, including consciousness that many believe is the ultimate and exclusive human characteristic? Can AI "jump over" evolutionary steps in the development of cognition and decouple from metabolism and reproduction?* From what we know today, it seems that the answer is positive. In a similar way as machines move, or airplanes fly based on mechanisms different from living beings, cognitive computing is being developed that implements cognitive functions on a basis different from the biological one. Is cognition or consciousness essentially different from movement? This remains to be seen. The development indicates that it might be the case *that cognition is necessary for life, but life is not necessary for cognition.*

Finally, an argument is advanced that computational cognition by no means implies panpsychism. We argue that computing nature/pancomputationalism is a sound scientific approach based on insights from physics, chemistry, biology, cognitive science (including distributed/social cognition) and theories of information and computation, which is opposite to panpsychism.

For the future, a lot of work remains to be done, especially on the connections between the low-level cognitive processes and the high-level ones. It is important to find relations between cognition and consciousness and the detailed picture of info-computational mechanisms behind cognitive phenomena in the developmental and evolutionary time frame.

Acknowledgements The authors want to thank the reviewers Jan van Leeuwen, Marcin Schroeder, Matej Hoffmann, Raffaela Giovagnoli and Tom Froese for their constructive and very helpful comments.

References

1. Horsman, C., Stepney, S., Wagner, R., Kendon, V.: When does a physical system compute? Proc. R. Soc. A **470**(2169), 20140182 (2014)
2. Burgin, M., Dodig-Crnkovic, G.: A taxonomy of computation and information architecture. In: Galster, M. (ed.) Proceedings of the 2015 European Conference on Software Architecture Workshops (ECSAW' 15), ACM Press, New York (2015)
3. Dodig-Crnkovic, G.: Significance of models of computation, from turing model to natural computation. Mind Mach. **21**(2), 301–322 (2011)
4. Fodor, J., Pylyshyn, Z.: Connectionism and cognitive architecture: a critical analysis. Cognit. **28**, 3–71 (1988)
5. Maturana, H., Varela, F.: Autopoiesis and cognition: the realization of the living. D. Reidel, Dordrecht (1980)
6. Stewart, J.: Cognition = life: implications for higher-level cognition. Behav. Process. **35**, 311–326 (1996)
7. Lorenz, H.: Ancient theories of soul. In: The Stanford Encyclopedia of Philosophy (2009)
8. Varela, F., Thompson, E., Rosch, E.: The Embodied Mind: Cognitive Science and Human Experience. MIT Press (1991)
9. Vernon, D.: A survey of artificial cognitive systems: implications for the autonomous development of mental capabilities in computational agents. IEEE Trans. Evol. Comput. **11**(2), 151–180 (2007)
10. Fresco, N.: Physical Computation and Cognitive Science. Springer (2014)
11. Bishop, M.: A cognitive computation fallacy? cognition, computations and panpsychism. Cognit. Comput. **1**, 221–233 (2009)
12. Dodig-Crnkovic, G.: Information, computation, cognition. Agency-based hierarchies of levels. In: Müller, V.C. (ed.) Fundamental Issues of Artificial Intelligence (Synthese Library). Springer (2014)
13. Fernandez, B., Xabier, E., Paolo, D., Rohde, M.: Defining agency: individuality, normativity, asymmetry, and spatio-temporality in action. Adapt. Behav. **17**(5), 367–386 (2009)
14. Cruse, H.: The evolution of cognition—a hypothesis. Cognit. Sci. **27**, 135–155 (2003)
15. Froese, T., Ziemke, T.: Enactive artificial intelligence: investigating the systemic organization of life and mind. Artif. Intell. **173**, 466–500 (2009)
16. Kauffman, S.: Origins of Order: Self-Organization and Selection in Evolution. Oxford University Press (1993)
17. Deacon, T.: Incomplete Nature. How Mind Emerged from Matter. W. W. Norton (2011)

18. Ben-Jacob, E., Shapira, Y., Tauber, A.I.: Smart bacteria. In: Margulis, L., Asikainen, C.A., Krumbein, W.E (eds.) Chimera and Consciousness. Evolution of the Sensory Self. MIT Press (2011)
19. Ben-Jacob, E., Shapira, Y., Tauber, A.I.: Seeking the foundations of cognition in bacteria. Phys. A **359**, 495–524 (2006)
20. Maturana, H.: Autopoiesis, structural coupling and cognition: a history of these and other notions in the biology of cognition. Cybern. Hum. Knowing **9**(3–4), 5–34 (2002)
21. Witzany, G.: Biocommunication and Natural Genome Editing. Springer (2010)
22. Ng, W., Bassler, B.L.: Bacterial quorum-sensing network architectures. Annu. Rev. Genet. **43**, 197–222 (2009)
23. Pratt, S.C.: Quorum sensing by encounter rates in the Ant Temnothorax-Albipennis. Behav. Ecol. **16**(2), 488–496 (2005)
24. Doya, K.: Metalearning and neuromodulation. Neural Netw. **15**(4–6), 95–506 (2002)
25. Dodig-Crnkovic, G.: Investigations into information semantics and ethics of computing. Zhurnal Eksperimental'noi I Teoreticheskoi Fiziki. Mälardalen University Press, Västerås, Sweden (2006). http://scholar.google.com/scholar?hl=en&btnG=Search&q=intitle:No+Title#0. http://www.diva-portal.org/smash/record.jsf?pid=diva2:120541
26. Bull, L., Holley, J., Costello, B.D.L., Adamatzky, A.: Computing Nature, vol. 7. Springer, Berlin, Heidelberg (2013)
27. Dodig-Crnkovic, G.: Physical computation as dynamics of form that glues everything together. Information **3**(4), 204–218 (2012)
28. Landauer, R.: Computation: a fundamental physical view. Phys. Scr. **35**, 88–95 (1987)
29. Landauer, R.: Information is physical. Phys. Today **44**, 23–29 (1991)
30. Landauer, R.: The physical nature of information. Phys. Lett. A **217**, 188 (1996)
31. Clark, A.: Being There: Putting Brain, Body and World Together Again. Oxford University Press (1997)
32. Pfeifer, R., Bongard, J.: How the Body Shapes the Way We Think—A New View of Intelligence. MIT Press (2006)
33. Hewitt, C.: What is computation? actor model versus Turing's model. In: Zenil, H. (ed.) A Computable Universe, Understanding Computation & Exploring Nature as Computation. World Scientific Publishing, Imperial College Press (2012)
34. Castelfranchi, C.: Guarantees for autonomy in cognitive agent architecture. Intel. Agent **890**, 56–70 (1995)
35. Putnam, H.: Mathematics, Matter and Method. Cambridge University Press (1975)
36. Ganti, T.: The Principles of Life. Oxford University Press (2003)
37. Luisi, L.: Autopoiesis: a review and a reappraisal. Naturwissenschaften **90**, 49–59 (2003)
38. Ben-Jacob, E.: Bacterial complexity: more is different on all levels. In: Nakanishi, S., Kageyama, R., Watanabe, D. (eds.) Systems Biology—The Challenge of Complexity, pp. 25–35. Springer (2009)
39. Ben-Jacob, E.: Learning from bacteria about natural information processing. Ann. N. Y. Acad. Sci. **1178**, 78–90 (2009)
40. Adamatzky, A.: Physarum Machines. Computers from Slime Mould. World Scientific (2010)
41. Pombo, O., Torres J.M., Symons J., Rahman S., (eds.): Special Sciences and the Unity of Science. Springer (2012)
42. Rosen, R.: Anticipatory Systems. Pergamon Press (1985)
43. Popper, K.: All Life Is Problem Solving. Routledge (1999)
44. Bohan B., Paul: On communication and computation. Mind Mach. **14**(1), 1–19 (2004)
45. Shapiro, J.A.: Evolution: A View from the 21st Century. FT Press Science, New Jersey (2011)
46. Kauffman, S., Clayton, P.: On emergence, agency, and organization. Biol. Philos. **21**(4), 500–520 (2006)
47. Crutchfield, J., Ditto, W., Sinha, S.: Introduction to focus issue: intrinsic and designed computation: information processing in dynamical systems-beyond the digital hegemony. Chaos **20**(3), 037101–037106 (2010)

48. Crutchfield, J., Wiesner, K.: Intrinsic quantum computation. Phys. Lett. A **374**(4), 375–380 (2008)
49. Clark, A.: Microcognition: Philosophy, Cognitive Science, and Parallel Distributed Processing. MIT Press (1989)
50. Rozenberg, G., Kari, L.: The many facets of natural computing. Commun. ACM **51**, 72–83 (2008)
51. Rozenberg, G., Bäck, T., Kok, J.N. (eds.): Handbook of Natural Computing. Springer (2012)
52. Denning, P.: Computing is a natural science. Commun. ACM **50**(7), 13–18 (2007)
53. Wang, Y.: On abstract intelligence: toward a unifying theory of natural, artificial, machinable, and computational intelligence. Int. J. Softw. Sci. Comput. Intel. **1**(1), 1–17 (2009)
54. Dodig-Crnkovic, G., Müller, V.: A dialogue concerning two world systems: info-computational vs. mechanistic. In: Dodig-Crnkovic, G., Burgin, M. (eds.) Information and Computation, pp. 149–184. World Scientific (2011)
55. Pauen, M.: Reality and representation qualia, computers, and the 'explanatory gap'. In: Riegler, A., Peschl, M., von Stein, A. (eds.) Understanding Representation in the Cognitive Sciences. Kluwer Academic/Plenum Publishers, New York (1999)
56. Turing, A.M.: The chemical basis of morphogenesis. Philos. Trans. R. Soc. Lond. **237**(641), 37–72 (1952)
57. Kampis, G.: Self-Modifying Systems in Biology and Cognitive Science: A New Framework for Dynamics, Information, and Complexity. Pergamon Press, Amsterdam (1991)
58. Kantor, F.W.: An informal partial overview of information mechanics. Int. J. Theor. Phys. **21**(6–7), 525–535 (1982)
59. Sloman, A.: Meta-morphogenesis: evolution and development of information-processing machinery. In: Cooper, S.B., van Leeuwen, J. (eds.) Alan Turing: His Work and Impact. Elsevier, Amsterdam (2013)
60. Valiant, L.: Probably Approximately Correct: Nature's Algorithms for Learning and Prospering in a Complex World. Basic Books (2013)
61. Fisher, J., Henzinger, T.A.: Executable cell biology. Nat. Biotechnol. **25**(11), 1239–1249 (2007)
62. Chaitin, G.: Epistemology as information theory: from Leibniz to Ω. In: Dodig Crnkovic, G. (ed.) Computation, Information, Cognition—The Nexus and The Liminal, pp. 2–17. Cambridge Scholars Publishing (2007)
63. Dodig-Crnkovic, G.: Information and energy/matter. Information **3**(4), 751 (2012)
64. Floridi, L.: A defense of informational structural realism. Synthese **161**(2), 219–253 (2008)
65. Sayre, K.M.: Cybernetics and the Philosophy of Mind. Routledge & Kegan Paul (1976)
66. Zeilinger, A.: The message of the quantum. Nature **438**(7069), 743 (2005)
67. Vedral, V.: Decoding Reality: The Universe as Quantum Information. Oxford University Press (2010)
68. Shannon, C.: Mathematical theory of the differential analyzer. J. Math. Phys. **20**, 337–354 (1941)
69. Dodig-Crnkovic, G.: Epistemology naturalized: the info-computationalist approach. APA Newslett. Philos. Comput. **06**(2), 9–13 (2007)
70. Dodig-Crnkovic, G.: Knowledge generation as natural computation. J. Syst. Cybernet. Inf. **6**(3), 12–16 (2008)
71. Dodig-Crnkovic, G.: Info-computationalism and morphological computing of informational structure. In: Plamen, L. Simeonov, L., Smith, S., Ehresmann, A.C. (eds.) Integral Biomathics, pp. 97–104. Berlin, Heidelberg (2012)
72. Dodig-Crnkovic, G.: Dynamics of information as natural computation. Information **2**(3), 460–477 (2011)
73. Bateson, G.: Steps to an ecology of mind: collected essays in anthropology, psychiatry, evolution, and epistemology. In: Adriaans, P., Benthem van, J. (eds.). University of Chicago Press, Amsterdam (1972)

74. McGonigle, D., Mastrian, K.: Introduction to information, information science, and information systems. In: Nursing Informatics and the Foundation of Knowledge, vol. 22. Jones & Bartlett (2012)
75. Hewitt, C.: What is commitment? physical, organizational, and social. In: Noriega, P., Vazquez-Salceda, J., Boella, G., Boissier, O., Dignum, V. (eds.) Coordination, Organizations, Institutions, and Norms in Agent Systems II, pp. 293–307. Springer, Berlin, Heidelberg (2007)
76. Hauser, H., Füchslin, R.M., Pfeifer, R.: Opinions and Outlooks on Morphological Computation. e-book (2014)
77. Maturana, H., Varela, F.: The Tree of Knowledge. Shambala (1992)
78. Schroeder, M.: Dualism of selective and structural manifestations of information in modelling of information dynamics. In: Dodig-Crnkovic, G., Giovagnoli, R. (eds.) Computing Nature, SAPERE 7, pp. 125–137. Springer (2013)
79. Grossberg, G.A., Carpenter, S.: ART 2: self-organization of stable category recognition codes for analog input patterns. Appl. Opt. **26**(23), 4919–4930 (1987)
80. Wheeler, J.A.: Information, physics, quantum: the search for links. In: Zurek, W.(ed.) Complexity, Entropy, and the Physics of Information. Addison-Wesley (1990)
81. Kantor, F.W.: Information Mechanics. Wiley-Interscience, New York (1977)
82. Dodig-Crnkovic, G., Giovagnoli, R.: Natural/unconventional computing and its philosophical significance. Entropy **14**(12), 2408–2412 (2012)
83. Tononi, G.: An information integration theory of consciousness. BMC Neurosci. **5**(42), 1–22 (2004)
84. Tononi, G.: Consciousness as integrated information: a provisional manifesto. Bio. Bulletin **215**(3), 216–242 (2008)
85. Tononi, G.: The integrated information theory of consciousness: an updated account. Arch. Ital. Biol. **150**(2/3), 290–326 (2012)
86. Koch, C.: Consciousness—Confessions of a Romantic Reductionist. MIT Press (2012)
87. Piccinini, G.: Computation in physical systems. In: The Stanford Encyclopedia of Philosophy (2012)
88. von Uexküll, J.: Theoretical Biology. Harcourt, Brace, New York (1926)
89. Putnam, H.: Representation and Reality. MIT press (1988)
90. Maturana, H.: Biology of Cognition. Research Report BCL 9. 1970, Biological Computer Laboratory: Urbana, IL. Defense Technical Information Center, Illinois (1970)
91. van Dijk, J., Kerkhofs, R., Van Rooij, I., Haselager, P.: Special section: can there be such a thing as embodied embedded cognitive neuroscience? Theory Psychol. **18**, 297–316 (2008)
92. Dodig-Crnkovic, G.: Modeling life as cognitive info-computation. In: Beckmann, A., Csuhaj-Varjú, E., Meer, K. (eds.) Computability in Europe 2014. LNCS, pp. 153–162. Springer, Berlin, Heidelberg (2014)
93. Ehresmann, A.C.: MENS, an info-computational model for (neuro-)cognitive systems capable of creativity. Entropy **14**, 1703–1716 (2012)
94. Ghosh, S., Aswani, K., Singh, S., Sahu, S., Fujita, D., Bandyopadhyay, A.: Design and construction of a brain-like computer: a new class of frequency-fractal computing using wireless communication in a supramolecular organic, inorganic system. Information **5**(1), 28–100 (2014)

Part IV
Philosophical Perspectives

The Relevance of Language for the Problem of Representation

Raffaela Giovagnoli

Abstract This chapter deals with the relationship between representation and language, which becomes more relevant if we do not intend the process of forming internal representations of reality but rather the representative function of language. Starting from some Fregean ideas, we present the notion of representation theorized by Searle. According to Searle, a belief is a "representation" (not in the sense of having an "idea") that has a propositional content and a psychological mode: the propositional content or intentional content determines a set of conditions of satisfaction under certain aspects and the psychological mode determines the direction of fit of the propositional content. We draw attention to some very interesting ideas proposed by Brandom in response to the challenge of Searle to AI, as they propose formal aspects of representation that rely on the use of ordinary language while avoiding the psychological order of explanation.

1 Introduction

The notion of "representation" has a medieval origin and indicates an "image" or "idea" that is similar to the object represented. Aquinas thinks that "to represent something is to include the similitude to a thing". The Scholastic thought introduces the interpretation of representation as the meaning of a word. Ockam distinguishes between three different meanings: (1) We intend with this term the mean to know something and this is the sense in which knowledge is representative and to represent means to be the entity by which something is known. (2) We intend representation as to know something through which we know something else; in this sense, the image represents the thing in the act of remembering. (3) We intend representing as causing knowledge, namely the way in which the object causes knowledge. Descartes and Leibniz have an original interpretation of the term to indicate the "picture" or "image" of the thing. Leibniz thinks that the monad is a

R. Giovagnoli (✉)
Faculty of Philosophy, Lateran University, Vatican City, Italy
e-mail: raffa.giovagnoli@tiscali.it

© Springer International Publishing AG 2017
G. Dodig-Crnkovic and R. Giovagnoli (eds.), *Representation and Reality in Humans, Other Living Organisms and Intelligent Machines*, Studies in Applied Philosophy, Epistemology and Rational Ethics 28, DOI 10.1007/978-3-319-43784-2_11

representation of the universe. Wolf uses rather the Cartesian notion of *Vorstellung* as an image of the thing. Kant introduces a broader meaning of the term as the genus of all acts or manifestations of knowledge, which overcomes the traditional sense of image or similitude.

The notion of representation plays an important role in the problem of meaning. Medieval logic (Ispano, Ockam, Buridan and Albert of Saxony) distinguishes between meaning and *suppositio*: meaning is the representation or concept that we use for objective reference and the very objective reference is defined as *suppositio*. Aquinas follows this distinction while changing the terminology. He thinks that meaning and *suppositio* overlap in the use of singular terms but not in the general ones where meaning is their essence. Leibniz and Mill continue this tradition even though they introduce respectively the pairs: comprehension and extension, intension and extension; connotation and denotation.

The sense we use to represent the object is well interpreted in a logical sense by Frege. A sign that can be a name or a nexus of words or a single letter entails two distinguished things: the designed object or meaning (*Bedeutung*) and the sense (*Sinn*) that denotes the way in which the object is given to us. The absence of a psychological characteristic in Frege's philosophy of language is inherited by Carnap who maintains that to understand a linguistic expression means to grasp its sense and to investigate the state of affairs it refers to. The intentional meaning or sense can be applied to humans and robots. Church refers to Frege by distinguishing two dimensions of the use of a name: a name indicates its denotation and expresses its sense. The sense determines the denotation or it is a "concept" of the denotation. As we will see, pragmatism shares with formal semantics the view that we must overcome psychologism, but it shows original interpretations of the notion of representation, which has fruitful results for AI.

2   A Fregean Background

To start with a philosophical account of the use of language that matters for representing reality, I recall some ideas from Frege [1]. Frege inherits the Kantian conception according to which there could not be any combination of ideas unless there were already an original unity that made possible such a combination [2]. According to Sluga [3], Kant anticipates the Fregean doctrine of concepts: "Concepts, as predicates of possible judgments, relate to some representation of a not yet determined object". Thus the concept of body means something, for instance, metal, which can be known by means of the concept. It is therefore a concept solely by virtue of its comprehending other representations, by means of which it can relate to objects. It is therefore the predicate of a possible judgment, for instance, "every metal is a body". In the *Begriffsschrift* we can find an anticipation of the notion of judgment presented in the later work *Function and Concept*. The main point is that functions, concepts and relations are incomplete and require variables in their expression to indicate places of arguments. To establish a correspondence between function and concept Frege maintains that the linguistic form of an equation or identity is an assertoric sentence. It embeds a thought as its sense or, more precisely, we can say that it "raises a claim" to have one. Generally speaking, the thought is true or false, namely it possesses a truth-value that could be considered as the meaning of the sentence, like the number 4 is the meaning of the expression "2 + 2" and London is the meaning of the expression "the capital of Britain".

The assertoric sentences can be decomposed into two parts: the "saturated" and the "unsaturated" one. For instance, in the sentence "Caesar conquered Gaul" the second part is unsaturated and it must be filled up with a proper name (in our case Caesar) to give the expression a complete sense. In *Concept and Object* Frege clearly describes the nature of the denotation of a predicate. A concept is the denotation of a predicate.

Searle's account presents a step beyond Frege's descriptivism because in order to give weight to propositions and their intentional contents we must distinguish them from the sense [4]. The sense of a referring expression is given by the descriptive general terms entailed by that expression but the Fregean notion of sense is often not sufficient to communicate a proposition. Consequently, it is the utterance of the expression "in a certain context" (namely a pragmatic context) that communicates a proposition. For example, the expression "the dog" has the descriptive content entailed by the simple term "dog"; this very content is not sufficient for a successful reference which also requires the communication or the possibility to communicate a uniquely existential proposition (or "fact", e.g. "There is one and only one dog barking on the right of the speaker and it is in the field of vision of both speaker and hearer"). The classical formalization (x fx) could be used to mean that "the predicate f has at least one instance" instead of "Some object is f". The meaning of this option does not establish a correspondence between the original proposition and its revised existential formulation; rather it says that the circumstances in which one option is true are identical with the circumstances in which the other is true.

According to Searle, a belief is a "representation" (not in the sense of having an "idea") that has a propositional content and a psychological mode: the propositional content or intentional content determines a set of conditions of satisfaction under certain aspects and the psychological mode determines the direction of fit of the propositional content. "Conditions of satisfaction are those conditions which, as determined by the intentional, must obtain if the state must be satisfied" [5].

In this context, it is crucial to distinguish between the *content* of a belief (i.e. a proposition) and the *objects* of a belief (i.e. the ordinary objects). For instance, the content of the statement or belief that de Gaulle was French is the proposition that de Gaulle was French. The statement or belief is not directed at the proposition but is about de Gaulle. It represents him as being French by virtue of the fact that it has "propositional content" and "direction of fit".

The process of representation functions because of a Network of other intentional states and against a Background of practices and pre-Intentional assumptions that are neither themselves intentional states nor are they parts of the conditions of satisfaction of Intentional states [6]. The intentionality of mental states represents an original interpretation of the Fregean account of beliefs. A further step in the analysis is the distinction between Intentionality-with-a-t and Intentionality-with-an-s. In this case a belief in Intentional-with-an-s does not permit us to determine its extension, i.e. substitution *salva veritate*. For instance, if I say "Vic believes that Rossella is an Irish setter" I simply report Vic's belief but I cannot commit myself to its truth, namely the fact that Rossella is an Irish setter. Obviously, Vic's belief is extensional and Vic is committed to its truth (it is intentional-with-a-t).

Let's now briefly refer to a difference between Frege's and Searle's accounts of belief. Standardly, beliefs are introduced by a "that" clause as in our example (1) "Vic believes that Rossella is an Irish setter". This report is different from the statement (2) "Rossella is an Irish setter": (1) is intensional whether or not (2) is extensional. A fundamental difference between the two forms of sentence is that in a serious literal utterance (2) is asserted, while in a serious literal utterance of (1) the proposition is not asserted.

Searle sets up conditions for the adequacy of intensional reports of intentional states [7]:

1. The analysis should be consistent with the fact that the meanings of the shared words in pairs such as (1) and (2) are the same, and in serious literal utterances of each they are used with these same meanings.
2. It should account for the fact that in (1) the embedded sentence does not have the logical properties it has in (2), viz., (2) is extensional, (1) is intensional.
3. It should be consistent with the fact that it is part of the meanings of (1) and (2) that, in serious literal utterances of (1), the proposition that Rossella is an Irish setter is not asserted, whereas in (2) it is.
4. The analysis should account for other sorts of sentences containing "that" clauses, including those where some or all of the logical properties are preserved, such as "It is a fact that Rossella is an Irish setter".
5. The analysis should apply to other sorts of reports of intentional states and speech acts which do not employ "that" clauses embedding a stance but use infinitives, interrogative pronouns, the subjunctive, change of tense, etc. Furthermore the analysis should work not just for English but for any language containing reports of intentional states and speech acts (as for example "Tess wants Rossella be an Irish setter").

The first condition could not be accepted from a Fregean perspective because, according to Searle, when we have sentences containing "that" clauses we have always the same meanings of the shared words and a variation of the illocutive act ("to believe that", "to say that", etc.). Nevertheless, Searle's account respects Frege's notion of belief. In the case of "to believe that…" sentences do not have the so-called "direct Bedeutung" but they have "indirect Bedeutung". This fact means that the truth-value can be assigned only to the second thought, i.e. the thought of the subordinate sentence.

The background as Searle describes it and the very notion of intentionality of beliefs have no normativity in establishing whether an individual belief is "true" in a strong sense, as something that can be shared by different people. According to this thesis, our second claim is that common beliefs as true beliefs are possible only in an intersubjective context in which individual descriptions can overlap by referring to the same object under precise substitutional rules.

According to Frege, the same object or *Bedeutung* can be thought of in different ways, namely the same object can have different "senses". Frege's famous example of the proper names or descriptions such as "Venus", "Morning Star" and "Evening Star" is however to be considered as valid for the explanation of common beliefs as "true" beliefs. For instance, A thinks Venus is the Morning Star and B thinks Venus is the Evening Star with the resulting communication problem; the problem is solved because the two descriptions have the same meaning, i.e. Venus, and surely it is possible to establish whether the two descriptions work for the object to which they refer.

Thoughts can be true or false but sentences do not express them "randomly". Sentences express thoughts as related to contexts of use in which they acquire their truth-value, i.e. they are true or false. For instance the sentence "That is a funny play" can be true or false depending on the context of use. We can grasp thoughts but Frege does not present an analysis of the "grasping" because he thinks that this implies a psychological order of explanation. Searle rather gives an account of the grasping through his brilliant account of the functioning of background based on intentionality. We can therefore show the complementarity between the description of the functioning of the cognitive grasping of the content of beliefs and the "normative" objective content that represents the ground of shared beliefs.

3   An Interpretation of Concepts Beyond Cognitive Science

There is a different interpretation of the Fregean semantics, which is bound to the concept's use in ordinary language along the lines of Davidson, Dummett and Sellars [8]. This theoretical option cannot be discussed in the ambit of cognitive sciences (in particular cognitive psychology, developmental psychology, animal psychology and artificial intelligence) because [9]:

Each of these disciplines is in its own way concerned with how the trick of cognition is or might be done. Philosophers are concerned with what counts as doing it—with what understanding, particularly discursive, conceptual understanding consists in, rather than how creatures with a particular contingent constitution, history, and armamentarium of basic abilities come to exhibit it. I think Frege taught us three fundamental lessons about the structure of concepts, and hence about all possible abilities that deserve to count as concept-using abilities. The conclusion we should draw from his discoveries is that concept-use is intrinsically stratified. It exhibits at least four basic layers, with each capacity to deploy concepts in a more sophisticated sense of "concept" presupposing the capacities to use concepts in all of the more primitive senses. The three lessons that generate the structural hierarchy oblige us to distinguish between:

    I. *concepts that only label and concepts that describe,*
   II. *the content of concepts and the force of applying them, and*
  III. *concepts expressible already by simple predicates and concepts expressible only by complex predicates.*

AI researchers and cognitive, developmental and animal psychologists need to take account of the different grades of conceptual content made visible by these distinctions, both in order to be clear about the topic they are investigating (if they are to tell us how the trick is done, they must be clear about exactly which trick it is) and because the empirical and in-principle possibilities are constrained by the way the abilities to deploy concepts in these various senses structurally presuppose the others that appear earlier in the sequence.

Concepts are acquired through the use of language and provide the classification of reality, i.e. shared knowledge. Classification is the traditional goal of classical philosophy, and starting from Ancient Philosophy it seems to be investigated beyond the mere exercise of reliable responsive dispositions to respond to environmental stimuli, even though we find very fruitful investigations in the natural sciences [9]. But, the conceptual classification is better explained by intending the application of a concept to something as *describing* it. One thing is to apply a label to objects, another is to describe them. In Sellar's words [10]:

It is only because the expressions in terms of which we describe objects, even such basic expressions as words for perceptible characteristics of molar objects, locate these objects in a space of implications, that they describe at all, rather than merely label.

Moving from the Fregean difference between judgeable content and judgment we considered above, we can isolate the semantic content of the descriptive concept (the ones that do not label) from the act of describing or the pragmatic force of describing by applying those concepts. In the case of compound sentences formed by the use of conditionals there are differences between, for instance, denial (as a kind of speech act) and negation (a kind of content). So, one thing is to say "I believe that labeling is not describing" and another is to say "If I believe that labeling is not describing, then labeling is not describing". In the first case, I am denying something, in the second not. The endorsement of judgeable content is the capacity to endorse conditionals, i.e. to explore the descriptive content of proposition, their inferential circumstances and consequence of application, which characterize a sort of "semantic self-consciousness". The higher capacity to form conditionals makes possible a new sort of hypothetical thought that seems to appear as the most relevant feature of human rationality because chimps or African grey parrots or other non-human animals just use concepts to describe things but are not able to discriminate the contents of those concepts from the force of applying them.

Complex concepts can be thought of as formed by a four-stage process [11]:

- First, put together simple predicates and singular terms, to form a set of sentences, say (Rab, Sbc, Tacd).
- Then apply sentential compounding operators to form more complex sentences, say (Rab → Sxc, Sbc&Tacd).
- Then substitute variables for some of the singular terms (individual constants), to form complex predicates, say (Rax → Sxy, Sxy&Tayz).
- Finally, apply quantifiers to bind some of these variables, to form new complex predicates, for instance the one-place predicates (in y and z) {∃x[Rax → Sxy], ∀x∃y[Sxy&Tayz]}.

The process is repeatable to form new sentences from the complex predicates playing the role that simple predicates played at the first stage like, for instance {∃x [Rax → Sxd], ∀x∃y[Sxy{&Taya}.

One fundamental difference to explain the role of conditionals for human logic is between "ingredient" content and "free-standing" content. The former belong to a previous stage in which it becomes explicit only through the force of sentence (query, denial, command, etc., that are invested in the *same* content). The latter is to be understood in terms of the contribution it makes to the content of compound judgments in which it occurs, consequently only indirectly to the force of endorsing that content. The process of human logical self-consciousness develops in three steps:

1. We are able to "rationally" classify through inferences, i.e. classifications provide reasons for others.
2. We form synthetic logical concepts formed by compounding operators, paradigmatically conditionals and negation.
3. We form *analytical* concepts, namely, sentential compounds are *decomposed* by noting invariants under substitution.

The third step gives rise to the "meta-concept" of ingredient content, i.e. we realize that two sentences that have the same pragmatic potential as free-standing, force-bearing rational classifications can nonetheless make different contributions to the content (and hence force) of compound sentences in which they occur as unendorsed components. It happens when [12]:

we notice that substituting one for the other may change the free-standing significance of asserting the compound sentence containing them. To form complex concepts, we must apply the same methodology to sub-sentential expressions, paradigmatically singular terms, that have multiple occurrences in those same logically compound sentences. Systematically assimilating sentences into various equivalence classes accordingly as they can be regarded as substitutional variants of one another is a distinctive kind of analysis of those compound sentences, as involving the application of concepts that were not components out of which they were originally constructed. Concepts formed by this sort of analysis are substantially and in principle more expressively powerful than those available at earlier stages in the hierarchy of conceptual complexity. (They are, for instance, indispensable for even the simplest mathematics.)

4 Analytic Pragmatism and the Problem
of Representation

Making It Explicit aims at describing the social structure of the game of giving and asking for reasons, which is typical of human beings. *Between Saying and Doing* has a different task: it pursues the pragmatic end to describe the functioning of autonomous discursive practices (ADPs) and the use of vocabularies [13]. ADPs start from basic practices that give rise to different vocabularies and the analysis is extended to nonhuman intelligence.

The so-called "analytic pragmatism" (AP) represents a view that clarifies what abilities can be computationally implemented and what are typical of human reasoning [14]. First, Brandom criticizes the interpretation of the Turing Test given by strong artificial intelligence or GOFAI, but he accepts the challenge to show what abilities can be artificially elaborated to give rise to an autonomous discursive practice (ADP). What is interesting to me is that AI-functionalism or "pragmatic AI" simply maintains that there exist primitive abilities that can be algorithmically elaborated and that are not themselves already "discursive" abilities. There are basic abilities that can be elaborated into the ability to engage and ADP. But these abilities need not be discovered only if something engages in any ADP, namely they are sufficient to engage in any ADP but not necessary. Brandom's view could be seen as a philosophical contribution to the discussion about how to revisit some classical questions: the role of symbols in thought, the question of whether thinking is just a manipulation of symbols and the problem of isomorphism as sufficient to establish genuine semantic contentfulness.

The strategy of AP is based on a "substantive" decomposition that is represented in algorithms. Any practice-or-ability P can be decomposed (pragmatically analyzed) into a set of primitive practices-or-abilities such that:

1. they are PP-sufficient for P, in the sense that P can be algorithmically elaborated from them (that is, that *all* you need in principle to be able to engage in or exercise P is to be able to engage in those abilities plus the algorithmic elaborative abilities, when these are all integrated as specified by some algorithm); and
2. one could have the capacity to engage or exercise *each* of those primitive practices-or-abilities without having the capacity to engage in or exercise the target practice-or-ability P.

For instance, the capacity to do long division is "substantively" algorithmically decomposable into the primitive capacities to do multiplication and subtraction. Namely, we can learn how to do multiplication and subtraction without yet having learned division. On the contrary, the capacities to differentially respond to colours or to wiggle the index finger "probably" are not algorithmically decomposable into more basic capacities because these are not things we do *by* doing something else. We can call them *reliable differential capacities to respond to environmental stimuli* but these capacities are common to humans, parrots and thermostats.

Along the line introduced by Sellars, we can intend ADP typical of human practices in an "inferential" sense and strictly correlated with capacities to deploy an autonomous vocabulary (namely a vocabulary typical of human social practices) [15]. They are grounded in the notion of "counterfactual robustness" that is bound to the so-called "frame problem". It is a cognitive skill, namely the capacity to "ignore" factors that are not relevant for fruitful inferences. The problem for AI is not *how* to ignore but *what* to ignore. Basic practices that provide the very possibility to talk involve the capacity to attend to complex relational properties lying within the range of counterfactual robustness of various inferences.

It is very interesting to see the new version of intentionality as a pragmatically mediated relation which departs from a specific account of human discursive practices while showing the connection between modal and normative vocabularies: normative vocabulary essentially addresses acts of committing oneself, and modal vocabulary essentially addresses the contents one thereby commits oneself to. We can consider the following example. Imagine a non-autonomous vocabulary focused on the use of the term "acid". In this make-believe instance, if a liquid tastes sour one is committed and entitled to apply the term "acid*" to it. And if one is committed to calling something "acid*", then one is committed to its turning phenolphthalein blue. In this community there is agreement, under concurrent stimulation, about what things are sour and what things are blue and it has experts certifying some vials as containing phenolphthalein. Moving from this background, the community implicitly endorses the propriety of the material inference from a liquid's tasting sour to its turning phenolphthalein blue. If a practitioner comes across a kind of liquid that tastes sour but turns phenolphthalein red, he "experiences" materially incompatible commitments. To repair the incompatibility he is obliged either to relinquish the claim that the liquid tastes sour, or to revise his concept of an acid* so that it no longer mediates the inference that caused the problem. In this case, he can restrict its applicability to clear liquids that taste sour, or restrict the consequence to turning phenolphthalein blue when the liquid is heated to its boiling point. This move clearly shows how difficult it is to undertake new commitments since the practitioner may discover that he is not entitled to them. The lesson we learn from this example is that the world can alter the "normal" circumstances and consequences of application embedded in our concepts. The concept acid* entails that it is not *necessary* that sour liquids turn phenolphthalein blue but is *possible* that a liquid both be sour and turn phenolphthalein red.

I would like to point out that we meet an interesting reformulation of the classical Kantian notion of representation of objects. The transcendental apperception is replaced by a kind of synthesis based on incompatibility relations [16]:

In drawing inferences and "repelling" incompatibilities, one is taking oneself to stand in representational relations to objects that one is talking about. A commitment to A's being a dog does not entail a commitment to B's being a mammal. But it does entail a commitment to A's being a mammal. Drawing the inference from a dog-judgment to a mammal-judgment is taking it that the two judgements represent one and the same object. Again, the judgment that A is a dog is not incompatible with the judgment that B is a fox. It is incompatible with the judgment that A is a fox.

Taking a dog-judgment to be incompatible with a fox-judgement is taking the to refer or represent an object, the one object to which incompatible properties are being attributed by the two claims.

5 Conclusion

Starting from Frege's inheritance we can offer a history of concept's formation and use, which shows the peculiarity of human cognition. Differences among human, non-human animals and machines arise only if we consider corresponding differences and relations among fundamentally different kinds of concepts. A serious theoretical consideration is exemplified by the four-membered Chomsky hierarchy that describes kinds of grammar, automaton, and syntactic complexity of languages in an array from the most basic (finite state automata computing regular languages specifiable by the simplest sort of grammatical rules) to the most sophisticated (two stack pushdown automata computing recursively enumerable language specifiable unrestricted grammatical rules).

I think that we can observe the contribution that the philosophical analysis brings to the clarification of conceptual hence "representational" human activity. And this task means to consider all the grades of it from the simpler and less articulated sorts to the more complex and sophisticated kinds of concepts. Following the lesson of Analytic Pragmatism, we can enrich the research in the field of the phylogenetic development of sapience especially because we do not know about a corresponding process in non-human creatures. In a different way, Searle describes the use of ordinary language in all its dimensions in order to enrich mere empirical research. Here some interesting questions to which Brandom draws our attention: Human children clearly cross that boundary, but when, and by what means? Can non-human primates learn to use conditionals? Has anyone ever tried to teach them?

Another problem is addressed to AI, proposing very interesting varieties of possible implementation of sentential compounding like connectionism and parallel distributed processing systems. The problem is to capture the full range of concepts expressed by complex predicates as they lack the syntactically compositional explicit symbolic representations. As we have seen, it moves from the substitutional decomposition of such explicit symbolic representations.

Acknowledgements I would like to thank the referees for fruitful criticisms and suggestions, and the participants at the AISB Convention 2014 and the UNILOG 2015 World Congress for thoughtful comments.

References

1. Giovagnoli, R.: Why the Fregean "square of opposition" matters for epistemology. In: Beziau, J.-Y, Jaquette, D. (eds.) Springer, Basel (2012)
2. Macbeth, D.: Frege's Logic. Harvard University Press, Cambridge (2005)
3. Sluga, H.D.: Gottlob Frege. Routledge & Kegan, London (1980)
4. Searle, J.: Speech Acts. Cambridge University Press, Cambridge (1969)
5. Searle, J.: Intentionality. Cambridge University Press, Cambridge (1983)
6. Searle, J.: Speech Acts, chap. 5. Intentionality. Cambridge University Press, Cambridge (1983)
7. Searle, J.: Speech Acts. Intentionality, pp. 182–183. Cambridge University Press, Cambridge (1983)
8. Brandom, R.: How analytic philosophy has failed cognitive science. In: Brandom, R. (ed.) Reason in Philosophy. Animating Ideas. Harvard University Press, Cambridge (2012)
9. Brandom, pp. 2–3 (2012). (online pdf)
10. Sellars, W.: Counterfactuals, dispositions, and causal modalities. In: Feigl, H., Scriven, M., Maxwell, G. (eds.) Minnesota Studies in the Philosophy of Science, Volume II: Concepts, Theories and the Mind-Body Problem, pp. 306–307. University of Minnesota Press (1958)
11. Brandom, R.: How analytic philosophy has failed cognitive science. In: Brandom R. (ed.) Reason in Philosophy. Animating Ideas, cfr. pp. 26–29. Harvard University Press, Cambridge (2012)
12. Brandom, R.: How analytic philosophy has failed cognitive science. In: Brandom R. (ed.) Reason in Philosophy. Animating Ideas, pp. 32–33. Harvard University Press, Cambridge (2012)
13. Brandom, R.: Between Saying & Doing. Oxford University Press, Oxford (2008); Giovagnoli, R.: Representation, analytic pragmatism and AI. In: Dodig-Crnkovic G., Giovagnoli, R. (eds.) Computing Nature. Springer (2013); Giovagnoli, R. (ed.): Prelinguistic practice, social ontology and semantic. Etica & Politica/Ethics & Politics (2009). www.units.it/etica/; Penco, C., Amoretti, C., Pitto, F. (ed.). Towards an analytic pragmatism. In: CEUR Proceedings (2009)
14. Brandom, R.: Between Saying & Doing. Oxford University Press, Oxford (2008)
15. Kibble, R.: Reasoning, Representation and Social Practices. Giovagnoli, R.: Computational aspects of autonomous discursive practices. Giovagnoli, R., Dodig-Crnkovic, G., Erden, Y. (eds.): Proceedings of the Symposium Social Aspects of Cognition and Computation. http://www.cs.kent.ac.uk/events/2015/AISB2015/proceedings.html
16. Brandom, R.: Between Saying & Doing, pp. 187–188. Oxford University Press, Oxford (2008)

Consciousness and Hyletics in Humans, Animals and Machines

Angela Ales Bello

Abstract This chapter aims to show that the scientific approach to nature, in particular to animals and human beings, is not sufficient to understand the sense of their organism, because it does not explain the sense of their life. Furthermore for the same reason it is not possible to affirm that the human being is a machine, or that a machine could develop so that it can become like—or sometimes as the same in— a human being. To support this assumption I assume a phenomenological attitude following the analyses proposed by Edmund Husserl and some of his scholars.

1 Introduction

Consciousness is a central object of investigation for Modern philosophy. The focus of the philosophical gaze on the human subject is certainly one of the key characteristics of the European Renaissance. One must not forget, however, that philosophy, from its beginnings in ancient Greece, even when it has fixed its attention on nature trying to discover its principles, has not neglected the human subject. From the "I have investigated myself" of Heraclitus to the "Know thyself!" of Socrates to the "Truth resides in the interiority of humans" of Augustine of Hippo to the "I think, therefore I am" of Descartes, it is clear that philosophers are investigating the human being. Hence, if their investigations are to yield results, the human being must understand him- or herself, that is, s/he must know his/her capacities and limits. This kind of knowledge constitutes the kernel of the philosophical problem of consciousness, which articulates itself in numerous ways, but which is also deepened by focused and sustained philosophical inquiry.

Translated by Antonio Calcagno

A.A. Bello (✉)
Pontifical Lateran University Rome, Vatican City, Italy
e-mail: alesbello@pul.it

© Springer International Publishing AG 2017
G. Dodig-Crnkovic and R. Giovagnoli (eds.), *Representation and Reality in Humans, Other Living Organisms and Intelligent Machines*, Studies in Applied Philosophy, Epistemology and Rational Ethics 28, DOI 10.1007/978-3-319-43784-2_12

I wish here to discuss a way to approach the question of consciousness that is the result of a long distillation process of western philosophy. I wish to launch a "new" challenge to those positions that ground themselves in the presumed "objectivity" of modern science, in the desire that everything be objective. In other words, these objectivist philosophers call for the disappearance of the subject, ultimately dismissing investigations of the subject as a form of "introspection" that cannot guarantee certainty.[1] This scientific way of thinking or, in other words, this scientist view, currently dominates not only philosophy but all areas of research. We have to remind those philosophers who want to eliminate the role of subject that also some scientists are in search of the subject of investigation and propose a probabilistic interpretation of nature. Thus, we have to ask: Can we give an account of reality, in all of its complexity, solely by following that kind of scientific approach? In fact, we are dealing here with science's battle against the very intention of western philosophy, that is, science struggles in many cases against a form of knowledge that is based on deep and far-reaching philosophical inquiry. One wonders, however, whether the aforementioned "scientific" view, in all of its various forms, is a preconceived and arbitrary "vision of the world" that is similar to the type of vision that philosophy is accused of defending? Given the presumption of that scientific viewpoint to account for reality as it is, it is necessary to begin to discuss the meaning of science and to do so we have to start from philosophy, which is from the perspective that can give reason for science itself. That is why I assume the phenomenological perspective which always seeks to "begin anew" and it does so by inviting us to participate in the study of its objects.

In order to combat the prejudices of some philosophers and scientists, we need first to discuss the major aspects of phenomenology, always with particular reference to the work of the founder of the phenomenological school, Edmund Husserl. We have to examine the essential elements of phenomenology in order to see how they have developed and how they stand in relation to various scientific views.

I propose that we understand phenomenological inquiry though a metaphor, namely, that of concentric circles, regarding in this case the topic of our inquiry: the human, the animal, and the machine. By employing this metaphor, we will clarify the object of research, the structure of consciousness, and the connections between different phenomenological realms.

In the phenomenological school one sees a great interest in the phenomena of animal and plant life, especially given that an understanding of the human being is arrived at, in part, through an understanding of these phenomena. This interest

---

[1]Shaun Gallagher and Dan Zahavi develop interesting positions on this debate in their work *The Phenomenological Mind* (London: Routledge, 2012). Both authors emphasize the importance of the phenomenological perspective, which, in their view, is neither purely psychological nor introspective; rather, they see the phenomenological mind as central to understanding the human being. Even though they acknowledge the contribution of the cognitive sciences and neurosciences, the nature of their work remains largely phenomenological and philosophical..

appears in the work of Hedwig Conrad-Martius and Edith Stein,[2] but it was Husserl who earlier began this kind of research.

At a first approach we can say that the difference between the human being and the animal can be located in an understanding of consciousness, as we will endeavor to show; we also discover in the phenomenological approach to consciousness the difference between human beings and machines. In what follows, I will discuss, first, the phenomenological description of the human being, then of animals and, third, the problem of consciousness and machines.

The leading thread will be the so-called "hyletic dimension", which permits us to establish the relationship between human beings and animals. Then it is necessary to explain what is "hyletics".

2   The Role of Hyletics in the Description of the Human Being

Through the epoché of the natural attitude—that is the operation of putting into parenthesis all that we already know in order to start anew—that characterizes the philosophical approach according to Husserl, he discovers a new territory inside the human being, that is the territory of the lived experiences (*Erlebnisse*) and we have consciousness of them.[3]

Husserl's analysis of lived experiences highlights the two-fold nature of their noetic moment—that is the intention to grasp the sense—and their hyletic or material moment. In what follows I will try to explain the meaning of the two aspects.

The term hyletics does not indicate matter in the traditional sense, but a new type of materiality that Husserl proposes in § 85 of the first volume of the *Ideas*.[4] He was looking for a new term and thought he found it in the Greek word *hylé*. It was a question of identifying what had never before been clearly delineated, and for this reason there was also no existing term to describe it. The description of this sphere

---

[2]I examined these phenomenologists' positions in *"Edith Stein's Contribution to Phenomenology,"* in *Analecta Husserliana*, vol. 80, ed. Anna Teresa Tymieniecka (Dordrecht: Kluwer, 2002), 232–240; *"Edith Stein: Phenomenology, the State and Religious Commitment"*, 648–656; and *"Hedwig Conrad-Martius and the Phenomenology of Nature"*, 210–232.

[3]Regarding the meaning of *epochè* and of the lived experiences I have described it in my book *The Sense of Things. Toward a Phenomenological Realism*, Translated by Antonio Calcagno, "Analecta Husserliana," vol. cxviii, Springer 2015.

[4]Edmund Husserl *Ideen zu einer reinen Phänomenologie und einer phänomenologischen Philosophie*, transcribed by Edith Stein between 1916 and 1918, revised by Ludwig Landgrebe from 1924–25 and by Husserl himself until 1928, and finally edited by Marly Biemel in 1952 as volume IV of *Husserliana*. Volumes I and II of the *Ideen* were edited by Karl Schumann as volumes III-I and III-2 of *Husserliana*..

is further developed in the second volume of the *Ideas*[5] with the analysis of the living body (*Leib*), which experiences both the localization of bodily sensations [*Empfindnisse*] important for the constitutive function of objects that appear in space and a completely different group of sensations that Husserl calls sensorial sensations [*die sinnlichen Empfindungen*], for example, the sensations of pleasure and pain, of bodily well-being or discomfort deriving from the body being physically unwell.[6]

This line of description continues to be present in a large number of manuscripts, especially the C and D groups of writings from the 1930s. The function of hyletics within the field of the sensations is particularly investigated in Ms. Trans. D 18, which is dedicated to a discussion of the formation of the kinaesthetic system. This system accounts for the relationship between one's own body and the changes of the surrounding world within the framework of the ocular-motor field. In Ms. Trans. D 10 I, Husserl specifies that the kinaesthetic system becomes constituted in relation to the constitution of hyletic objects,[7] but it is in Ms. C 10 that one grasps the connection between hyletic units and affectivity, because even though the hyletic universe is a non-egological universe that becomes constituted without the intervention of the I, "*das Ich ist immer 'dabei'*," the I is always present as the locus of affectivity and it is always active in some fashion.[8]

Let us examine further Husserl's reference to the two groups of localized sensations, which serve as a material similar to that of the primary sensations' intentional *Erlebnisse*, for example, hardness, whiteness etc. These groups of sensations, according to Husserl, are immediately, somatically localized: in every human being, these sensations manifest themselves in an immediate intuitive manner in the lived body (*Leib*) inasmuch as it is the person's "own body," a body that is a subjective objectivity that is different from the experience of the body as a purely material thing extended in space.[9] The experience of one's *own* body, "difficult to analyse and illustrate," is linked up with the sensations of tension and relaxation of energy, in the sensations of internal inhibition, of paralysis, of liberation, which forms the base of the life of desire and will.[10] But connected with this stratum of sensations are "intentional" functions. The material of the aforementioned sensations of ownness assumes a spiritual function much like that of the primary sensations that come to form part of the perceptions on which constitutive judgments become constituted.[11] Hence, stratification occurs on two levels: first, on the level of cognition, formed by primary sensations, perceptions, and perceptive judgments, and,

---

[5]*Ideen*, III-2.

[6]Ibid. § 39.

[7]Ms. Trans. D 10, *Zur Konstitution der physischen Natur. Zuerst Leib—Aussendung; dann rückführend auf Hyle und Kinästhese*, 23.

[8]Ms. Trans., C 10, *Das gehört zum Komplex der urtümlichen Gegenwart!*, 25.

[9]See Edmund Husserl, *Ideen* II, §39.

[10]Ibid.

[11]Ibid.

second, on a psycho-reactive level, formed by sensorial sentiments and valuing. The perceptive, judicative and valuing levels are noetic.

The relationship between hyletics and noetics is thus clearly delineated, but the hyletic moment seems to drag the noetic one behind it, and hence Husserl's peremptory affirmation: "… a man's entire consciousness is in a certain way with his body through its hyletic base".[12] But the duality is not eliminated. Indeed, the intentional *Erlebnisse*, as such, are not localized and do not constitute a stratum of one's body. The autonomy of the intentional moment with respect to the material one is in this way confirmed and corroborated. In fact, inasmuch as it is a tactile grasping of form, perception is neither in the finger that touches nor in the tactile sensations that are localized in the finger. Thought is not really localized intuitively in the head, as are the localized sensations of tension.[13] Husserl notes that we often express ourselves in this way, and one may wonder why we do so. One can reply by saying that the attractive force of the hyletic localization makes us concentrate our attention on our body.

Turning to the discussion of animals, though it is impossible for us to live what animals live, it is possible to perform an act of empathy [*Einfühlung*] that consists in grasping their lives and the acts lived by them insofar as these acts are similar to those lived by us. Recall that what Husserl said about the animal world and instinct was situated within Husserl's discussion (in *Ideas II*) of the human capacities to grasp that world.

There is a level of likeness between humans and animals: we can grasp the bodily sensations and reactions of animals through their psychic acts, especially when we are in contact with more highly evolved animals. The difference and the disparity emerges in the fact that animals cannot perform certain acts that we define as "spiritual," including intellectual comprehension and elaboration, acts of will, and motivated decisions, acts that lie at the very foundation of the construction of the human world. We are aware that we cannot establish empathy at the level of these spiritual acts, and that is why we cannot consider animals to be really "like us." If the aforementioned difference between animals and humans is correct, we cannot fully grasp the mechanism through which animals know the world at a perceptive level. Even if a great variety of modes of cognition exist, especially those researched by ethologists, human perception with its passive processes certainly shares certain similarities with animal perception, but there are also differences as well, especially at the spiritual level. Yet, in terms of psychic reactions, including the embodied, localized sensations and the expressions of satisfaction or disgust, attraction or repulsion, humans and certain animals seem to be quite similar.

---

[12]Ibid.
[13]Ibid.

3   The Animal and Its Instinctive Life

In Husserl's A, C, and E manuscripts, one finds a treatment of instinct in both the human and animal worlds. That this theme was not tangential in Husserl's reflections is evidenced by the presence of the manuscripts that were eventually to form the second volume of the *Ideas Pertaining to a Pure Phenomenology and a Phenomenological Philosophy*. Here, we find an explicit reference to the psychic constitution of animals vis-à-vis humans.[14] The fact that he later returned to the topic shows that the attention Husserl paid to the animal world was not by any means fleeting; rather, his discussion of animals is important, especially for our purposes here.

In Ms. E III 10, Husserl employs his study of the pre-given world from the viewpoint of impulse and instinct as the starting point for tackling the investigation of knowledge of the human and animal worlds. The text opens with one of the very few passages where Husserl refers to Freud's analyses and seems to share Freud's conclusion: Husserl accepts the possibility of the existence of "repressed" affects, of unsatisfied desires, which are relegated to the level of the unconscious and generate an "illness" of the soul. "Everything that is removed, everything that is of value, but remains hidden, continues to function in an associative and apperceptive manner, something that the Freudian method deems possible and presupposes."[15] Starting from this consideration, Husserl examines the dynamics of the special intentionality that characterizes instincts. The desire for food, for example, can be described by using an approach that is valid for certain modes of cognition but in this case it is linked up not with cognition, but with instinct: there is a tending toward a fulfillment that finds its realization in an object, which, in this case, is the act of eating. In actual fact, hunger helps Husserl to understand the dimension of instinct, for the I is always hungry: hunger is its habitual condition that is only temporarily satisfied or fulfilled by eating food.

The analysis of instinctual life in human beings leads Husserl to establish that, first, it is precisely thanks to habits that the unity of the I already constitutes itself at the level of instinct such that the unity of the subject, though recognized by consciousness, is of a prior origin; second, habits themselves influence and, in some cases, even determine the direction of the will, hence, passivity plays an important role within the sphere of the human will. For example, the need for walking becomes transformed into a decision: "I want to go out." One can also trace what is typical in the fundamental structure of needs that become articulated at different levels and that constitute the structural form of all life, thereby making it possible for the I to possess a systematic structure of its orientations of will. In a broader sense, we can consider the modes of the will and of originary instinctual life to be a *Vorgestalt*, that is, a form that precedes other forms.

---

[14]See Sect. 2, Chap. 4, §45.

[15]Edmund Husserl, Ms. Trans. E III 10, *Vorgegebene Welt, Historizität, Trieb, Instinkt*, January 1930, 3.

Husserl examines the "vital" instinct of animals in essential terms and not from the viewpoint of the natural sciences, which study only animals' physical aspects. Ethologists and animal psychologists, though they seek to penetrate the "inside" of psychic life, do not, according to Husserl, possess adequate instruments for going deeper into what they live. The vital instincts of animals can only be interiorly understood through transcendental phenomenological analysis: "... in this way we have the animal subject as subject of its pregiven world, of its acquired orientations and correlates, a world in which one always finds the same objects."[16]

Husserl begins his analysis by highlighting two particular instincts, namely, survival, which is linked with food, and that of generation, which is connected with a dimension of togetherness. These two instincts, of course, do not exhaust the description of instinctual life. The instinct of fear, for example, is also of considerable importance for survival. Moreover, pleasure and non-pleasure, attraction and repulsion through the sense of smell or vision are also connected with both the instincts of survival and togetherness.

If instinct is understood as internal, always in the phenomenological transcendental sense, how can a species preserve itself? An animal is born into the world and leaves it because of its natural death due to old age, illness, or as a result of chance events. Is it possible, therefore, to understand what happens in the psyche and consciousness of an animal, a highly evolved animal, of course, when it confronts death? The animal knows death through the deaths of its companions, but does it make sense to speak of companions, family relations, and education? Is there some relation between the I, the You and, therefore, the We of the animal world? Husserl's answer is affirmative, even though the intersubjective world of animals is characterized by a basic form of relations between males and females, between father, mother, and "offspring," between friends and enemies, or by the struggle for life or death. But what is the animals' level of awareness of the very individuality of each animal?

The answer to this question is found in the central part of the manuscript, which bears the subtitle, "*Das Tier und das Wissen von seinem Tode. Höhere Tier und tierische Ich und Wir in Umwelt (Begriff des höheren Tieres)*" [The Animal and the Knowledge of Its Own Death: Higher-Order Animals, the Animal-I and the We in Their Surrounding World: (The Concept of Higher-Order Animals).]"[17] Life preserves itself in a continuous development of its own actualizations or growth that begin for the individual at birth and terminate with death, but birth implies also generation, and thus survival of the species, and yet even species come to an end. The examination of the animal world, therefore, requires a comparison with the human world, which ultimately brings out both the similarities and the differences between both the animal and human worlds. And if we can carry out a non-anthropomorphizing comparison, we find that the human spirit is certainly absent in the animal. Nevertheless, we can understand the psychic life of animals

---

[16]Ibid., 10.

[17]See pages 12–17 in Ms. Trans. E III 10.

that we are able to understand without projecting onto the animal our own human traits. One may certainly and rightfully ask oneself whether the individual animal has consciousness of its own death, but it would certainly be nonsensical to wonder whether or not the animal is conscious of the end of its species. Yet, this latter concern is present for humans, albeit we find it in different grades of awareness in different people.

> Human life is explicitly connected with its own death, but it is also connected with its own human history and, more precisely, with the future of humanity and thus also with the life and death of humanity as such and with the surrounding human world as a cultural world. This is true for the highest levels of development achieved by humanity so far — in this case, once again, there are different real and possible degrees of awareness.[18]

It is clear that we can definitely speak about a hyletic dimension within the animal world, though the awareness of animals' own interiority, of their own consciousness, remains mysterious.

4 What Is Consciousness? Consciousness and Machines

If the hyletic dimension is shared by both humans and animals, we have to ask whether or not another, more specific dimension, rich with implications, exists. What is consciousness? Where can we locate it? Does it have a specific place or location?

The mental organizing that leads us to imaginatively locate things in space leads us to find a place for anything, for things that do not require a space, for things that are not spatial. For example, God, the gods who are in the heavens, the devil, the damned at the center of the earth, the ideas of a heavenly world, etc. Spatiality is linked to corporeity or embodiment. This is why space is important for human beings. But not everything in human beings is spatial, especially consciousness. Consciousness is the way in which the human being is aware of what he/she lives, that is of the lived experiences, through which one knows himself as body, as psyche and as spirit, i.e. an intelligent being. Intelligence and will are defined as spiritual, in the sense that they belong to humans and distinguish them from the other beings in nature.

Consciousness is then very important for the us. But to what extent can we recognize that animals and machines have consciousness? And if so, which kind of consciousness?

The special edition of the journal *ParadoXa*[19] dedicated to consciousness can assist us to understand the non-spatiality of consciousness. In this issue, we find examples of contemporary views of consciousness. The articles contained in it deal

---

[18]See page 17 in Ms. E III 10.

[19]*ParadoXa*, *What is Consciousness*, Fondazione Internazionale Nuova Spes, Roma, October–December 2009.

with the "ancient" problem of the relation between the body and the soul, which today is described as the mind-body problem. Three solutions are offered to this problem: a monistic account in which only the body exists; a dualistic account of two different and heterogeneous realities; and a dual account in which there is a stratified and complex unity.

The monistic or physicalist view is more prevalent, not only for those who work directly in the biological sciences but also for those working in psychology. Physicalists will often use the results of the sciences, especially the neurosciences, to justify their positions. Less diffuse is the view advocated by those who distinguish between the natural and human sciences—a view that was widely held until the end of the 19th century by such thinkers as Dilthey, Husserl and Stein in Germany and Henri Bergson in France. The acceptance of the more prevalent view can be seen as a victory for the positivist model. The physicalist or scientific viewpoint can largely be designated by the following traits: a rigid adherence to traditional evolutionary theory, the validity of the neurosciences, and the identification of consciousness with brain activity.

If we examine the articles contained in the aforementioned journal, on one hand we obtain a wide panorama of contemporary views, and on the other hand through the citations within the articles, we encounter various researchers who have opened up or helped develop new fields of research. We also find a divided field of inquiry.

If we focus our attention on what is said about consciousness in the journal, we find recognition, albeit superficial, of what Husserl and phenomenology claimed about consciousness: "Phenomenology is a study of consciousness said to have been founded by the German philosopher Edmund Husserl, who defined it (in 1901) as: "The reflective study of the essence of consciousness as experienced from the first-person point of view."[20]

We find here an interest in research carried out from the first-person perspective as opposed to inquiry carried out from a third-person perspective, which is largely practiced by the majority of scientists who describe themselves as carrying out "objective" research.

If we admit the viability of the first-person viewpoint, which undoubtedly signals a remarkable methodological turn, one notes the characteristics of consciousness described by the authors of the articles in *ParadoXa*: we find the I, the sense of the I, its identity, free deliberation of the will, and the will. Even if these concepts are considered with reference to the functions of consciousness, we do not find an answer to the question "What is consciousness?" When we do find intimations of what consciousness could be, we find polemical critiques of the Cartesian *cogito* that maintain "that such a spontaneous and approximate "explanation" of the I expressed from a physical, material mechanism in my brain gives to itself other

---

[20]Igor Aleksander, "*What Computation Can Tell Us About Consciousness,*" in *ParadoXa*, *op. cit.*, 77. Igor Aleksander is Emeritus Professor of Neural Systems at Imperial College, London, UK.

physical mechanisms necessary for the performances of the free deliberation of the will, which, again, happen "there" in my (amazingly physical!) brain."[21]

For those who maintain that the brain must be understood from the perspective of the neurosciences, one finds a functional definition of the brain. Mental processes, especially first-person ones discussed by phenomenology, constitute, in their totality, the mind, and can be employed in Virtual Machine Functionalism, understood as the "state structure of a system." This is Igor Aleksander's position, which proposes to construct not real, but virtual machines, for virtualism allows for the possibility of overcoming the difficulty of physicalism (monism) and the mind-body dualism by means of a "flexible" relation between structures and functions.

This argument could also be understood in terms of the identification of the brain with a neuronal machine. Naturally, given that the machine is not a living organism, it has no relation to the sense of the hyletic discussed above, but it certainly is related to consciousness. According to this hypothesis, if the brain possesses a neural network, it becomes possible, then, to construct conscious machines or thinking robots. This thesis is maintained by Domenico Parisi, who argues that if we wish to understand what consciousness is, we must be capable of building robots that have a mental life.[22] The construction of robots, which are seen to be different to living organisms, allows one to reproduce the different phenomena of consciousness that would allow us to understand the very nature of consciousness itself.

Aleksander does not accept the foregoing thesis and moves the discussion of machines to the virtual plain, but he also wishes to understand what consciousness is through "computational machines." The problem, however, remains: What is the relation between the brain (organism) and the machine (real or virtual)? In both cases, do we still not find ourselves lapsing into mechanism? We have to admit, then, that what is new with respect to the mechanistic model is the first-person perspective and the ability to examine whether or not the things of the mind are independent of the brain, as Aleksander maintains. He ultimately wishes to uphold the "comparison" between virtual machines and the brain. This is also confirmed in the essay by Lorenzo Magnani when he affirms, "Computers possess, for example, developed decisional capacities (even though now they do not have programs and structures that allow them to experience the conscious sensation of the will and the deliberation of the will)."[23]

Even if we admit that it is true, can we conclude that the human being is like a machine?

At this point, Martin Heidegger's objection concerning technology arises: Can technology explain the structures of human beings? Can technology do likewise for

---

[21]Lorenzo Magnani, "*L'evoluzione della coscienza e del libero arbitrio*," in *ParadoXa*, op. cit., 51. Lorenzo Magnani is Professor of the Philosophy of Science at the University of Pavia.

[22]Domenico Parisi, "*Robot che "hanno la coscienza"*," in *ParadoXa*, 62..

[23]Lorenzo Magnani, *L'evoluzione della coscienza e del libero arbitrio*, op. cit., 56.

other living organisms? This is also Heidegger's question is more complicate than the one posed by evolutionary theory, which maintains that consciousness is the product of evolution. In this case, biological processes of organisms are separated from the processes of artificially constructed machines (virtual or real). On the contrary, the tendency of contemporary biological studies, which we referred to above, is to link consciousness with the brain and to almost exclusively focus their scientific investigation on the brain, ultimately reducing it to a machine; even if a machine could be a highly complex system, it is not an organism. In fact the complexity cannot reach the qualitative level of an organism.

Not all researchers working in the aforementioned fields are reductionists. Tito Arecchi, for example, rejects the thesis of the so-called "eliminativists," who maintain that "every act is determined by the initial conditions of our atoms and molecules; we are automatons."[24] Critical of those who uphold the Strong Artificial Intelligence [*Intelligenza Artificiale Forte*] position, Arecchi argues that the adaptability of organisms—here, we see a strong connection with evolutionism—exceeds "the power of an algorithm that we can install in a computer."[25] We can admit that the DNA of an organism could be explained through an algorithm but its own structure or sense is more than an algorithm, it is life.

The picture that Arecchi paints of the evolutionary process permits us to overcome absolute determinism, thereby introducing adaptive, non-algorithmic leaps that exceed efficient causality and that allow for final causality. His argument is based on Gödel's incompleteness theorem. Hence, Arecchi is able to account for "leaps outside of a particular species," which lie outside the realm of algorithmic processes and which are based on "other enunciations that are compatible with axioms, but which are not demonstrable by a pre-selected algorithm."[26]

According to Arecchi, his thesis allows us to understand how mammals and birds replaced dinosaurs. His theory is very similar to that of the phenomenologist Hedwig Conrad-Martius, who posed the same question of the transition of early sauria into mammals. In order to justify this transition, she introduced her theory of qualitative leaps.[27]

The theories of Arecchi and Conrad-Martius highlight the insufficiency of a purely "scientific" approach, understood in the abovementioned sense, in order to understand nature. Both thinkers introduce a qualitative element into the discussion because they are convinced of the necessity for both biology and physics to be open to metaphysics. Is the insufficiency born from the practice of science itself or does science already contain such an insufficiency within itself? In part, science is concerned with truth and requires the obtainment of what is "most true." It also

---

[24]Tito Arecchi, "*Fenomenologia della coscienza: complessità e creatività,*" in *ParadoXa, op. cit.,* 38.

[25]Ibid.

[26]Ibid., 39.

[27]Hedwig Conrad-Martius, *Ursprung und Aufbau des lebendiges Kosmos* (Salzburg-Leipzig: Otto Müller, 1938).

needs to present truths that are coherent with their presuppositions. But are these presuppositions true? Can we even avoid the radical nature of the foregoing question?

Psychological and psychiatric studies note the opposition between those who follow a naturalist, organicist view and those who wish to examine the mental phenomena of the mind, albeit from a different perspective. The latter do not reduce mental phenomena to brain function. One can think here of psychoanalysis or phenomenological psychopathology.

An example of the first type of view can be found in the research of scholars and scientists who draw upon anatomy, neurology, and cognitive and experimental psychology. They examine psychic disturbances as based on neuronal activity. Through advanced technologies, including positron emission tomography, scientists can establish the difference between a vegetative state or state of minimal consciousness and states of awareness or wakefulness. Even functional magnetic resonance imaging applied to "resting cases," that is to subjects who are in a state of rest, can demonstrate the existence of functional networks important for researchers, including the "intrinsic self-reference of thought" and "conscious access to external stimuli." Both of these can, "in other words, "embody" the difference between "awareness of oneself" and the "surrounding environment."[28]

The term "embodiment" is particularly significant. Are these researchers looking for a substrate, base, source or even a locus in which the traditional, specific activities of human beings are grounded? Substrate, base, and locus are not equivocal terms. To affirm that consciousness is based in brain activity is different to saying that the brain is the locus of consciousness or is consciousness. The problem lies in the very functions of the "base." Certainly, without a brain the human being cannot live and cannot even be called a human being. But what is the brain the base of? So-called higher-order activities are only one manifestation of its functions. But is this function of the brain activated by something else, as was mentioned earlier?

The question of monism and dualism come to the fore again, but it is the third possibility that appears to be important as well, a view we termed dual. The dualist view, which inevitably leads to Platonism, maintains that body (lower-order activity) and soul (higher-order activity) stem from two different sources. The high-order activity is characteristic of human beings and distinguishes them from other animals that do not possess the same capacities of cognition, elaboration, and transformation as human beings. It is precisely because body and soul are qualitatively different that their interaction is only temporary and not posited before us.

---

[28]See *I disturbi degli stati di coscienza*, in *ParadoXa*, *op. cit.*, 115. It is interesting to note that the text cited here stems from the collective work of various authors: Andrea Bosco (Researcher in Experimental Psychology), Giulio Pancioni (Professor of Psychology), Marta Olivetti Belardinelli (Lecturer of Cognitive Psychology), Michele Papa (Professor of Human Anatomy). Mario Stanziano, Andrea Soddu, and Quentin Noirhomme (researchers at the Hospital of Liège) work with Papa. Working also at the Hospital of Liège is Steven Laureys (Professor of Neurology).

Hence, this is why it is possible to speak of the experience of the liberation from one's "body."

The traditional conception of the relation between body and soul has resulted in a completely different anthropology than the one advanced by science. The human being was created by God as a unified being that derives from a unique source; God wished the human being to be unified in all of his or her complexity and stratification while still possessing distinct and differentiated aspects that are both true and necessary. Western medieval philosophy has closely focused on investigating the unity of the human being and it could not ignore the insight of the revelation that came from theology, whose concepts were mined in order to gather indubitable "evidence." In order to have viable evidence, what was affirmed as true had to be irrefutably true. Hence, we can understand the articulation of criteria designed to guarantee the validity of knowledge, which were always present but which were most explicitly theorized by both Descartes and Husserl: evidence was seen as veridicality. It is *evident* that the human being is stratified and complex.

Descartes argued that the body is part of nature, which, in his age, began to be understood through the lens of a mathematically informed mechanistic science. Husserl was more prudent, for he conceived of nature in terms of causality, understood in qualitative, non-mathematical terms, a causality that was teleological and neither quantitative nor mechanistic. Certainly, the natural sciences arose because they made evident such causality, but is such an interpretation of nature sufficient to account for nature itself?

Of all of Edmund Husserl's students, Edith Stein maintains, in a less problematic way than did Husserl, that nature can be understood in terms of causality and, therefore, is capable of being understood by the sciences. Other sciences, however, can be employed to understand the human being, including physiology and anatomy, for example. The results obtained from such sciences can be valid, even though they may not be complete. Following Husserl, Stein argues that one finds in human beings both psyche and spirit. In the former, once finds causality, albeit not necessarily in the same form advocated by the natural sciences. In the latter, one finds the acts of the intellect and the will, which are inscribed within a framework of "motivation."[29]

Phenomenology maintains the view that the scientific reading of nature does not complete or exhaust our understanding of nature. A philosophy of nature is always required in order to describe the qualitative elements and aspects of nature. Hedwig Conrad-Martius, who knew the mathematics, physics and biology of her day, affirms the trans-physical elements of nature, which ultimately allow her to offer not only a quantitative understanding of nature but also a qualitative one.[30]

---

[29]Edith Stein, *Der Aufbau der menschlichen Person*, in *Edith Stein Gesamtausgabe*, vol. 14 (Freiburg: Herder 2004).

[30]Hedwig Conrad-Martius, *Naturwissenschaftlich Metaphysische Perspektiven* (Heidelberg: F. H. Kerle Verlag, 1949).

Today, a position akin to that of Hedwig Conrad-Martius seems to be impossible. An opening, however, has appeared within the neurosciences. The quantitative and the qualitative encounter one another in the relation between brain and mind. Is it possible to overturn the localizing of consciousness as an epiphenomenon of the brain? I believe this is possible: the brain is only the instrument of consciousness, and this claim is justified through the very complexity and stratification of the human being.[31] In the end, we do not have a dualism, but a duality in which an autonomous psycho-spiritual element is present. This aspect is not verifiable in strictly empirical terms through the use of machines; rather, such verification is the work of qualitative research.

5   Conclusion

I can conclude underscoring the difference between an organism and a machine. In order to grasp their difference it is necessary to reflect on the meaning of "life". We find life when an organism develops itself from inside and potentially can generate other organisms. Machines must be constructed by human beings in an artificial way and even if they reach a great complexity, they are not living beings. It is true that we can produce, using what nature offers to us, something that is more and more similar to us, but this production is not "generation" that is life transmission.

---

[31]This position could be considered as the same as classical functionalism, but it is different from it as far as it is based on a complex anthropology. Consciousness can show itself as a function, but it refers to psychic and spiritual aspects of the human being ultimately opening the way to a metaphysical analysis of it..

Matter, Representation and Motion in the Phenomenology of the Mind

Roberta Lanfredini

Abstract Not only the classical cognitive pattern but also the classical phenomenological pattern gives rise to a problem concerning the qualitative dimension.
This problem is essentially related to the notion of matter, conceived as residual
with respect to the notion of form: the sensorial hyle is residual with respect to the
intentional form; *plena* are residual with respect to the extension, and physical
matter (its resistance, its non-undifferentiation, its endurance) is also residual with
respect to the broad ensemble of connections where the physical thing is inscribed.
The residual component which characterizes the notion of matter is simply the other
side of the absolute predominance of form (representational form in the specific
context of mental phenomena). This predominance gives rise to the same problem
in the context of phenomenology and in philosophy of mind: the problem of the
ontological status of qualitative states. This issue is a crucial one and, in order to be
solved, requires a radical change of perspective. In the context of phenomenology
this change depends on the concept of enactive, embodied and situated mind. This
notion implies a temporal paradigm; it alludes to a dynamic, non-static pattern; that
is, it alludes not to projective notions (as in the representational model), but to
notions which are agentive and, ultimately, evolutive.

1 Representation and Embodiment

In the context of the contemporary philosophy of mind, we run into an increasing
number of arguments claiming a crisis of the classical cognitive and representational theory of mental events [1–7]. We must acknowledge that a large number of
homogeneous notions come into play here; and those notions are all meant to
provide an utterly new way of conceiving the mental dimension. Notions such as
extended or ecological mind, as well as *enactive, embodied,* and *situated* mind, are
all instruments which depict the human mind not precisely, or not only, as an

R. Lanfredini (✉)
Dipartimento di Lettere e Filosofia, VIa Bolognese, 52, Florence 50139, Italy
e-mail: lanfredini@unifi.it

© Springer International Publishing AG 2017   261
G. Dodig-Crnkovic and R. Giovagnoli (eds.), *Representation and Reality in Humans,*
Other Living Organisms and Intelligent Machines, Studies in Applied Philosophy,
Epistemology and Rational Ethics 28, DOI 10.1007/978-3-319-43784-2_13

ensemble of functions, but rather as something which takes shape by means of vibrant drives, a progressive integration with the surrounding environment, a tenacious aiming process and the safeguarding of our own movements within the space.

In this respect, we are witnessing the arising of a different paradigm which emphasizes a close connection with instincts and vital drives, at the expense of the paradigm which confines the mind to its projective and representational function. At the same time, the classical notions of mind, consciousness and representation start to be replaced by the more physical notions of organism, life and moving body. In this new scenario, cognitive processes are understood as dynamically adaptive, incarnated, deep-rooted in sensory and motor processes and situated each time in a given context. The sensory and motor paradigm, which replaces the representational one, aims at a minimal image of the *I* [8], which takes the form of a Bodily Self [9]. The latter consists in the possibility of recognizing ourselves as integrated subjectivities, endowed with sensory and motor functions [10], but also with bodily, and environmental (or ecological) capabilities; and, lastly, as subjectivities capable of carrying out a flexible integration with the environment by virtue of an incessant developmental process [11]. The paradigm emerging from the crisis of the classical cognitive system is, therefore, that of Embodied Cognition [12–15]. More precisely, it takes shape as a system which consistently diminishes transparency and lack of friction, distinctive characteristics of cognition understood as representative form, in favour of the *opacity* and *resistance* of cognition conceived as an ensemble of active opportunities and perceptive-motor capabilities.

According to this paradigmatic hypothesis, perception, instead of being conceived as a process which takes place in the brain, is to be understood as an activity which features the animal in its entirety and is, therefore, related to the exploration of its environment; the latter activity being characterized by a systematic interdependence between the sensory information at our disposal (*affordances*), as well as by a crucial pre-reflective and sub-categorial dimension [16–18]. Working in this way, Embodied Cognition claims to account, from a new standpoint, for one of the fundamental issues of the classical cognitive pattern, namely the issue of *qualia* or, as Nagel puts it, of the way *it feels* to have mental states [19]. In the following pages I will endeavor to argue that this particular issue, along with many others emerging from the cognitive and representational model, primarily depends on a certain way of conceiving the mind, and on a certain way of conceiving matter. Once a specific theory of matter is provided, it will give rise, in a perfect mirroring, to a specific theory of mind. This repercussion is crystal clear in the Cartesian framework. Descartes' notion of 'thought' is to be read as a direct consequence of his adherence to the mechanistic notion of the universe.

The fundamental idea is as follows: the physical world is a great mechanism whose functioning is mathematically depicted by Galileo's physics. In this context, *thinking* stands out as a notable exception, for it cannot be reduced to that mechanism and is, therefore, unrelated to the natural and material world. Accordingly, we may assert that the quantitative (not qualitative) and mathematical accounts apply to everything but the human mind. This explanatory gap is the fundamental source of the conflict between the non-mechanistic notion of mind and the mechanistic notion of body. In philosophical terms, that explanatory gap takes the shape of the ontological dualism of *res cogitans* and *res extensa*. This dichotomy is the ultimate result of a twofold theoretical movement, which can be traced back to a precise notion of matter. To clarify: if the world is a pure mechanism whose functioning can be wholly described by quantitative physics, it will necessarily lose every qualitative feature; on the other hand (and bear in mind that this second movement is the consequence and not the origin of the first one), subjectivity, in order to keep liberty of thought and action, will have to shed all its material attributes, thus withdrawing from the natural world.

This sacrifice of the natural component gives rise to the notion of a pure thinking subject, a disembodied mind placed outside materiality, an irreducible, autonomous and spiritual substance. This notion of the mental universe is not the source, but rather the consequence of the quantitative and mechanistic notion of body and matter.

The classical cognitive pattern and the significance ascribed to the notion of representation find here a crucial theoretical confirmation. One of the fundamental results of this conceptual framework is the so-called dismissal of the qualitative, along with the rejection of a certain epistemology, and, by and large, of the underlying anthropology, the main core of which can be traced back on one hand to the reduction of the living body to pure extension and movement (receptive surface and membrane interposed between inside and outside), and on the other hand to the notion of sensory experience as rough and formless matter, which needs a structure (or a form) in order to be organized, shaped and, hence, able to reasonably represent something.

That implies a notion of sensibility as pure passiveness, indifference, non-orientation, solitude and ineffability. Furthermore, it entails a notion of representation as empty structure, as transparent, pure and easily expressible form.

In this essay, I will take into account the dualism of *feeling subject* and *thinking subject*, and I will depict it as prior and founding, far more subtle and complex than the more renowned dualism of *res cogitans* and *res extensa*. Concurrently, I will endeavor to argue that we may find the first kind of dualism in the context of the Husserlian phenomenology. That kind of dualism gives rise to a broad ensemble of issues, which are all related to the classical cognitive pattern; first and foremost, the issue of *qualia*. This issue has been depicted as so *hard* that some scholars ended up conceiving it as absolutely unsolvable.

2 The *Hard Problem* of Matter

Whenever we employ the expression the "hard problem of consciousness" [20–27], we are referring to a problem of irreducibility: irreducibility of the qualitative, phenomenal features of a broadly constructed mental experience (expressible in

R. Lanfredini

terms of first-person perspective) to its functional and cognitive features (which are, on the contrary, expressible in terms of a third-person perspective). The word 'problem' testifies to a certain refractoriness or, more precisely, the residuality of something that possesses some sort of priority and definitional status, namely the conceptual definition of the mental universe.

Let us consider the following problem: does a falling tree make noise, even though there is nobody around who can hear it? The prompt answer is oftentimes taken for granted: certainly the tree produced a sound, although there was nobody in the surroundings listening to it. It could not be otherwise. But everybody knows that this naive answer, although quite understandable if we adopt a natural attitude, can be replaced by a less naive answer, which is nonetheless as plausible as the first one: the fall of the tree emits sound waves that, departing from a source, radiate outwards like concentric circles in water. If the waves are intercepted by a human ear, they are processed as the sound of a fall. If the sound waves are not intercepted, the sound is not perceived. Therefore, the effective emission of sound by the falling tree depends on what we mean by 'sound'. If we mean a heard noise, then the tree falls silently. On the contrary, if we mean "a distinctive spherical pattern of impact waves in the air, open to public inspection and measurement, given the right instruments" [28], then the falling tree emits a sound. While mentioning 'sound' in this second sense, we are referring to the *physical sound*; whereas, when speaking about 'sound' in the first sense, we are referring to *the experience of hearing a sound*. The latter has three components: (1) It depends essentially on the observer, namely it is reached by means of a private access [19]; (2) It essentially possesses a qualitative or phenomenal character (not just occasionally) [26]; (3) It is, in itself, not measurable. The behavioral responses to the sound experience can be measured, not the experience itself. Of course the hearing of the sound can be enhanced to include all the other kinds of experience: experience of colors, smells, tactile qualities and so on. All the problems that the philosophy of mind focuses on arise from these three components. Qualia are "those properties of mental states that type those states by what it is like to have them", and so "correspond to mental state-type" [21].

They give rise to the so-called *hard problem* of the philosophy of the mind: the problem of *qualia*. That is to say, the problem of how subjective states, which are qualitative and essentially tied to the subjectivities and their response (and, therefore, distinct from *quanta*, which are measurable, quantifiable and expressible in the third person), could spring from something that is no longer qualitative in its own nature, but quantitative and material. The geography of *qualia*, as outlined above, is wide and complex. It involves tactile, visual and auditory properties, as well as taste and olfaction; experiences of heat and cold; affective sensations such as pain, pleasure and other bodily sensations (itches, tickles); mental imagery, the sense of self and so on. But all these states have the same common denominator: privileged access, qualitative character and non- measurability. The problem of *qualia* and the incompatibility between description in first and third person is founded on a crucial hypothesis which, in the context of epistemology, has the value of an auxiliary methodological hypothesis. On the one hand, we have the 'external' or 'real' world, (sound waves, light radiations and so on) and the properties of its components (mass, shape, size, surface, motion) [29–31]. On the other hand, we have the 'internal' or 'phenomenal' world and its properties: touched objects, tasted flavors, heard sounds, seen colors and so on. In this light, we may assert that the problem of *qualia* perfectly resembles the traditional distinction between primary properties and secondary properties, between the way objects actually are and the way they are experienced. As Dennett puts it: "qualia is an unfamiliar term for something that could not be more familiar to each of us: the ways things seem to us. As is so often the case with philosophical jargon, it is easier to give examples than to give a definition of the term. Look at a glass of milk at sunset; the way it looks to you—the particular, personal, subjective visual quality of the glass of milk is the quale of your visual experience at the moment. The way the milk tastes to you then is another, gustatory quale and *how it sounds to you* as you swallow is an auditory *quale*; these various 'properties of conscious experience' are prime examples of *qualia*. Nothing, it seems, could you know more intimately than your own qualia" [32].

In spite of a widely held agreement, phenomenology itself formulates problems of residuality. These issues revolve around two different hard problems: the first one is that of transcendental ego, whereas the second one is that of matter. The latter provokes two distinct problems: the problem of *sensible matter* and that of *physical matter*. Lastly, we must consider that also the problem of sensible matter is a composite one, for it comprehends the problem of *sensorial hyle* (or the problem of *feeling*), and the problem of the *felt property*. When we talk about the problem of matter within the phenomenological tradition we then have three different concepts coming into play:

1. matter as material *hyle*, or material *content*
2. matter as felt *property*
3. inanimate matter (or physical matter)

The problem of matter, in its constitutive variety [33], is precisely the problem of qualia in philosophy of mind. It is the consequence of the fundamental distinction between what is prior and what is residual.

Let us start from the first problem, namely matter as material hyle. The problem of the sensorial *hyle* springs out of the distinction, clearly present in Husserl [34]: Fifth Investigation; [35], between intentional experience and feeling (the experience of feeling something). Mental experience is, according to Husserl, essentially if not exclusively, an intentional or cognitive experience [27, 36–42]. Intentional structure does not feature relational states of affairs—think, in this regard, of an asteroid falling on earth—nor directional yet non-intentional acts, such as the placing of a book on a table. Furthermore, that structure does not shape those experiences which are not directed towards anything, such as the pain caused by a fire or a panic attack. Intentionality is that property of the mind by means of which it is able to direct itself towards different objects, and this by virtue of its own structure and contents. It is precisely through that content that consciousness 'departs' from itself.

And it does this in a twofold way. On the one hand, directing itself towards its contents as in self-reflection, the mind considers itself as an object. On the other hand, directing itself towards objects which are not among its *Erlebnisse*, it directs itself towards something which is not actually contained in itself, namely the perceived object, or even the imagined object. In the latter case, the object is always given prospectively. Consciousness, with its intentional stance, is never a 'look from nowhere' [25], a bare and plain perception of the given object, but is always a perspectival look. As such, it is always incomplete. Conceiving intentionality as the conceptual core of the phenomenological inquiry implies the following aspects: (i) to focus the research primarily on the notion of *essence*; (ii) to individuate the essence of subjectivity in the notion of representation; (iii) to identify the essence of representation in the notion of determination. "Every act is either a representation or is grounded in a representation": this is considered by Husserl, and Brentano before him, the indisputable core of every phenomenological analysis.

Nevertheless, intentionality and its representational stance do not exhaust the notion of experience. The answer to the question about whether all conscious states are intentional states is, according to Husserl and contrary to Brentano, a negative answer.

3   Res Cogitans, Res Extensa, Res Viva

The Husserlian distinction between intentional content (structured and representational) and non- intentional content (hyletic and qualitative) echoes the Cartesian distinction between *intellectual cogitationes* (thinking, judging, desire) and *sensible cogitationes* (feeling, imagination). Descartes, and Husserl along with him, conceive imagination and sensation as clearly distinct from those cogitations that, adopting a phenomenological terminology, we might define as purely intentional (such as intellect, will and judgment). Furthermore, according to both Descartes and Husserl, the sensible cogitations convey *something more*. This expression is fundamental, for sensibility and imagination crucially depend on something that is both not essential and distinct from myself. This *something more* is, according to Descartes, the body. There is no feeling without a feeling body. And, without feeling, the thinking substance would be mutilated, merely reduced to a function.

> And, in doing so, I notice quite clearly that imagination requires a peculiar effort of mind which is not required for understanding: this additional effort of mind clearly shows the difference between imagination and pure understanding [*intellectionem puram*]. Besides, I consider that this power of imagining [*vis imaginandi*] which is in me, differing as it does from the power of understanding [*vis intelligendi*], is not required for my essence, that is, the essence of my mind. For if I lacked it, I should undoubtedly remain the same thing that I am now [*ille idem qui nunc sum*] (Descartes [], 2: 51).

Yet there is another irreducible entity beyond *res cogitans* and *res extensa*. We may call this entity *res viva*. In his account of the *res viva*, Descartes does not talk about essence, but rather about *human nature*. This nature is not the disembodied dimension of pure thought, but rather an inextricable bundle of material and spiritual components: "Nature also teaches me, through these sensations [*sensus*] of pain, hunger, thirst, etc., that I am not merely present to my body as a sailor is present to a ship, but most closely [*arctissime*] (*joined*) to it, and as if intermingled [*permixtum*] with it, [*and on that account I with it compose one thing*]. For otherwise, when the body is hurt [*laeditur*], I, who am nothing other than a thinking thing, would not on that account sense pain, but would perceive this injury [*laesionem*] with the pure intellect, as a sailor perceives by sight whether something is broken in his boat; and when the body needs food or drink, I would expressly understand this thing itself, and I would not have the confused sensation [*sensus*] of hunger and thirst. For surely these sensations of thirst, hunger, pain and so on are nothing other than certain confused modes of cognition, arising from the union of, and the as it were intermingling of, the mind with the body" (Descartes [43], 2: 56).

After having established a distinction between material and spiritual substance, Descartes comes to realize that the distinction gives rise to a substantial difficulty. The living body, namely the animated and pulsating matter—which remains separated—is the privileged place of interaction between different ontological dimensions of matter. Descartes himself acknowledges a considerable anomaly in his theoretical framework: human nature, which is, at the same time, union and fusion of mind and body, a human nature that we must account for in order to explain the two fundamental cogitations, namely feeling and imagining.

What is a thinking substance? Descartes' answer consists of a distinction between the formal, structural and functional dimension, on the one hand, and the material and hyletic dimension, on the other hand. The first dimension, which constitutes the very essence of the thinking substance, plays a definitory role; the merging of that formal feature with the material aspect corresponds to the nature of the thinking substance. The former (essence) is conceived as *disembodied*. The latter (nature) fully realizes itself only in its embodiment. The former keeps the distinction between immanence and transcendence, interior and exterior: the latter undermines such distinction. The former keeps the transparency of functionality, while the latter introduces considerable elements of opacity into the notion of subject. Both Descartes and Husserl strive towards the very same result: the individuation of a dimension of the cogito—or consciousness—which can always be objectively detached and be the subject of some sort of distant self-reflection. This possibility is grounded in the fact that this dimension does not have any traces of material existence. Descartes himself singles out quite lucidly the controlling function that reflective thought performs upon the more tacit one, in particular in the possibility of converting the feeling (the qualia, as we would call them) into *thinking of feeling*, or even *pretending to feel*.

As he says:

> And finally it is the same I who sense, or who observes corporeal things as if through the senses. For example, I am now seeing light, hearing a noise, feeling heat [*colorem sentio*]. But I am asleep, so all this is false [*falsa*]. Yet I certainly seem to see, to hear, and I to

become warm [*calescere*]. This cannot be false [*falsum*]; and this is properly [*proprie*] what is called sensing in me [*quod in me sentire appellatur*]; and this, precisely so taken [*praecise sic sumpum*], is nothing other than thinking (Descartes [43], 2: 19)

Since [*Cum*] I now know that even bodies are not strictly [*proprie*] perceived by the senses or the faculty of imagination but by the intellect alone [*solo intellectu*], and are not perceived from that they are touched or seen, but only from that they are understood [*intelligantur*], I clearly know [*aperte cognosco*] nothing more easily or evidently that can be perceived by me than my own mind (Descartes [43], 2: 22–23)

In conclusion, we may assert that feeling is, for Husserl as it was for Descartes, residual with respect to the intentional structure. If the phenomenological inquiry tends towards essence, it is clear that only the *structure* (we could even say the intentional architecture) can meet the requirements of this essentiality. Matter, understood in terms of material *hyle*, is the component which resists to the intentional *morphè*. In the context of this resistance, it limits the transparency and the representability of things.

4 Qualia and Plena

Let us take into account the second feature of phenomenological matter, namely matter as felt property, relative to the intended object. In this regard, the phenomenological counterpart of the term *quale*, or (the properties which are felt), is *plenum* (or *plena*) [45] Sects. 8 and 9. These are similar terms which, nevertheless, are the result of very different philosophical movements, so to speak. The term 'qualia' refers to everything which can be qualified. Traditionally, it corresponds to the so-called secondary properties which, unlike the primary ones, are subjective, dependent upon a particular situation and point of view, resistant to quantification and measurement. The phenomenological term 'plena' refers to the *filling* of the intuitive act which fills an empty act every time we say that we perceived something, imagined something, or even daydreamed something.

Plena are filling properties. Husserlian *plenum* is, however, similar to the *quale*, for a substantial undifferentiation characterizes both of them. There is clearly a descriptive difference between visual and musical *qualia*. It is a difference that, according to Husserl, can be reduced to different material, or better, regional, ontologies. For this reason, converting a scarlet red into a violin sound turns out to be impossible, whereas it is entirely possible to convert a scarlet red into a crimson one, or even an ocean blue. Evidently the point is not just avoiding confusion between qualitative differences: there are different regional boundaries that cannot be crossed qualitatively. However, according to Husserl, they share a common ground. And it is exactly this common ground that allows both phenomenologists and philosophers of mind to talk about sensible qualities, although with different aims. This common ground has, as far as I can see, two facets: on the one hand, both *plena* and *qualia* (*qualia* as opposite to *quanta*, *plena* as opposite to emptiness), thus underlying the priority of the structural, functional and quantifiable dimension. On the other hand (and this second facet is related to the first one), both plena and qualia need something else to manifest themselves, and this something else is *extension*.

Each plenum, according to Husserl, fills an extension. Surely this is the case for colors (a color without extension is simply impossible, this being the most infamous example of material a priori). But it is also the case for sounds and tactile qualities. Both in phenomenology and philosophy of mind qualia are those properties that qualify an extension, this phenomenal aspect which emerges from an extension which is not sensibly qualified. It is a conception that we may call, once again, residual, for it depends crucially upon a substantive spatial conception. Furthermore even for Husserl, in the foundation relation existing between extension and plena and in its consequent constitution of an independent part, the latter acquires from extension that divisibility which they cannot have in themselves (it is impossible to divide 'the red'). This is testimony, once again, to the priority and the fundamental character of extension, just as quanta are prior to qualia in the philosophy of mind.

In more than one setting, Husserl refers to the intuitive properties as a veil which covers an extension, providing it with a certain qualification. According to Husserl, plena are fundamental (if there were no plena we would only have some sort of empty spectral phenomena) but they are not essential: only extension is truly essential. If the function and role of qualia is that of qualifying an extension, the function and role of plena is that of filling one. Without plena the phenomenon would be destined to lose its boundaries, its shape. Once again the notion of matter is conceived as residual with respect to a functional architecture or map, in this case the extension which the plenum fills. Matter, in this context (that is, when it is understood as a material plenum), is conceived as something more, something inessential that is added to what essentially constitutes the object: it is true that, without any plena, the object would be 'an empty something'; nevertheless, it is precisely that 'being an empty something' which defines it as an object.

5 Space and Time Metaphors

At this stage of our analysis, I would like to consider a final notion of matter, namely inanimate or physical matter. In this sense, materiality, despite its having spatial extension as an essential feature—as it was for Descartes—cannot be reduced to it. There are two essential components of materiality: duration and relational context. Duration is illustrated by Husserl as a relation between the temporal determination and a real feature which fills the duration and extends itself through it [44, 46]. In this sense, and only in this sense, each feature, when it comes to its content, is necessarily subject to change throughout its duration in an a priori way: things change when the temporal plena of their duration change either discretely or continuously. Whereas, when this condition does not occur, things stay the same.

The fact that the materiality of things depends upon the circumstances and the context in which they are placed cannot be disputed. If we consider a thing by itself, distinguishing between something and its phantom, i.e. its pale, empty and ghostly counterpart, becomes virtually impossible. The ghost of a certain thing has all the essential features that render that thing exactly what it is and nothing else: essential features that are dispersed throughout an extension. In this sense, we would see rainbows and blue skies but we could not define them as material things. On the other hand, if we consider the thing within a given context, the thing and the ghost of that thing cannot be regarded as the same element. Things exist, are real, substantial and causal (these terms are all synonyms) when they behave in a certain way. In this sense, real (or material) properties are, *ipso facto*, functional links: for example causal links. In order to get to know the reality of a given thing, we must be able to predict its behavior under a certain force, pressure, when it is smashed up, cooled down, heated up and so on. In the multiplicity of its dependence relations the real thing will retain its own identity. Therefore, inanimate or physical matter is once again, according to Husserl, a purely relational and functional concept; and, in this context, this aspect is even clearer.

The entire phenomenological analysis of consciousness and matter is pervaded and justified by an ensemble of *spatial metaphors*: the psychic life is described, indeed, as a *stream* or *field* of consciousness; Husserl refers to a pure *dimension* of immanence; the ontology of psychic experience is described (employing a geographical analogy) as a *region*; the notion of *stratum* turns out to be absolutely crucial; and, lastly, the notions of *perspective*, *adumbration* and *backcloth* (which are employed in this context) are all *spatial notions*. In this respect, the phenomenological notion of matter is not an exception: Husserl refers, indeed, to a *space* of understanding in relation to its contents, to a *space* of extension with regard to the intended object, and to the *space* of causal networks while considering physical matter. This space is filled each time by a material content: the material *hyle* in the case of the experience, the plenum in the case of the intended object, and the causal links for the physical matter.

It is exactly because of the spatial metaphor that the intuitive notion of matter collapses (through its own disembodiment and through its own taming, so to speak) into the notion of material thing. The material thing is reduced to its essential features, in their twofold facets of determinations and qualifications. Furthermore, the material thing is constituted by different layers: the layer of the material, inanimate thing is essentially distinct from the layer of the spiritual thing, mostly because of its fragmentability or divisibility—the latter being a disposition which matter acquires by means of spatial extension. Nevertheless, we must consider that matter is not entirely reducible to extension. On the contrary, matter is *something more* than its extension, something more than its being simply a thing.

A systematic account of this *something more* eludes Husserl. He does not get a clear and thorough grasp on the primary and pre-categorical character that characterizes our intuitive notion of matter; its resistance and independence. According to Husserl, matter remains (almost paradoxically) an unessential residual of the definition of material thing (extension on the one hand, function on the other hand).

In this sense, according to Husserl, Descartes and the contemporary philosophers of mind, matter actually turns out to be an actual problem, precisely because phenomenology does not capture the intuitive aspect of this notion, namely its being flesh and blood.

To sum up, the problem of matter underlines what we might call a *logic of residuum*, which involves a profound, paradigmatic distinction between primary and essential aspects (structure, function, form and so on) and secondary or inessential aspects (fullness, pre-cognitive, pre-categorical, embodiment, materiality, vagueness and so on). A possible solution to the problem of matter and to the problem of consciousness requires some sort of overturning, a paradigm shift, precisely as stated in Kuhn's framework [47, 48]. This shift reintroduces a material, tacit, unexpressed, but also temporal, active, integrated and intertwined dimension within the phenomenological perspective. This new paradigm is founded on a new metaphor that we might call a *temporal metaphor*.

6 Sensing, Flesh and Motion

The new approach mentioned above calls into question three crucial notions: the concept of sensing, the concept of flesh and, above all, the concept of motion. The first notion, that of sensation, replaces the original notion of consciousness. Sensation, in this context, is neither a *quale* nor an immanent content (*hyle*), but rather an integrated and intertwined unity of matter and spirit. As stated by Merleau-Ponty:

> There are two ways of being mistaken about quality: one is to make it into an element of consciousness, when in fact it is an object for consciousness, to treat it as an incommunicable impression, whereas it always has a meaning; the other is to think that this meaning and this object, at the level of quality, are fully developed and determinate [49].

And again:

> The pure quale would be given to us only if the world were a spectacle and one's own body a mechanism with which some impartial mind made itself acquainted. Sense experience, on the other hand, invests the quality with vital value, grasping it first in its meaning for us, for that heavy mass which is our body, whence it comes about that it always involves a reference to the body (Merleau-Ponty 46).

The notion of flesh replaces, here, the original notion of body. In phenomenology, the concept of body as *Körper* is tightly connected with the notion of extension whereas the concept of body as *Leib* is tightly connected with the notion of kinesthesis. Once again, and in both cases, we are dealing with a notion that underlines (contemplates, expresses and conveys) a formal, and functional aspect of body. On the contrary, the notion of flesh (*chair* in Merleau-Ponty) guarantees the presence of immanence and transcendence both in the stream of consciousness and in the matter. Several times Merleau-Ponty asserts that the notion of flesh is an "ultimate notion"; a "concrete emblem of a general manner of being, which provides access both to subjective experience and objective existence" [50].

The phenomenon he focuses on is that of 'touching one hand with the other hand' (Merleau-Ponty 133–4). This phenomenon, he suggests, reveals to us the two dimensions of our 'flesh', namely that it is both a form of experience (tactile experience) and something that can be touched. It is both 'touching' and 'tangible'.

Furthermore, the relationship is reversible: the hand that touches can be felt as touched, and vice versa, and according to Merleau-Ponty, this 'reversibility' is precisely what constitutes the essence of flesh. This component is crucial, for it shows the ambiguous status of our bodies as both subject and object. Husserl's influence is evident here. Actually in *Ideas II* he states that "the Body as Body presents, like Janus, two faces" [33]. But Husserl considers these two faces of the body as *strata*. On the contrary, in the example of the handshake, the crisscrossing of touching and the tangible incorporate themselves into the same world: the two systems—Merleau-Ponty says, "are applied upon one another, as the two halves of an orange" and "we say therefore that our body is a being of two leaves", "because the body belongs to the order of things as the word is universal flesh". By and large, Merleau-Ponty transforms the *correlative analysis* (typical of the Husserlian phenomenology, in which the structure of consciousness is the basic element) into a *bilateral analysis* according to which both the subjective and objective poles require a foundational priority. Accordingly, he extends the methodological approach from a perspective that privileges the *external frame* of the experience and its sensory fulfilment to a perspective that privileges the *actual content* and its interlacing with the material body.

This bilateral perspective aims at overcoming the distinction between intentional (or representational) structure and sensible matter and, more generally, the epistemological model based on a face-to-face pattern: mind-body, spirit-matter, outer-inner, active-passive, organism- environment and so on. Merleau-Ponty's proposal takes the shape of an *intertwined paradigm*, namely circular and reversible. Within this paradigm sensibility functions as an intersection between form (mind) and matter (body), whereas the notion of flesh aims at rejecting the distinction between extension (essence) and *plena* (accident). In the first case, the weakened notion is that of representation, in the second case that of represented object. In both cases Merleau-Ponty's paradigm undermines the very core of the representational pattern, namely the absolute centrality of the notion of determination. In this respect, Heidegger writes:

> The question asks about being. What does being mean? Formally, the answer is: Being means this and this. The question seeks an answer which determines something which is somehow already given in the very questioning. The question is what I called *a question of definition*. It does not ask whether there is anything like being at all, but rather what is meant by it, what is understood under it, under "being". When we thus ask about the sense of being, then being, which is to be determined, is in a certain way already understood. In a certain way: here this means according to a *wholly indeterminate pre-understanding*, an indeterminacy whose character can however be phenomenologically grasped. [51].

Overcoming the determinative (or attributive) conception, in favor of plasticity and notions such as reversibility, environmentality, pre-categoriality, implies that the foundation of the mental is not to be sought in the notion of representation, but rather in that of motion and, therefore, of time.

Merleau-Ponty's perspective about matter does not disclose the priority of structure (structure of the representation, structure of extension, structure of relations) but rather the priority of motion, which is indivisible in itself. The operation of building movement out of immobility is completely justified when it comes to action, but is rather misleading from a theoretical perspective. The radical overturning we just mentioned consists in emphasizing the component that, so far, had been regarded as residual: *hyle*, namely the plenum, the impact, so to speak, of matter. We might say, quoting Gibson [10], that the objects of the physical world are not sets of properties but a cluster of affordances, opportunities, motions. Nevertheless the concept of motion is not conceived as *displacement*, as movement from one place to another, but as moving *process* [7].

This shifts the focus of the philosophical analysis to the notion of duration (a temporal metaphor rather than a spatial one) and, along with that, to notions such as difference, embodiment, context. All these notions hint at the abandonment of the distinction between form and content, interior and exterior, inside and outside in favor of their integration and fusion. The result is the rejection of a dualistic paradigm, and the adoption of an interactive model, which in turn leads to an alternative conception of matter.

In conclusion, we may assert that the classical representational and cognitive pattern (but also the classical phenomenological pattern) gives rise to a problem concerning the qualitative dimension. This problem is essentially related to the notion of matter, conceived as residual with respect to the notion of form: the sensorial *hyle* is residual with respect to the intentional form; *plena* are residual with respect to extension, and physical matter (its resistance, its non-undifferentiation, its endurance) is also residual with respect to the broad ensemble of connections in which the physical thing is inscribed. The residual component which characterizes the notion of matter is simply the other side of the absolute predominance of form (*representational form* in the specific context of mental phenomena). This predominance necessarily gives rise to a problem (both in the context of phenomenology and in philosophy of mind): the problem of the ontological status of the qualitative states.

This issue is a crucial one and, in order to be solved, requires a radical change of perspective. In the context of phenomenology this change takes the shape of specific notions, such as enactive, embodied and situated mind [12–14, 52–54].

In conclusion: conceiving the concept of matter in this way requires the adoption of a paradigm that is no longer spatial but temporal [55–57]. And it is certainly not by chance that, in the course of his reflections on nature during his final years, Merleau-Ponty and Séglard [58] turned with renewed attention to the scientific revolutions concerning the concept of time in physics and, above all, in contemporary biology. In biological time the primacy accorded to impressional consciousness (and in consequence, to the notions of *datum*, fixity, immobility, arrest) is replaced by the primacy accorded to that *"masse intérieurement travaillée"* offered by the flesh (and in consequence, to notions of tendency, movement, duration). All these notions imply a temporal paradigm (and not a spatial one); they allude to a dynamic, non-static pattern, to a paradigm that, instead of being projective (as in the representational model), is *agentive* and, ultimately, evolutive.

References

1. Clark, A.: Supersizing the Mind. Oxford University, Oxford (2008)
2. Noë, A.: Vision and Mind. Cambridge University Press, Cambridge (2010)
3. Bitbol, M.: Science as if situation mattered. Phenomenol. Cognit. Sci. **1**, 181–224 (2002)
4. Bitbol, M.: Is consciousness primary? NeuroQuantology **6**(1), 53–71 (2008)
5. Bitbol, M.: Neurophenomenology, an ongoing practice of/in consciousness. Constr. Found. **7** (3), 165–173 (2012)
6. Noë, A., Thompson, E.: Vision and Mind: Selected Readings in the Philosophy of Perception. MIT Press, Cambridge, Mass. (2002)
7. O'Regan, J.K.: Sensorimotor approach to (phenomenal) consciousness. In: Baynes, T., Cleeremans, A., Wilken, P. (eds.) Oxford Companion to Consciusness, pp. 588–593. Oxford University Press, Oxford (2009)
8. Blackmore, S.J.: The question is: who am I? J. Am. Soc. Psych. Res. **96**, 143–151 (2002)
9. Legrand, D.: The bodily self: the sensory-motor roots of pre-reflexive self-consciousness. Phenomenol. Cognit. Sci. **1**, 90–135 (2006)
10. Gibson, J.J.: The Ecological Approach to Visual Perception. Houghton Mifflin Heil, Boston (1979)
11. Beer, R.D.: Dynamical approaches to cognitive sciences. Trends Cognit. Sci. **4**(3), 91–99 (2000)
12. Thompson, E., Varela, F.J.: Radical embodiment: neural dynamics and consciousness. Trends Cognit. Sci. **5**(10), 418–425 (2001)
13. Varela, F.J.: Neurophenomenology: a methodological remedy for the hard problem. J. Conscious. **3**(4), 330–349 (1996)
14. Varela, F.J., Thompson, E., Rosch, E.: The Embodied Mind: Cognitive Science and Human Experience. MIT Press, Cambridge (1991)
15. Weber, A., Varela, F.J.: Life after Kant: natural purposes and the autopoietic foundations of biological individuality. Phenomenol. Cognit. Sci. **1**, 97–125 (2002)
16. Froese T.: Breathing new life into cognitive science. J. Philos.—Interdisc. Vanguard, **2**(1), 95–111 (2011a)
17. Froese, T.: From second order cybernetics to enactive cognitive science: Varela's turn from epistemology to phenomenology. Syst. Res. Behav. Sci. **28**, 631–645 (2011)
18. Hutto, D., Myin, E.: Radicalizing Enactivism: Basic Minds Without Content. The MIT Press, Cambridge (2013)
19. Nagel, T.: What Is It Like to Be a Bat? Philos. Rev. **83**(4), 435–450. In: Noë, A. (2010). Vision and Mind, Cambridge, Cambridge University Press (1974)
20. Chalmers, D.J.: Facing up to the problem of consciousness. J. Conscious. Stud. **2**(3), 200–219 (1995)
21. Chalmers, D.J.: The Conscious Mind. Oxford University Press, Oxford (1996)
22. Chalmers, D.J.: Moving forward on the problem of consciousness. J. Conscious. Stud. **4**, 3–46 (1997)
23. Chalmers, D.J.: Consciousness and its place in nature. In: Stich, S., Warfield, F. (eds.) Blackwell Guide to Philosophy of Mind, pp. 1–46. Blackwell, UK (2003)

24. Crane, T.: The origins of qualia. In: Crane, T., Patterson, S.A. (eds.) The History of the Mind-Body Problem, pp. 169–194. Routledge, London (2000)
25. Crane, T.: Elements of Mind. Oxford University, Oxford (2001)
26. Jackson, J.: Epiphenomenal qualia. J. Philos. **83**, 127–136 (1982)
27. Searle, J.: The Rediscovery of the Mind, Massachussetts: Massachussetts Institute of Technology Seron S. (2010). Perspectives récentes pour une phénoménologie de l'intentionnalité, Bulletin d'analyse phénoménologique, **6**(8), 162–191 (1992)
28. Heil, J.: Philosophy of Mind. A Contemporary Introduction, Routledge, NY (2013)
29. Churchland, P.M.: Eliminative materialism and the propositional attitudes. J. Philos. **78**(2), 67–90 (1981)
30. Churchland, P.M.: Reduction, qualia, and direct introspection of brain sciences. J. Philos. **82** (1), 8–28 (1985)
31. Churchland, P.M.: The Engine of Reason, the Seat of the Soul. The MIT Press, Cambridge (1995)
32. Dennett, D.C.: Quining Qualia. In: Marcel, A.J., Bisiach, E. (eds.) Consciousness in Contemporary Sciences. Oxford University Press, Oxford (1988)
33. Husserl, E.: Ideen zu einer reinen Phänomenologie und einer phänomenologischen Philosophie, Zweites Buch, Phänomenologischen Untersuchungen zur Konstitution, Husserliana IV. Martinus Nijhoff, Den Haag (1929)
34. Husserl, E.: Logische Untersuchungen. Halle, Max Niemeyer (1901)
35. Husserl, E.: Ideen zu einer reinen Phänomenologie und einer phänomenologischen Philosophie: Allgemeine Einführung in die reine Phänomenologie, Husserliana, III/1 e III/2, p. 1976. Martinus Nijhoff, Den Haag (1913)
36. Horgan, T., Tienson, J.: The intentionality of phenomenology and the phenomenology of intentionality. In: Chalmers, D. (ed.) Philosophy of Mind: Classical and Contemporary Readings, pp. 520–533. Oxford University Press, Oxford (2002)
37. Horgan, G., Tienson, J.: Consciousness and intentionality. In: Velmans, M., Schneider, S. (eds.) The Blackwell Companion to Consciousness, pp. 468–484. Blackwell, UK (2007)
38. Kriegel, U.: Subjective Consciousness: A Self-representational Theory. Oxford University Press, Oxford (2009)
39. Kriegel, U.: The phenomenal intentionality research program. In: Kriegel, U. (Ed.), Phenomenal Intentionality, Oxford University Press, Oxford (2013)
40. Loar, B.: Phenomenal intentionality as the basis of mental content. In: Hahn, M., Ramberg, B. (eds.) Reflections and Replies: Essays on the Philosophy of Tyler Burge. MIT Press, Cambridge (2003)
41. Siewert, C.: Phenomenality and Self-consciousness. In: Kriegel, U. (ed.) Phenomenal Intentionality. Oxford University Press, Oxford (2013)
42. Zahavi, D.: Intentionality and phenomenality. A Phenomenological Take on the Hard Problem. In Thompson, E. (Ed.), The Problem of Consciousness: New Essays in Phenomenological Philosophy of Mind. Canadian Journal of Philosophy, Supplementary vol. 29, pp. 63–92 (2003)
43. Descartes R.: The Philosophical Writings of Descartes (1641). In: J. Cottingham, R. Stoothoff, D. Murdoch and A. Kenny, (eds.), 3 vols. Cambridge University Press, Cambridge (1991)
44. Husserl, E.: Analysen zur Passiven Synthesis. Kluwer Academic Publishers Dordrecht (1926), 1966
45. Husserl E.: Die Krisis der europäischen Wissenschaften und die transzendentale Phänomenologie, Husserliana VI, 1959 (1937)
46. Husserl, E.: Zur Phänomenologie des Inneren Zeitbewusstseins. Martinus Nihoff, The Hague (1917)
47. Kuhn, T.: The Structure of Scientific Revolutions. University of Chicago Press, Chicago (1962)
48. Kuhn, T.: Logic of discovery or psychology of research? In: Schilpp, P.A. (Eds.), The Philosophy of Karl Popper, vol. II, pp. 798–819. The Open Court Publishing Company, La Salle (1974)

49. Merleau-Ponty, M.: Phenomenology of Perception. Routledge, London (1945)
50. Merleau-Ponty, M.: The Visible and the Invisible. Northwestern University Press, Evanston (1964)
51. Heidegger, M.: History of the Concept of Time: Prolegomena. Indiana University Press, Indiana, 2009 (1975)
52. Hanna, R., Thompson, E.: The mind-body problem. Theoria et Historia Scientiarum **7**(1), 23–42 (2003)
53. Thompson, E.: Mind in Life: Biology, Phenomenology, and the Sciences of Mind. Harvard University Press, Harvard (2010)
54. Thompson, E.: Précis of mind in life: biology, phenomenology and the Sciences of mind. J. Conscious. Stud. **18**(5–6), 1–13 (2011)
55. Longo, G., Montévil, M.: Protention and retention in biological systems. Theory Biosci. **130** (2), 107–117 (2011)
56. Longo, G., Montévil, M.: Perspectives on Organisms: Biological Time. Springer, Symmetries and Singularities, Berlin (2014)
57. Longo, G., Montévil, M., e Pocheville A.: From bottom-up approaches to levels of organization and extended critical transitions. Front. Physiol. **3**, 232 (2013)
58. Merleau-Ponty, M., Séglard, D.: Nature: Course Notes from the Collège de France. Northwestern University Press (2003)

Part V
Logical Perspectives

From the Structures of Opposition Between Similarity and Dissimilarity Indicators to Logical Proportions

A General Representation Setting for Capturing Homogeneity and Heterogeneity

Henri Prade and Gilles Richard

Abstract Comparative thinking plays a key role in our appraisal of reality. Comparing two objects or situations A and B, described in terms of Boolean features, may involve four basic similarity or dissimilarity indicators referring to what A and B have in common (positively or negatively), or to what is particular to A or particular to B. These four indicators are naturally organized into a cube of opposition, which includes two classical squares of opposition, as well as other noticeable squares. From the knowledge of one situation A, it is possible to recover the description of another one, B, provided that we have enough information about the comparison between A and B. Then comparison indicators between A and B can be equated with comparison indicators between two other situations C and D. A conjunction of two such comparisons between pairs (A, B) and (C, D) gives birth to what is called a logical proportion. Among the 120 existing logical proportions, 8 are of particular interest since they are independent of the encoding (positive or negative) used for representing the situations. Four of them have remarkable properties of homogeneity, and include the analogical proportion "A is to B as C is to D", while the four others express heterogeneity by stating that "there is an intruder among A, B, C and D, which is not X" (where X stands for A, B, C or D). Homogeneous and heterogeneous logical proportions are of interest in classification, anomaly detection tasks and IQ test solving.

H. Prade (✉) · G. Richard
IRIT, Université Paul Sabatier, 118 route de Narbonne, 31062
Toulouse Cedex 09, France
e-mail: prade@irit.fr

G. Richard
e-mail: richard@irit.fr

© Springer International Publishing AG 2017
G. Dodig-Crnkovic and R. Giovagnoli (eds.), *Representation and Reality in Humans,
Other Living Organisms and Intelligent Machines*, Studies in Applied Philosophy,
Epistemology and Rational Ethics 28, DOI 10.1007/978-3-319-43784-2_14

1 Introduction

The role of comparison in our perception of reality has been recognized for a long time. Making comparison is closely related to similarity judgment [24] and analogy making [9]. In particular, comparison applied to numerical quantities is a matter of differences and ratios, which, by equating differences or ratios, leads to the idea of a proportion. Ancient Greek mathematicians, such as Archytas of Tarentum, or his follower Eudoxus of Cnidus, already studied mathematical proportions, including arithmetic, geometric and harmonic ones.

Numerical proportions involve four terms A, B, C and D. Arithmetic (resp. geometric) proportions state the equality of two differences (resp. ratios) between two ordered pairs (A, B) and (C, D) of numbers. Harmonic proportions combine geometric and arithmetic comparisons by stating that $A/D = (A - B)/(C - D)$. Associated with each type of proportion is a particular *mean* operation, obtained by taking $B = C$ as an unknown number in the equality stating the proportion.

Aristotle, following Eudoxus of Cnidus, does not only refer to geometric proportions, but also considers comparative relations between four symbolic (i.e. nonnumerical) terms that form in modern words an *analogical proportion*, namely a statement of the form "A is to B as C is to D" expressing an identity of relation between the ordered pairs (A, B) and (C, D) of symbols (where now A, B, C and D refer to objects or situations). The use of pictorial tests based on analogical proportions has often been used in psychological studies on the development of human thought [15].

However, it is only recently that analogical proportions have been cast in a logical setting, when A, B, C and D refer to four situations described in terms of Boolean features [12]. The logical expression of an analogical proportion exactly states that "A differs from B as C differs from D, and B differs from A as D differs from C". Thus an analogical proportion equates the dissimilarity of A with respect to B (resp. of B with respect to A) with the dissimilarity of C with respect to D (resp. of D with respect to C). This modeling provides a way of extrapolating D from A, B and C, as in the numerical setting, and to validate a computational procedure first discovered by Sheldon Klein [11], where the analogical proportion is supposed to hold, in a pointwise manner, for each feature, between the four Boolean values corresponding to A, B, C and D. However, A, B and C cannot always be completed with a D in order to make a valid analogical proportion. But if it exists, D is unique.

As already mentioned, completing an analogical proportion where the last item is missing is the basis of psychological tests. This mechanism may also be useful for solving more sophisticated IQ tests such as Raven's progressive matrices tests. Such tests are built from a 3×3 array with eight pictures in the first eight cells while the ninth picture has to be completed. The analogical proportion mechanism enables us to directly build the missing picture from the eight given pictures [3], rather than choosing it from among a set of candidate pictures as in the real test.

Then, the logical expression of an analogical proportion has been shown to be an important particular case of the remarkable notion of *logical proportion* which has been recently introduced [17, 19, 20]. However, the phrase "logical proportion" was first coined by [14] in a more restrictive sense for naming a quaternary relation between Boolean propositions, inspired from geometrical proportions. It turns out that the relation defined by Piaget is one of the noticeable expressions of an analogical proportion [19], a fact remained unnoticed by Piaget. The general form of a logical proportion is a conjunction of two equivalences between *comparison indicators* expressing similarities or dissimilarities pertaining to pairs of Boolean variables encoding the values of a considered feature for the pairs (A, B) and (C, D).

A form of comparative reasoning, simpler than the extrapolation of D from A, B and C, amounts to reconstruct, feature by feature, of the description of an object B from the value of some comparison indicators with respect to another object A, knowing A. This latter type of reasoning can be used as a starting point for providing a new introduction to the idea of a logical proportion, from the notion of comparison indicators. Besides, a natural question is then to identify the logical proportions that have extrapolative power, i.e. which lead to a unique solution for the fourth situation, knowing the three others, and assuming that a particular logical proportion holds for each feature between the four situations.

The paper is organized as follows: In Sect. 2, we investigate how we can compare an object A with an object B, thanks to four types of comparative information. We show that this gives birth to a cube of opposition, which includes two classical squares of opposition, and exhaustively exhibits the relations between the possible values of the four comparison indicators. These indicators are also shown to be necessary and sufficient for describing the respective configurations of two sets. Moreover, we determine, by easy Boolean calculations, in what cases the description of an object A may be recovered from the description of another object B and the knowledge of the values of two comparative indicators relating them. This establishes a link with logical proportions where two comparison indicators pertaining to a pair (A, B) are respectively equated with two comparison indicators pertaining to a pair (C, D). In Sect. 3, we provide a short background on logical proportions, highlighting some of their remarkable properties, and identifying eight logical proportions of particular interest: four homogeneous ones that include the analogical proportion and are based on equivalences between indicators of the same nature (referring either to similarity or dissimilarity), and four heterogeneous ones that encode the presence of an intruder among four items. In Sect. 4, we investigate the way to solve Boolean equations involving logical proportions, and identify the proportions suitable for extrapolation. It turns out that this is exactly the eight previously mentioned ones. In Sect. 5, after listing and discussing problems of interest in relation to extrapolation, we present a generic transduction rule, which, when combined with the equation solving process, provides a general basis for extrapolation. Lastly, we briefly discuss the respective merits of homogeneous and heterogeneous logical proportions for classification and prediction, or intruder detection. The contents of this article is partly based on a workshop paper [21].

2 Indicators as Comparative Descriptors

Given a collection $\mathcal{U}$ of Boolean properties, it is common practice to describe an object A by a set a of properties that this object satisfies. Thus, a is a representation of A, where $a \subseteq \mathcal{U}$. Note that $a = \emptyset$ or $a = \mathcal{U}$ are not forbidden, since A may have none of the properties in $\mathcal{U}$, or all of them. Having all the characteristics of A w.r.t. $\mathcal{U}$ means that we have complete knowledge of a (w.r.t. the conceptual space induced by $\mathcal{U}$).

When comparing objects A and B, one looks for their similarities and dissimilarities. More precisely, there are only four types of comparative information between two objects A and B with respect to $\mathcal{U}$ (in the following $\overline{a}$ denotes the set complement of set a):

- The set of properties that A and B share: $a \cap b$
- The set of properties that A has but that B does not have: $a \cap \overline{b}$
- The set of properties that B has but that A does not have: $\overline{a} \cap b$
- The set of properties that neither A nor B have: $\overline{a} \cap \overline{b}$

Thus, these four expressions provide comparative information regarding A and B, and we call them *set indicators* or *indicators* for short. We have two types of indicators:

- $a \cap b$ and $\overline{a} \cap \overline{b}$, telling us about properties that both A and B have, or that both A and B do not have; they are called *similarity indicators*.
- $a \cap \overline{b} = a \setminus b$ and $\overline{a} \cap b = b \setminus a$, telling us about properties that only one among A and B has; they are called *dissimilarity indicators*.

We notice that the union of the four indicators is just $\mathcal{U}$ and the intersection of any two indicators is empty. This is reminiscent of the well-known work of Amos Tversky [24], taking into account the common features, the specificities of A w.r.t. B and the specificities of B w.r.t. A, respectively, modeled by $a \cap b$, $a \setminus b$ and $b \setminus a$ in order to define a global measure of similarity. However, here, we are rather interested in keeping track of in what respect items are similar and in what respect they are dissimilar using Boolean indicators in a logical setting.

2.1 The Cube of Opposition of Comparison Indicators

Aristotle and his followers noticed that universally and existentially quantified statements, which are encountered in syllogisms, can be organized into a *square of opposition*. More precisely, consider a statement (**A**) of the form "all P's are Q's", which is negated by the statement (**O**) "at least a P is not a Q", together with the statement (**E**) "no P is a Q", which is clearly in even stronger opposition to the first statement (**A**). These three statements, together with the negation of the last statement, namely (**I**) "at least a P is a Q", give birth to the square of opposition [13], traditionally

Fig. 1 Square of opposition

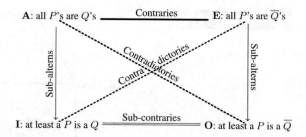

denoted by the letters **A**, **I** (affirmative half) and **E**, **O** (negative half), pictured in Fig. 1 (where $\overline{Q}$ stands for "not Q").

As can be checked, noticeable relations hold between the four vertices of the square:

1. **A** and **O** are the negation of each other, as are **E** and **I**;
2. **A** entails **I**, and **E** entails **O** (we assume that there is at least a P to avoid existential import problems);
3. **A** and **E** cannot be true together, but may be false together;
4. **I** and **O** cannot be false together, but may be true together.

Negating the predicates, i.e. changing P into $\overline{P}$ and Q into $\overline{Q}$, leads to another similar square of opposition **a**, **i**, **e**, **o**, based on "not-P's" assumed to constitute a non-empty set. Then the eight statements, **A**, **I**, **E**, **O**, **a**, **i**, **e**, **o** may be organized in what may be called a *cube of opposition* [6] as in Fig. 2. The front facet and the back facet of the cube are traditional squares of opposition, where the thick nondirected segment relates the contraries, the double thin nondirected segments the subcontraries, the diagonal nondirected segments the contradictories and the vertical unidirected segments point to subalterns, and express entailments.

Assuming that there are at least a P and at least a not-P entails that there are at least a Q and at least a not-Q. Then, we have that **A** entails **i**, **a** entails **I**, **e** entails **O**, and **E** entails **o**. Note also that the vertices **a** and **E**, as well as **A** and **e**, cannot be

Fig. 2 Cube of opposition

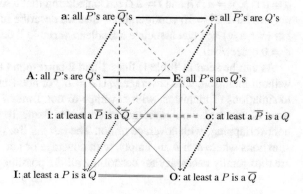

Fig. 3 Cube of opposition
of comparison indicators

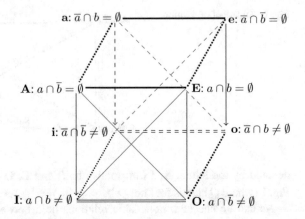

true together, while the vertices **i** and **O**, as well as **I** and **o**, cannot be false together.
Lastly note that there is no logical link between **A** and **a**, **E** and **e**, **I** and **i**, or **O** and **o**.

Going back to the set of properties a and b describing two situations A and B, the
cube of opposition may be rewritten in terms of the four set comparison indicators,
which may be empty or nonempty. The four indicators appear in the side facets of
the cube. The nonclassical squares of opposition of the side facets (with a different
display of the vertices) were already discussed in [19], without referring to any cube.
See Fig. 3. Note that we assume $a \neq \emptyset$, $\overline{a} \neq \emptyset$, $b \neq \emptyset$ and $\overline{b} \neq \emptyset$ here to avoid the
counterpart of the existential import problems, since now the situations A and B play
symmetric roles in the statements associated with the vertices of the cube.

2.2 Possible Configurations of Two Subsets in Terms
of Set Indicators

Let us here denote by s, t, u, v the four set indicators pertaining to a and b: $s = a \cap b$,
$u = a \cap \overline{b}$, $v = \overline{a} \cap b$ and $t = \overline{a} \cap \overline{b}$. Considering that a set indicator can be empty or
not, we have $2^4 = 16$ configurations that we describe in Table 1. To each indicator
$i \in \{s, t, u, v\}$, we can associate a Boolean variable i' defined as: $i' = 1$ if $i \neq \emptyset$ and
$i' = 0$ otherwise.

As can be seen in Table 1, lines 1 and 2 correspond to situations of overlapping
without inclusion, with coverage ($a \cup b = \mathcal{U}$) or not. Lines 3, 4, 5 and 6 correspond
to situations of inclusion, with coverage or not. Lines 7 and 8 correspond to situations of equality, with coverage or not. Lines 9 and 10 correspond to situations of
nonoverlapping, with coverage or not. The last six lines correspond to pathological
situations where a or b are empty, with coverage or not. The four indicators s, t, u, v
are thus jointly necessary for describing all the possible situations pertaining to the

Table 1 Respective configurations of two subsets

	Configuration	$s = a \cap b \neq \emptyset$	$u = a \cap \bar{b} \neq \emptyset$	$v = \bar{a} \cap b \neq \emptyset$	$t = \bar{a} \cap \bar{b} \neq \emptyset$
1	$a \cap b \neq \emptyset; a \not\subseteq b;$ $b \not\subseteq a; a \cup b \neq \mathcal{U}$	1	1	1	1
2	$a \cap b \neq \emptyset; a \not\subseteq b;$ $b \not\subseteq a; a \cup b = \mathcal{U}$	1	1	1	0
3	$b \subset a \subset \mathcal{U}$	1	1	0	1
4	$b \subset a; a = \mathcal{U}$	1	1	0	0
5	$a \subset b \subset \mathcal{U}$	1	0	1	1
6	$a \subset b; b = \mathcal{U}$	1	0	1	0
7	$a = b \subset \mathcal{U}$	1	0	0	1
8	$a = b = \mathcal{U}$	1	0	0	0
9	$a \cap b = \emptyset;$ $a \cup b \neq \mathcal{U}$	0	1	1	1
10	$a \cap b = \emptyset;$ $a \cup b = \mathcal{U}$	0	1	1	0
11	$a \subset \mathcal{U}; b = \emptyset;$	0	1	0	1
12	$a = \mathcal{U}; b = \emptyset$	0	1	0	0
13	$a = \emptyset; b \subset \mathcal{U}$	0	0	1	1
14	$a = \emptyset; b = \mathcal{U}$	0	0	1	0
15	$a = b = \emptyset;$ $\mathcal{U} \neq \emptyset$	0	0	0	1
16	$a = b = \emptyset = \mathcal{U}$	0	0	0	0

relative position of two subsets a and b, possibly empty, in a referential. Then the following properties can also be checked in Table 1: $\max(s', t', u', v') = 1$.

This means that the four set indicators cannot be simultaneously empty, except if the referential $\mathcal{U}$ is empty, which corresponds to line 16 in Table 1. In logical terms we have: $s' \vee t' \vee u' \vee v' \equiv \top$. If $a \neq \emptyset, b \neq \emptyset, a \neq \mathcal{U}, b \neq \mathcal{U}$, we have

$$\max(n(u'), n(v')) \leq \min(s', t'),$$

where $n(x) = 1$ if $x = 0$ and $n(x) = 0$ if $x = 1$. A counterpart of this property holds in possibility theory and has also been noticed in formal concept analysis [7]. This corresponds to lines 1, 2, 3, 5, 7, 9 and 10 in Table 1. This also corresponds to the five possible configurations of two nonempty subsets a and b (lines 1, 3, 5, 7, 9), first identified by Gergonne [8, 10] when discussing syllogisms, plus two configurations (lines 2, 10) where $a \cup b = \mathcal{U}$ (but where $a \neq \mathcal{U}, b \neq \mathcal{U}$).

It should also be noticed that a subset a can be defined from another subset b, with the help of some of its positive and negative similarities s and t to a, and its differences u and v, in many different ways, where b appears or not:

$$a = s \cup u; \qquad a = \overline{v} \cup t;$$

$$a = s \cup (\overline{b} \cap \overline{t}); \qquad a = \overline{v \cup (\overline{b} \cap \overline{u})} = \overline{v} \cap (b \cup u) = u \cup (b \cap \overline{v}).$$

2.3 Recovering a Subset from Another Subset and Comparative Information

It may be the case that we have no direct information at our disposal about A, but only some comparative information with respect to another known object B (i.e. where b is known w.r.t. $\mathcal{U}$). Obviously, if we know that A is identical to the known object B, then $a = b$ and we are done. On the other hand, if we know that A is the exact opposite of B, then we are done again with $a = \mathcal{U} \setminus b = \overline{b}$. But, in general, we may only have information about partial match and/or partial mismatch. In the following, we assume that the information is complete (with respect to $\mathcal{U}$) about some of the above comparison indicators. For instance, we know *all* the properties in $\mathcal{U}$ that A and B share. But, we may also take into account dissimilarities as well when the information is available. We now discuss how to get accurate information regarding A starting from B and from some of the previous types of comparative information.

It is quite clear that knowing b and only one comparison indicator is not enough to recover complete knowledge of a. For instance, there is not a unique subset a such that $s = a \cap b$ and b are given. We only know that $s \subseteq a \subseteq s \cup \overline{b}$.

In order to get an accurate view of A starting from comparative information w.r.t. B, we need to know at least two comparison indicators. So our problem can now be stated as follows: what are the pairs of indicators which could lead to complete knowledge of A? More formally, what are the pairs of indicators (i_1, i_2) such that a is a function of b, i_1, i_2?

As we have four indicators, we have four choices for i_1 and then three remaining choices for i_2, leading to a total of 12 options. But as the problem is entirely symmetrical with respect to i_1 and i_2, we only have six cases to consider:

1. $i_1 = a \cap b$ and $i_2 = a \cap \overline{b}$: in that case, we compute $a = (a \cap b) \cup (a \cap \overline{b}) = i_1 \cup i_2$ as the unique solution.
2. $i_1 = a \cap b$ and $i_2 = \overline{a} \cap b$: in that case, the solution is not unique for a, i.e. we do not have enough information for computing a. We only know that $i_1 \subseteq a \subseteq \overline{i_2}$.
3. $i_1 = a \cap b$ and $i_2 = \overline{a} \cap \overline{b}$: we start from $a = (a \cap b) \cup (a \cap \overline{b}) = i_1 \cup (a \cap \overline{b})$, but $a \cap \overline{b} = \overline{(a \cup b)} \cap \overline{b} = \overline{i_2} \cap \overline{b} = \overline{i_2 \cup b}$, leading to a unique $a = i_1 \cup \overline{i_2 \cup b}$.
4. $i_1 = a \cap \overline{b}$ and $i_2 = \overline{a} \cap b$: starting from $a = (a \cap b) \cup (a \cap \overline{b}) = (a \cap b) \cup i_1$, since $b = (a \cap b) \cup (\overline{a} \cap b) = (a \cap b) \cup i_2$, we have $(a \cap b) = b \setminus i_2$, leading to the unique solution $a = i_1 \cup (b \setminus i_2)$.
5. $i_1 = a \cap \overline{b}$ and $i_2 = \overline{a} \cap \overline{b}$: this case is the dual of case 2 where b is replaced with $\overline{b}$. Similarly, the solution is still not unique. We still have $i_1 \subseteq a \subseteq \overline{i_2}$.
6. $i_1 = \overline{a} \cap b$ and $i_2 = \overline{a} \cap \overline{b}$: obviously $\overline{a} = i_1 \cup i_2$ then $a = \overline{i_1 \cup i_2}$.

Table 2 a is not computable

b	1	0	0	1	1	0
$b \cap a$	1	0	0	1	0	0
$b \cap \overline{a}$	0	0	0	0	1	0
a	1	?	?	1	0	?

Table 3 A unique a is computable

b	1	0	0	1	1	0
$b \cap a$	1	0	0	1	0	0
$\overline{b} \cap \overline{a}$	0	1	0	0	0	0
a	1	0	1	1	0	1

As we can see, there are only four pairs of indicators leading to a unique a (we recognize the expressions of the previous subsection), while the two remaining pairs are not informative enough to uniquely determine the set a.

Let us consider an example where $\mathcal{U}$ contains six properties such that every object can be represented as a six digit Boolean vector. For instance, object B is represented by the vector $b = 100110$, which means that B has the first, the fourth and the fifth property and not the others. Another object A is unknown, but we have some comparative information with respect to B: for instance we know $b \cap a$, i.e. the properties that B and A share: $b \cap a = 100100$. We also know $b \cap \overline{a} = 000010$. We are exactly in case 2 above, and we cannot recover A, as three components are not deducible, as shown in Table 2. This results from the fact that we have no information for A regarding the properties that B does not have. But in the case where we know $\overline{b} \cap \overline{a} = 010000$ (in place of $b \cap \overline{a}$), we can recover A as in Table 3, where the set d is just $(b \cap a) \cup (\overline{b} \cap \overline{a}) \cup b$ (as in case 3 above). It could also happen that the given disjoint subsets $i_1 \subset \mathcal{U}$ and $i_2 \subset \mathcal{U}$ are themselves indicators for other objects C and D different from A and B. A related question then arises: given two disjoint subsets i_1 and i_2, can we always find a pair of objects (C, D) such that i_1 and i_2 are the values of indicators for the pair (C, D)?

The answer to this question is positive and can be checked as follows: Considering only similarity indicators for instance, let us look for a pair (C, D) such that $c \cap d = i_1$ and $\overline{c} \cap \overline{d} = i_2$, knowing that $i_1 \cap i_2 = \emptyset$. The second equation is equivalent to $c \cup d = \overline{i_2}$, and the problem can be restated as an equivalent one: Knowing that $i_1 \subseteq \overline{i_2}$, find two sets c and d such that: $c \cap d = i_1$ and $c \cup d = \overline{i_2}$. It is well known that this problem has solutions (generally not unique). This remark allows us to change the setting of our initial problem, since it appears that having the values of two indicators for a pair (A, B) is equivalent to having two equalities between indicators for the pair (A, B) and indicators for another pair (C, D).

Since the power set $2^{\mathcal{U}}$ is a Boolean lattice, we can reword the problem in a Boolean setting where $\cap$ is translated into $\wedge$, $\cup$ into $\vee$, and $=$ into $\equiv$. We keep the complementation overbar, as a compact notation for negation. Then, equivalences between indicators lead to the notion of logical proportions that we investigate in the next section.

3 Brief Background on Logical Proportions

Adopting a propositional logic view, for a given pair of Boolean variables (a, b), we have four distinct indicators:

- $a \wedge b$ and $\overline{a} \wedge \overline{b}$, called *similarity indicators*
- $a \wedge \overline{b}$ and $\overline{a} \wedge b$, called *dissimilarity indicators*

Following what we did in the subset-based analysis above, comparing a pair of variables (a, b) with another pair (c, d) is done via a pair of equivalence between indicators.

Note that a, b, c and d no longer denote subsets of properties that respectively hold in situations A, B, C and D. Rather, they encode Boolean variables pertaining respectively to situations A, B, C and D, which refer to one particular property on the basis of which these situations are compared. In other words, variables a, b, c and d encode respectively if a particular considered property holds true, or not, in situations A, B, C and D. So we need as many 4-tuples of Boolean variables (a, b, c, d) of this kind as there are properties used for describing situations A, B, C and D.

3.1 The 120 Logical Proportions

As shown in the introductory discussion, it makes sense to consider the values of two indicators to be in a position to compute a Boolean variable related to them. As a consequence, it is legitimate to consider all the conjunctions of two equivalences between indicators: such a conjunction is called a *logical proportion* [16, 17]. More formally, let $I_{(a,b)}$ and $I'_{(a,b)}$[1] (resp. $I_{(c,d)}$ and $I'_{(c,d)}$) denote two indicators for (a, b) (resp. (c, d)).

Definition 1 A logical proportion $T(a, b, c, d)$ is the conjunction of two distinct equivalences between indicators of the form $(I_{(a,b)} \equiv I_{(c,d)}) \wedge (I'_{(a,b)} \equiv I'_{(c,d)})$.

It can be easily checked that there are 16 distinct equivalences relating a comparison indicator pertaining to a and b to a comparison indicator pertaining to c and d. Their truth tables are given in Table 4, exhibiting, in each case, the ten patterns of truth values of a, b, c and d that make the corresponding equivalence true. It is false

---

[1]Note that $I_{(a,b)}$ (or $I'_{(a,b)}$) refers to one element in the set $\{a \wedge b, \overline{a} \wedge b, a \wedge \overline{b}, \overline{a} \wedge \overline{b}\}$, and should not be considered as a functional symbol: $I_{(a,b)}$ and $I_{(c,d)}$ may be indicators of two different kinds. Still, we use this notation for the sake of simplicity.

Table 4 Patterns making true the 16 equivalences between indicators. The conjunctive combination of $\underline{X}$ and $\overline{X}$ yields the logical proportion X

$\underline{A}$	$\overline{A}$	$\underline{R}$	$\overline{R}$	$\underline{P}$	$\overline{P}$	$\underline{I}$	$\overline{I}$
$a\bar{b} \equiv c\bar{d}$	$\bar{a}b \equiv \bar{c}d$	$a\bar{b} \equiv \bar{c}d$	$\bar{a}b \equiv c\bar{d}$	$ab \equiv cd$	$\bar{a}\bar{b} \equiv \bar{c}\bar{d}$	$ab \equiv \bar{c}\bar{d}$	$\bar{a}\bar{b} \equiv cd$
0 0 0 0	0 0 0 0	0 0 0 0	0 0 0 0	0 0 0 0	0 0 0 0	0 0 1 1	0 0 1 1
1 1 1 1	1 1 1 1	1 1 1 1	1 1 1 1	1 1 1 1	1 1 1 1	1 1 0 0	1 1 0 0
0 0 1 1	0 0 1 1	0 0 1 1	0 0 1 1	0 1 0 1	0 1 0 1	0 1 0 1	0 1 0 1
1 1 0 0	1 1 0 0	1 1 0 0	1 1 0 0	1 0 1 0	1 0 1 0	1 0 1 0	1 0 1 0
0 1 0 1	0 1 0 1	0 1 1 0	0 1 1 0	0 1 1 0	0 1 1 0	0 1 1 0	0 1 1 0
1 0 1 0	1 0 1 0	1 0 0 1	1 0 0 1	1 0 0 1	1 0 0 1	1 0 0 1	1 0 0 1
0 0 0 1	0 0 1 0	0 0 1 0	0 0 0 1	0 0 0 1	0 1 1 1	0 0 0 1	0 1 0 0
0 1 0 0	1 0 0 0	0 1 0 0	1 0 0 0	0 0 1 0	1 0 1 1	0 0 1 0	1 0 0 0
0 1 1 1	1 0 1 1	0 1 1 1	1 0 1 1	0 1 0 0	1 1 0 1	0 1 1 1	1 1 0 1
1 1 0 1	1 1 1 0	1 1 1 0	1 1 0 1	1 0 0 0	1 1 1 0	1 0 1 1	1 1 1 0

$\underline{H_a}$	$\overline{H_a}$	$\underline{H_b}$	$\overline{H_b}$	$\underline{H_c}$	$\overline{H_c}$	$\underline{H_d}$	$\overline{H_d}$
$a\bar{b} \equiv cd$	$\bar{a}b \equiv \bar{c}\bar{d}$	$\bar{a}b \equiv cd$	$a\bar{b} \equiv \bar{c}\bar{d}$	$ab \equiv c\bar{d}$	$\bar{a}\bar{b} \equiv \bar{c}d$	$ab \equiv \bar{c}d$	$\bar{a}\bar{b} \equiv c\bar{d}$
0 0 0 1	0 0 0 1	0 0 0 1	0 0 0 1	0 0 0 1	0 0 0 1	0 0 1 0	0 0 1 0
0 0 1 0	0 0 1 0	0 0 1 0	0 0 1 0	0 1 0 0	0 1 0 0	0 1 0 0	0 1 0 0
0 1 0 0	0 1 0 0	1 0 0 0	1 0 0 0	1 0 0 0	1 0 0 0	1 0 0 0	1 0 0 0
1 0 1 1	1 0 1 1	0 1 1 1	0 1 1 1	0 1 1 1	0 1 1 1	0 1 1 1	0 1 1 1
1 1 0 1	1 1 0 1	1 1 0 1	1 1 0 1	1 0 1 1	1 0 1 1	1 0 1 1	1 0 1 1
1 1 1 0	1 1 1 0	1 1 1 0	1 1 1 0	1 1 1 0	1 1 1 0	1 1 0 1	1 1 0 1
0 0 0 0	0 0 1 1	0 0 0 0	0 0 1 1	0 0 0 0	0 1 1 0	0 0 0 0	0 1 0 1
0 1 0 1	1 0 0 1	1 0 0 1	0 1 0 1	0 0 1 1	1 0 1 0	0 0 1 1	1 0 0 1
0 1 1 0	1 0 1 0	1 0 1 0	0 1 1 0	0 1 0 1	1 1 0 0	0 1 1 0	1 1 0 0
1 1 0 0	1 1 1 1	1 1 0 0	1 1 1 1	1 0 0 1	1 1 1 1	1 0 1 0	1 1 1 1

for the six missing patterns. Note that conjunctions are omitted in the expressions of comparison indicators in Table 4 for the sake of a compact writing. The meaning of the names given to these 16 equivalences will be made clear in the next two subsections.

Since we have to choose 2 distinct equivalences among $4 \times 4 = 16$ possible ones for defining a logical proportion, we have $\binom{16}{2} = 120$ such proportions, and it has been shown that they are all semantically distinct. Consequently, if two proportions are semantically equivalent, they have the same expression as a conjunction of two equivalences between indicators (up to the symmetry of the conjunction). Then a logical proportion is just a particular Boolean formula involving four variables and as such, has a truth table with 16 lines. It has been shown [19] that an equivalence between 2 indicators has exactly ten lines leading to truth value 1 in its truth table, and that any logical proportion has exactly six lines leading to truth value 1 in its truth table.

First of all, depending on the way the indicators are combined, different types of logical proportions are obtained [16, 19]. There are:

- **4** *homogeneous* proportions that involve only dissimilarity, or only similarity indicators: they are called, *analogy* (*A*), *reverse analogy* (*R*), *paralogy* (*P*) and *inverse paralogy* (*I*) (see definitions below).
- **16** *conditional* proportions defined as the conjunction of an equivalence between similarity indicators and of an equivalence between dissimilarity indicators, as, e.g. $((a \wedge b) \equiv (c \wedge d)) \wedge ((a \wedge \overline{b}) \equiv (c \wedge \overline{d}))$, which expresses that the two conditional objects[2] $b|a$ and $d|c$ have the same examples (first condition) and the same counter-examples (second condition).
- **20** *hybrid* proportions obtained as the conjunction of two equivalences between similarity and dissimilarity indicators, as in the following example:
  $(a \wedge b \equiv \overline{c} \wedge d) \wedge (\overline{a} \wedge \overline{b} \equiv c \wedge \overline{d})$.
- **32** *semihybrid* proportions for which one half of their expressions involve indicators of the same kind, while the other half requires equivalence between indicators of opposite kinds, as, e.g. $(a \wedge b \equiv c \wedge d) \wedge (\overline{a} \wedge \overline{b} \equiv c \wedge \overline{d})$.
- **48** *degenerated* proportions whose definition involves three distinct indicators only.

3.2 The Four Homogeneous Logical Proportions

When we consider the logical proportions with homogeneous equivalences only, i.e. the ones that involve only dissimilarity, or only similarity indicators, we get the four logical proportions mentioned above, which are listed below with their expression:

---

[2]A conditional object $b|a$ can take three truth values: true if $a \wedge b$ is true, false if $a \wedge \neg b$ is true, not applicable if a is false; it may be intuitively thought of as representing the rule "if a then b" [5].

Table 5 Analogy, reverse analogy, paralogy and inverse paralogy truth tables

A	R	P	I
0 0 0 0	0 0 0 0	0 0 0 0	1 1 0 0
1 1 1 1	1 1 1 1	1 1 1 1	0 0 1 1
0 0 1 1	0 0 1 1	1 0 0 1	1 0 0 1
1 1 0 0	1 1 0 0	0 1 1 0	0 1 1 0
0 1 0 1	0 1 1 0	0 1 0 1	0 1 0 1
1 0 1 0	1 0 0 1	1 0 1 0	1 0 1 0

- *Analogy*: $A(a, b, c, d)$, defined by $((a \wedge \overline{b}) \equiv (c \wedge \overline{d})) \wedge ((\overline{a} \wedge b) \equiv (\overline{c} \wedge d))$
- *Reverse analogy*: $R(a, b, c, d)$, defined by $((a \wedge \overline{b}) \equiv (\overline{c} \wedge d)) \wedge ((\overline{a} \wedge b) \equiv (c \wedge \overline{d}))$
- *Paralogy*: $P(a, b, c, d)$, defined by $((a \wedge b) \equiv (c \wedge d)) \wedge ((\overline{a} \wedge \overline{b}) \equiv (\overline{c} \wedge \overline{d}))$
- *Inverse paralogy*: $I(a, b, c, d)$, defined by $((a \wedge b) \equiv (\overline{c} \wedge \overline{d})) \wedge ((\overline{a} \wedge \overline{b}) \equiv (c \wedge d))$

Reverse analogy expresses that "*a* differs from *b* as *d* differs from *c*", and conversely, paralogy expresses that "what *a* and *b* have in common, *c* and *d* have also". The inverse paralogy expresses a complete opposition between the pairs (a, b) and (c, d). The truth tables of these homogeneous proportions are recalled in Table 5, where we only show the lines leading to truth value 1. The next proposition, easily deducible from the definition, establishes a link between analogy, reverse analogy and paralogy (while inverse paralogy *I* is not related to the three others through a simple permutation):

Proposition 1 $A(a, b, c, d) \leftrightarrow R(a, b, d, c)$ *and* $A(a, b, c, d) \leftrightarrow P(a, d, c, b)$.

Clearly, the truth table of **A** (resp. **R**, **P** and **I**) is obtained as the conjunction of the truth tables of equivalences $\underline{\mathbf{A}}$ and $\overline{\mathbf{A}}$ (resp. $\underline{\mathbf{R}}$ and $\overline{\mathbf{R}}$, $\underline{\mathbf{P}}$ and $\overline{\mathbf{P}}$, and $\underline{\mathbf{I}}$ and $\overline{\mathbf{I}}$), given in Table 4. It is not the aim of this paper to investigate the whole set of properties of this class of Boolean formulas, nevertheless we recall the basic behavior that can be expected from these proportions, in Table 6.

To conclude this section, let us note that the analogical proportion is the Boolean counterpart of the numerical proportion $\frac{a}{b} = \frac{c}{d}$. As such, it is expected that this proportion will enjoy a property similar to the well-known rule of three for numerical proportions: knowing three of the variables, we can compute a 4th one in order to build up a proportion. This is obviously linked to our initial problem, where we want to derive some missing information. We will investigate this issue in Sect. 4.

Table 6 Properties of logical proportions

Property	Definition	# of proportions	Homogeneous
Full identity	$T(a,a,a,a)$	15	A, R, P
Reflexivity	$T(a,b,a,b)$	6	A, P
Reverse reflexivity	$T(a,b,b,a)$	6	R, P
Sameness	$T(a,a,b,b)$	6	A, R
Symmetry	$T(a,b,c,d) \rightarrow T(c,d,a,b)$	12	A, R, P, I
Central permutation	$T(a,b,c,d) \rightarrow T(a,c,b,d)$	16	A, I
All permutations	$\forall i,j, T(a,b,c,d) \rightarrow$ $T(p_{i,j}(a,b,c,d))$	1	I
Transitivity	$T(a,b,c,d) \wedge T(c,d,e,f) \rightarrow$ $T(a,b,e,f)$	54	A, P
Code independency	$T(a,b,c,d) \rightarrow T(\overline{a},\overline{b},\overline{c},\overline{d})$	8	A, R, P, I

3.3 The Four Heterogeneous Logical Proportions

A remarkable property shared by the four homogeneous logical proportions is their independency with respect to a positive or negative encoding of the features, namely $T(a,b,c,d)$ holds true if and only if $T(\overline{a},\overline{b},\overline{c},\overline{d})$ holds true. Indeed it is the same for the evaluation of these proportions to describe, e.g. the size of objects A, B, C and D by indicating if they are large or not, or by indicating if they are small (understood as "not large") or not. There are only 8 logical proportions among the 120 that are code independent [18]. The four others are the heterogeneous proportions that we present now.

While homogeneous proportions are defined through equivalences between comparison indicators of the same type (either similarity or dissimilarity), heterogeneous proportions are hybrid proportions defined between indicators of opposite types, while the code independency property is preserved. Their logical expression is given below:

$$\mathbf{H_a} : ((a \wedge \overline{b}) \equiv (c \wedge d)) \wedge ((\overline{a} \wedge b) \equiv (\overline{c} \wedge \overline{d})),$$

$$\mathbf{H_b} : ((\overline{a} \wedge b) \equiv (c \wedge d)) \wedge ((a \wedge \overline{b}) \equiv (\overline{c} \wedge \overline{d})),$$

$$\mathbf{H_c} : ((a \wedge b) \equiv (c \wedge \overline{d})) \wedge ((\overline{a} \wedge \overline{b}) \equiv (\overline{c} \wedge d)),$$

$$\mathbf{H_d} : ((a \wedge b) \equiv (\overline{c} \wedge d)) \wedge ((\overline{a} \wedge \overline{b}) \equiv (c \wedge \overline{d})).$$

Their truth tables are shown in Table 7. Clearly, the truth table of $\mathbf{H_x}$ is obtained as the conjunction of the truth tables of equivalences $\mathbf{H_x}$ and $\overline{\mathbf{H_x}}$ given in Table 4. It is stunning to note that these truth tables exactly involve the eight missing tuples of the homogeneous tables, i.e. those ones having an odd number of 0 and 1. The meaning of $\mathbf{H_x}$ is easy to grasp: "there is an intruder among the four values, which is not x".

Table 7 H_a, H_b, H_c, H_d: the six patterns that make them true

$\mathbf{H_a}$	$\mathbf{H_b}$	$\mathbf{H_c}$	$\mathbf{H_d}$
1 1 1 0	1 1 1 0	1 1 1 0	1 1 0 1
0 0 0 1	0 0 0 1	0 0 0 1	0 0 1 0
1 1 0 1	1 1 0 1	1 0 1 1	1 0 1 1
0 0 1 0	0 0 1 0	0 1 0 0	0 1 0 0
1 0 1 1	0 1 1 1	0 1 1 1	0 1 1 1
0 1 0 0	1 0 0 0	1 0 0 0	1 0 0 0

Heterogeneous proportions satisfy obvious permutation properties. For instance, in $\mathbf{H_a}$, b and c, b and d, and c and d can be exchanged. Besides, note that, if we change d into $\bar{d}$ (and vice versa) in $\mathbf{H_a}$, $\mathbf{H_b}$, $\mathbf{H_c}$ and $\mathbf{H_d}$, are changed in $\mathbf{A}$, $\mathbf{R}$, $\mathbf{P}$ and $\mathbf{I}$, respectively. It should come as surprise that they satisfy the same association properties as the homogeneous ones: for instance, any combination of two or three heterogeneous proportions is satisfiable by 4-tuples, and the conjunction $H_a(a,b,c,d) \wedge H_b(a,b,c,d) \wedge H_c(a,b,c,d) \wedge H_d(a,b,c,d)$ is not satisfiable. These facts contribute to make the heterogeneous proportions the perfect dual of the homogeneous ones.

3.4 Extension to $\mathbb{B}^n$

It is unlikely that we can represent objects in practice with only one property. In standard datasets, objects are represented by Boolean vectors, and it makes sense to extend the previous framework to $\mathbb{B}^n$. The simplest way to do this is to consider a definition involving componentwise proportions as follows (where T denotes any proportion and $\vec{a}, \vec{b}, \vec{c}$ and $\vec{d}$ are in $\mathbb{B}^n$):

$$T(\vec{a}, \vec{b}, \vec{c}, \vec{d}) \text{ iff } \forall i \in [1,n], T(a_i, b_i, c_i, d_i),$$

where $\vec{x} = (x_1, \dots, x_n)$ for $x = a, b, c$ and d. All the previous properties of Boolean proportions remain valid for their counterpart in $\mathbb{B}^n$: For instance, if $T = A$, then the *central permutation* property (see Table 6) is still valid:

$$T(\vec{a}, \vec{b}, \vec{c}, \vec{d}) \rightarrow T(\vec{a}, \vec{c}, \vec{b}, \vec{d}).$$

Table 8   A non-univocal proportion

$$
\begin{array}{cccc}
0 & 0 & 0 & 0 \\
0 & 1 & 0 & 1 \\
1 & 0 & 1 & 0 \\
1 & 0 & 1 & 1 \\
1 & 1 & 1 & 0 \\
1 & 1 & 1 & 1 \\
\end{array}
$$

4   Equation Solving Process

As said in the introduction, the main problem we want to tackle is to compute missing information starting from existing information. In the context of logical proportions, the equation solving problem can be stated as follows:

Given a logical proportion T and three Boolean values a, b, c, does a Boolean value x such that $T(a, b, c, x) = 1$ exist, and in that case, is this value unique?

This is an *equation solving problem*. First of all, it is easy to see that there are always cases where the equation has no solution. Indeed, the triple a, b, c may take $2^3 = 8$ values, while any proportion T is true for only six distinct valuations, leaving at least two cases with no solution. For instance, when we deal with analogy A, the equations $A(1, 0, 0, x) = 1$ and $A(0, 1, 1, x) = 1$ have no solution.

When the equation $T(a, b, c, x) = 1$ is solvable, a proportion T which has a unique solution will be said to be *d-univocal*. The following proportion is not *d-univocal*, as can be seen from its truth table (Table 8), despite the fact that it satisfies full identity, reflexivity, symmetry and transitivity: $((\overline{a} \wedge \overline{b}) \equiv (\overline{c} \wedge \overline{d})) \wedge ((\overline{a} \wedge b) \equiv (\overline{c} \wedge d))$.

We have the following result:

Proposition 2   *There are exactly 64 d-univocal proportions (including the homogeneous ones). They are the 4 homogeneous proportions, 8 conditional proportions, 12 hybrid proportions, 24 semihybrid proportions and 16 degenerated proportions.*

When moving to $\mathbb{B}^n$, the previous result remains valid; i.e. for 64 proportions, the equation $T(\vec{a}, \vec{b}, \vec{c}, \vec{x}) = 1$ has at most one solution.

In a similar manner, one may define proportions that are a-, b- or c-univocal. If we impose the proportion to be univocal with respect to two positions, we get 32 solutions. With respect to three positions, we get 16 solutions. Lastly, we have the following result:

Proposition 3   *There are only eight logical proportions that are a-, b-, c- and d-univocal: they are the four homogeneous and the four heterogeneous logical proportions.*

This result should not be a surprise, since a proportion that is not x-univocal is necessarily true both for a pattern with an odd number of 1 and 0, and for a pattern with an even number of 1 and 0. This excludes the homogeneous and the heterogeneous proportions. When considering only the homogeneous proportions A, R, P, I, we have a more detailed result [17–19]:

Proposition 4
The analogical equation $A(a, b, c, x)$ is solvable iff $(a \equiv b) \vee (a \equiv c)$ holds.
The reverse analogical equation $R(a, b, c, x)$ is solvable iff $(b \equiv a) \vee (b \equiv c)$ holds.
The paralogical equation $P(a, b, c, x)$ is solvable iff $(c \equiv b) \vee (c \equiv a)$ holds.
In each of the three above cases, when it exists, *the unique solution x is given by*
$c \equiv (a \equiv b)$ *(or equivalently $x = a \equiv b \equiv c$).*
The inverse paralogical equation $I(a, b, c, x)$ is solvable iff
$(a \not\equiv b) \vee (b \not\equiv c)$ *holds. In that case, the unique solution x is given by $c \not\equiv (a \not\equiv b)$.*

As we can see, the first three homogeneous proportions A, R, P behave similarly. Still, the conditions of their solvability differ. Moreover, it can be checked that at least two of these proportions are always simultaneously solvable. Besides, when they are solvable, there is a common expression that yields the solution. This again points out a close relationship between A, R and P. This contrasts with proportion I, which in some sense behaves in an opposite manner.

5 Inference Based on Logical Proportions

From an inference perspective, the equation solving property is the main tool: when we have three known objects A, B, C and another one D whose properties are unknown, but we have the information that $T(A, B, C, D)$ for a given d-univocal proportion, then we can compute D.

5.1 Existence of a Logical Proportion Linking Four Boolean Vectors

Now we are back to our initial framework of objects represented as a collection of Boolean properties. In that case, an object A is represented as a Boolean vector and belongs to $\mathbb{B}^n$, where n is just the cardinal of $\mathcal{U}$ (the whole set of properties). We can then provide a list of relevant problems we may need to solve:

1. Given four objects A, B, C, D, is there a proportion T, among the 120 proportions, such that $T(a, b, c, d)$?
2. If the answer to question 1 is "Yes", can we exhibit such a proportion?
3. If the answer to question 1 is "Yes", is such a proportion unique?

4. Given three objects A, B, C and a proportion T, can we compute an object D such that $T(A, B, C, D)$? (i.e. is T a d-univocal proportion?)

Reasoning about the first problem is relatively straightforward. As soon as we have the Boolean representation a, b, c, d of A, B, C, D, we have to consider the n valuations (a_i, b_i, c_i, d_i) corresponding to the components of a, b, c and d. Among these n valuations, some can be identical: so let us denote by m the number of distinct valuations among these n valuations. Obviously, if $m > 7$, there is no proportion such that $T(a, b, c, d)$, since only six valuations can lead to a given proportion being true.

If $m = 1$, we have exactly 45 candidate proportions. If $m = 2$, we have exactly 15 candidate proportions. If $m = 3$, we have three or six candidate proportions. If $m = 4$, the landscape is a bit different. Obviously, we cannot get more than six proportions, but some combinations lead to zero candidate proportions. In fact, we can have zero, one, three or six candidate proportions. For instance:

Lemma *A logical proportion cannot satisfy the class of valuation* $\{0111, 1011, 1101, 1110\}$ *or the class* $\{1000, 0100, 0010, 0001\}$.

Proof It is enough to show that this is the case for an equivalence between indicators. So let us consider such an equivalence $l_1 \wedge l_2 \equiv l_3 \wedge l_4$. If this equivalence is valid for $\{0111, 1011\}$, it means that its truth value does not change when we switch the truth value of the two first literals from 0 to 1: there are only two indicators for a and b satisfying this requirement: $a \wedge b$ and $\bar{a} \wedge \bar{b}$. On top of that, if this equivalence is still valid for $\{1101, 1110\}$, it means that its truth value does not change when we switch the truth value of the two last literals from 0 to 1: there are only two indicators for c and d satisfying this requirement: $c \wedge d$ and $\bar{c} \wedge \bar{d}$. Then the equivalence $l_1 \wedge l_2 \equiv l_3 \wedge l_4$ is just $a \wedge b \equiv c \wedge d$, $a \wedge b \equiv \bar{c} \wedge \bar{d}$, $a \wedge b \equiv \bar{c} \wedge \bar{d}$ or $\bar{a} \wedge \bar{b} \equiv \bar{c} \wedge \bar{d}$. None of these equivalences satisfies the whole class $\{0111, 1011, 1101, 1110\}$. The same reasoning applies for the other class. □

From a practical viewpoint, as soon as we observe this class appearing as a part of the n valuations (a_i, b_i, c_i, d_i), we know that there is no suitable proportion. If $m = 5$ or $m = 6$, we get zero or one candidate proportion.

5.2 Induction with Proportions

We can now adopt a viewpoint, similar in some sense to the k-nearest neighbors philosophy, where we infer unknown properties of a partially known object D starting from the knowledge we have about its other specified properties. This induction principle can be stated as follows (where J is a subset of $[1, n]$):

$$\frac{\forall i \in [1, n] \setminus J, T(a_i, b_i, c_i, d_i)}{\forall i \in J, T(a_i, b_i, c_i, d_i)}.$$

This can be seen as a continuity principle assuming that, if it is known that a proportion holds for some attributes, this proportion should still hold for the other attributes. It extends the inference principle defined in the case of the analogical proportion [22, 23], to d-univocal proportions.

The application of this inference principle to predict missing information for a given object D is straightforward. If we are able to find a triple of known objects A, B, C and a d-univocal proportion T such that $T(a_i, b_i, c_i, d_i)$ holds true for the attributes i belonging to $[1, n] \setminus J$, we then solve the equations $T(a_j, b_j, c_j, x_j)$ for every $j \in J$, and the (unique) solution is considered to be the value of d_j.

A particular case of this inductive principle has been implemented for classification purposes, i.e. when there is only one missing piece of information, which is the class of the object. Different approaches have been implemented, mainly using homogeneous proportions, all of them relaxing the induction principle to allow some flexibility: one may cite [1, 2] using analogical proportion only, using the four homogeneous proportions. In terms of accuracy, it appears that the results are similar when using $\mathbf{A}, \mathbf{R}, \mathbf{P}$, while $\mathbf{I}$ exhibits a different behavior. The prediction of missing values, using $\mathbf{A}$, has recently been experimented with successfully [4].

This leads us to the question of choosing the most suitable proportion for a given task. We can also consider other proportions among the 60 remaining univocal ones. These proportions express different kinds of regularities, which may be more or less often encountered in datasets. While the three homogeneous proportions $\mathbf{A}, \mathbf{R}, \mathbf{P}$, which play similar roles up to permutations, appear to be suitable for classification or for completing missing values, the involutive paralogy $\mathbf{I}$ seems to be of a different nature by expressing orthogonality between situations. The four heterogeneous proportions play similar roles with respect to the presence of intruder values. Clearly, if situations A, B and C are very close, enforcing $\mathbf{H_a}, \mathbf{H_b}$ or $\mathbf{H_c}$ for each feature will lead to a D very different from the three other situations. On the contrary, if A, B and C are quite different, D will not be more different from each of them. The respective roles of the univocal logical proportions is a topic for further research. Generally speaking, the proposed approach tends to indicate the possibility of transposing the idea of proportional reasoning from numerical settings to Boolean or nominal worlds.

6 Conclusion

We have outlined a systematic discussion of the interest of logical comparison indicators, and of their potential value for reasoning tasks such as classification or completion of missing information. We have emphasized the role of logical proportions in these reasoning tasks, exploiting similarity and dissimilarity indicators. In particular, we have identified the prominent place of homogeneous proportions and of heterogeneous proportions. They capture very different semantics, and appear to be basic tools for classification, intruder detection [20], and other reasoning tasks [3].

References

1. Bayoudh, S., Miclet, L., Delhay, A.: Learning by analogy: a classification rule for binary and nominal data. In: Proceedings of the International Joint Conference on Artificial Intelligence (IJCAI'07), pp. 678–683 (2007)
2. Bounhas, M., Prade, H., Richard, G.: Analogical classification: a new way to deal with examples. In: Schaub, T., Friedrich, G., O'Sullivan, B. (eds.) Proceedings of the 21st European Conference on Artificial Intelligence (ECAI'14), Prague, 18–22 Aug. Frontiers in Artificial Intelligence and Applications, vol. 263, pp. 135–140. IOS Press (2014)
3. Correa, W., Prade, H., Richard, G.: When intelligence is just a matter of copying. In: Proceedings of the 20th European Conference on Artificial Intelligence, Montpellier, 27–31 Aug, pp. 276–281. IOS Press (2012)
4. Correa Beltran, W., Jaudoin, H., Pivert, O.: Estimating null values in relational databases using analogical proportions. In: Laurent, A., Strauss, O., Bouchon-Meunier, B., Yager, R.R. (eds.) Proceedings of the 15th International Conference on Information Processing and Management of Uncertainty in Knowledge-Based Systems (IPMU'14), Part III, Montpellier, 15–19 July. Communication in Computer and Information Science, vol. 444, pp. 110–119. Springer (2014)
5. Dubois, D., Prade, H.: Conditional objects as nonmonotonic consequence relationships. IEEE Trans. Syst. Man Cybern. **24**, 1724–1740 (1994)
6. Dubois, D., Prade, H.: From Blanché's hexagonal organization of concepts to formal concept analysis and possibility theory. Logica Universalis **6**(1–2), 149–169 (2012)
7. Dubois, D., Prade, H.: Possibility theory and formal concept analysis: characterizing independent sub-contexts. Fuzzy Sets Syst. **196**, 4–16 (2012)
8. Faris, J.A.: The Gergonne relations. J. Symbolic Logic 207–231 (1955)
9. Gentner, D., Holyoak, K.J., Kokinov, B.N.: The Analogical Mind: Perspectives from Cognitive Science. Cognitive Science, and Philosophy, MIT Press, Cambridge, MA (2001)
10. Gergonne, J.D.: Essai de dialectique rationnelle. Ann. Math. Pures Appl. 189–228 (1817)
11. Klein, S.: Culture, mysticism & social structure and the calculation of behavior. In: Proceedings of the 5th European Conference in Artificial Intelligence (ECAI'82), Orsay, France, pp. 141–146 (1982)
12. Miclet, L., Prade, H.: Handling analogical proportions in classical logic and fuzzy logics settings. In: Proceedings of the 10th European Conference on Symbolic and Quantitative Approaches to Reasoning with Uncertainty (ECSQARU'09), Verona, pp. 638–650. Springer, LNCS 5590 (2009)
13. Parsons, T.: The traditional square of opposition. In: Zalta, E.N. (ed.) The Stanford Encyclopedia of Philosophy. Summer 2015 edn. (2015)
14. Piaget, J.: Logic and Psychology. Manchester University Press (1953)
15. Piaget, J.: The Development of Thought: Equilibration of Cognitive Structures (trans. Rosin, A.). Viking, Oxford, UK (1977)
16. Prade, H., Richard, G.: Logical proportions—typology and roadmap. In: Hüllermeier, E., Kruse, R., Hoffmann, F. (eds.) Computational Intelligence for Knowledge-Based Systems Design: Proceedings of the 13th International Conference on Information Processing and Management of Uncertainty (IPMU'10), Dortmund, 28 June–2 July. LNCS, vol. 6178, pp. 757–767. Springer (2010)
17. Prade, H., Richard, G.: Reasoning with logical proportions. In: Lin, F.Z., Sattler, U., Truszczynski, M. (eds.) Proceedings of the 12th International Conference on Principles of Knowledge Representation and Reasoning, KR 2010, Toronto, 9–13 May, pp. 545–555. AAAI Press (2010)
18. Prade, H., Richard, G.: Homogeneous logical proportions: their uniqueness and their role in similarity-based prediction. In: Brewka, G., Eiter, T., McIlraith, S.A. (eds.) Proceedings of the 13th International Conference on Principles of Knowledge Representation and Reasoning (KR'12), Rome, 10–14 June, pp. 402–412. AAAI Press (2012)
19. Prade, H., Richard, G.: From analogical proportion to logical proportions. Logica Universalis **7**(4), 441–505 (2013)

20. Prade, H., Richard, G.: Homogenous and heterogeneous logical proportions. If CoLog J. Logics Appl. **1**(1), 1–51 (2014)
21. Prade, H., Richard, G.: Proportional reasoning in a Boolean setting. In: Booth, R., Casini, G., Klarman, S., Richard, G., Varzinczak, I.J. (eds.) International Workshop on Defeasible and Ampliative Reasoning (DARe@ECAI'14), Prague, 19 Aug. CEUR Workshop Proceedings, vol. 1212 (2014)
22. Stroppa, N., Yvon, F.: An analogical learner for morphological analysis. In: Online Proceedings of the 9th Conference on Computational Natural Language Learning (CoNLL-2005), pp. 120–127 (2005)
23. Stroppa, N., Yvon, F.: Analogical learning and formal proportions: Definitions and methodological issues. Technical report, ENST, June 2005
24. Tversky, A.: Features of similarity. Psychol. Rev. **84**, 327–352 (1977)

A "Distinctive" Logic for Ontologies and Semantic Search Engines

Ferdinando Cavaliere

Abstract We present here the theoretical basis of the "Semantic Prompter Engine", a hypothetical program which, when applied to popular online search engines, enriches them with a semantics. Other applications lie in the field of translation and de/encoders. The query language of the "prompter" (realized through synonymy and new relationships) and the construction of its related ontologies (structured universes of discourse) are based on the "distinctive predicate calculus". This logic, designed by the author, is close to natural language and common intuition and is easily represented by means of Venn diagrams or numbered segments. It is a blend of past (quantification of the predicate, syllogistic, Vasil'ev's logic of notions) and contemporary (fuzzy logic, N-oppositional theory) logical systems. With this approach it is possible to build a bridge between artificial (machine) and natural (human or animal) representations of concepts and reality.

1 Introduction

The rock on which many online search procedures founder, with all attendant losses of time and information, is formed by the absence of "semantic" search machines operating on the basis of semantic similarities of the terms involved rather than on the basis of formal (phonological or spelling) similarities. For example, the word form "ship" is more similar to the form "slip" than to "vessel". Take the case of an official site of a public body. A citizen needs information about certain provisions and places a request for "administrative measures". If his search leads him to, say, "ordinance" or "specify", the poor citizen is not helped much, frustrating the efforts made by the public body to make the procedures simpler and more transparent.

Some progress has been made. Some search engines are now beginning to suggest synonyms, hyponyms or hyperonyms. However, the attempts have remained incoherent. Synonymy is understood in too vague a sense: if "horse" is

F. Cavaliere (✉)
Circolo Matematico Cesenate, Cesena, (FC), Italy
e-mail: cavaliere.ferdinando@gmail.com

© Springer International Publishing AG 2017
G. Dodig-Crnkovic and R. Giovagnoli (eds.), *Representation and Reality in Humans, Other Living Organisms and Intelligent Machines*, Studies in Applied Philosophy, Epistemology and Rational Ethics 28, DOI 10.1007/978-3-319-43784-2_15

synonymous with "equine", and even a zebra is called equine, are "horse" and "zebra" synonyms?

The difficulty lies in the development of semantics for the search engines. This semantics requires the setting up of an "ontology"—an organic network of semantic relations between words—in a logically coherent way, not, for example, on a purely statistical basis but on a formally checkable axiomatic-deductive basis (see [1]). This engine should, moreover, be economical and synthetic but maximally extended and "friendly" not only for the computer scientists who created the ontology but also, and especially, for the users of the search machine. If realized, a semantic search engine would certainly have vast application possibilities in many fields of knowledge, such as dictionaries, translators, library science (indexing systems) and thesaurus making.

To solve this type of problem, we propose the creation of a database program that can do the work of a semantic search engine, the Semantic Prompter Engine. The guidelines for the construction of the ontological database—a structured archiving system of notions, terms and data—will have to come from the definitions of notions such as synonym or antonym, based on a logico-linguistic approach closely following natural language and natural intuition. Querying the system, one will receive suggestions for further searches to be forwarded to the traditional engines. The output of the combined two systems will be equivalent to the output of a semantic search engine.

2 Synonymy and Distinctive Logic

The point of departure of our entire construction consists in the conception of a word/concept/term as a name for a nonnull aggregate or set of listable characteristics/qualities that are sufficient for a differentiation from other words. This nonnull aggregate is, moreover, distinct from a "universe of discourse" containing all possible characteristics of all concepts.

The setting off of the elements of two such "word aggregates" against each other will establish the semantic relations between the two words concerned,[1] showing whether they are fully or partially or not at all interchangeable, or whether they are similar or opposed. Or else, by choosing a term and a relational type, we can find out which other terms satisfy the relation. As such logics come to fruition, we will see to what extent we will be able to refine these semantico-linguistic considerations. For pragmatic and computational purposes, this approach to concepts is reductionist. What we wish to develop amounts, in fact, to a "calculus ratiocinator" that lends itself to applications such as search engines, dictionaries, catalogs or

---

[1]Such an operation uses an intensional definition of the concepts involved. This might look like a reversal to efforts at developing an intensional predicate calculus, frequently witnessed in the history of logic but always without success. Here, however, the qualities/characteristics are treated as mere elements and the concepts as sets; hence, the resulting calculus is extensional.

similar structures. It is obvious that we are thinking of a small-scale "ars characteristica universalis" as envisaged by Gottfried Wilhelm Leibniz.

At this point we must clarify a fundamental aspect of the extensions of sets and those of concepts. In our perspective, a concept/word/term is defined by a set whose elements are its (essential) attributes or features. A proper subset will thus contain a smaller number of those very features. As one goes from a set to one of its proper subsets, certain features will get lost, which means a *generalization* or *abstraction* of the initial set. Such a restriction of features will correspond to an increase of referential extensions. For example, the set of characteristics associated with the concept "lion", such as having retractable nails, feral canines and suckling their young, corresponds extensionally to all lions. The subset of these features containing the feature "suckling their young", corresponding to the term "mammal", has a larger extension comprising not only lions but also, for example, whales or elephants. The inverse procedure, the passing from a set to a superset, represents a specialization or particularization. Therefore, when one engages in the development of concrete applications, such as dictionaries, ontologies or databases, it will be necessary to specify whether the definitions or relations defined are to be interpreted at a connotational-intensional or a denotational-extensional level. And one will have to be consistent in this respect, on pains of a total collapse of the system. For example, when it is said that the step from mammal to vertebrate is restrictive, one is at the level of features, whereas when it is said that this step is expansive, one is at the level of extensional referents, that is, of individual entities carrying the features at issue.

Thus, taking an ordered pair of sets ba, we can use categorical predications to express the various possible cases. That is, we have:

every b is a = Aba universal affirmative (inclusion/to be) (e.g. the tables are furniture)
no b is a = Eba universal negative (denial/to be not) (e.g. the tables are not chairs)
only some b is a = Yba *distinctive* or *partial* or *exclusive* particular (intersection/to be in part = to be fuzzy) (e.g. only some tables are wooden objects).

The quantifier Y represents the intuitive natural language *some* and not the existential *some* of classical predicate logic. The latter means "at least some, perhaps all", not excluding the universal quantifier, whereas the former stands for "only some" or "at least one but not all", "all but for some", "neither all nor no".[2]

---

[2]The adjective "distinctive" alludes to the necessity to distinguish, in the subject of the particular, those elements for which the predicate delivers truth from those for which it does not. Partial *some* may be considered to be the logical product of the classical affirmative particular Iba (at least some a is b) and the classical negative particular Oba (at least some a is not b). Elsewhere [2] I have presented the "distinctive" predicate calculus. The partial one has been studied by many logicians, amongst whom I can mention N. A. Vasil'ev (1880–1940), A. Sesmat (1885–1957) and R. Blanché (1898–1975); see [2–9, 12, 27, 28]. The concept of a partial (ἐν μέρει) quantifier goes back to Aristotle [10] himself, even though he eventually chose for the rival concept of existential particular, thus conditioning logical research for many centuries to follow.

From a set-theoretical and diagrammatical point of view (Euler, Gergonne, Venn), there are five distinct possible cases of possible relations between two sets:

   I. Identity or equivalence
  II. Proper inclusion of the first in the second
 III. Proper inclusion of the second in the first
 IV. Mutual proper (nonnull and nonexhaustive) intersection
  V. Exclusion or incompatibility.

As shown in Fig. 1, owing to the three categoricals, we obtain an exhaustive tripartition of the five cases, something which would not be possible using the traditional partials (for example, the cases I or II can validate Aba as well as Iba). At the same time, there is a gain in economy.

From the classical point of view, the three categoricals are mutually *contrary*. Yet in the wake of certain observations made by Aristotle (Metaphysics X.4, 1055a19–23), we prefer to reserve this term (Greek: ἐνάντιαι) for the two universals. The term *intermediate* (μεταξύ, μέσον) seems to be more suitable for the "partial". The classic *square of oppositions* is thus replaced with the distinctive segment (see Fig. 1b, c). Figure 2 shows that the redundancy of the square of opposition (a) may be absorbed into the distinctive segment (b). The "U" quantifier means "all or no", and it is the negation of the "Y" quantifier.

On these bases we have built the Distinctive Predicate Calculus (DPC) (see [2]). This system reflects the need to follow as closely as possible natural language usage without any loss of logical rigor. And from a semantic point of view, it enables us to express intermediate or "fuzzy" concepts, essential for a proper account of linguistic communication, which is what our project is about. As shown elsewhere [2], it is in fact possible to interpret the distinctive system metalogically in terms of nonstandard logics. For example, a sentence like "Only some x is P" (where "some x" stands for a set of individuals and P for a predicate) can also be expressed as "It is

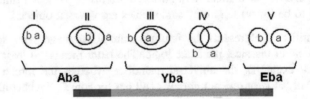

Fig. 1 Tripartition of the five cases

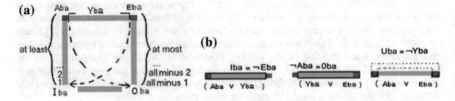

Fig. 2 Square of opposition and distintive segment

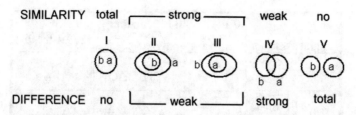

Fig. 3 Scales of similarity and difference

only partially true that the x's are P". In particular, the five cases lend logical support to the common notions of *similarity* (*affinity*) and *difference*. These may form two inversely proportional scales in the sense that the more two terms are similar, the less they are opposed, and vice versa. Using the five cases, we could thus establish the correspondences laid out in Fig. 3.

In [5] the different types or degrees of opposition[3] became a new logical-geometric theory: the *N-oppositional theory*, probably strongly connected with our DPC: see [12].

Further refinements of the analysis presented here can be realized by means of a predicate calculus with numerical quantifiers, sketched but not fully elaborated by the author, which may possibly bridge the gap between classical and fuzzy logics [2, 13].

We are thus in a position to trace a linguistic-semantic parallel (in boldface) of ordinary language usage (in italics) of the five logical relations:

  I. **Synonymous** or *Equivalent*
 II. **Hyponymous** or *Restriction* of the first in the second
III. **Hyperonymous** or *Expansion* of the first in the second
 IV. **Meronymous "sui generis"** or *Connected*
  V. **Antonymous (or Contrary) "sui generis"** or *Disconnected.*

The cases I, II and III are commonly known in linguistics and semantics. However, in these disciplines, the notions of *meronymy* (from the Greek word *méros* "part") and *antonymy* have a wider and vaguer meaning than is intended here. In particular, for *meronymy "sui generis"* of case IV, there is a special restriction: b is only partly a, but, symmetrically, a is only partly b—that is, we speak of mutual partial intersection, whereas in the literature the expression *b is meronymous to a* does not exclude the asymmetrical case II.

At this point it is necessary to establish a complete logical interpretation of these semantic/linguistic concepts.

---

[3]About this theme see [6, 7, 11].

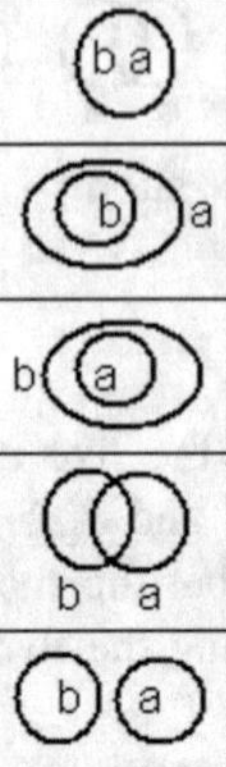

(a) SET RAPRESENTATIONS	(b) Gergonne's NOTATION	(c) CONJUNCTION OF INVERTED CATEGORICALS	(d) $_{\text{Л}}$CA QUANTIFICATION OF PREDICATE
I	b I a	Aba ∧ Aab	AbAa
II	b ⊂ a	Aba ∧ Yab	AbYa
III	b ⊃ a	Yba ∧ Aab	YbAa
IV	b X a	Yba ∧ Yab	YbYa
V	b H a	Eba ∧ Eab	EbEa

Fig. 4 Gergonne's five cases and their interpretations

It is to Joseph Diez Gergonne (1771–1859), in his study of 1816, that we owe the first creation of five distinct logical symbols for each of the five relations[4] (see column b of Fig. 4). From these five relations Gergonne distilled the first valid logical calculus exceeding the expressive and deductive power of classical syllogistic logic (see [14]). The only problem is that this symbolism takes us farther away from ordinary linguistic usage.

At the same time, as theorists of the "quantification of the predicate" (QP)[5] found out, the three categoricals are insufficient to pick out uniquely the five situation classes to which they can refer. For example, if it is true to say "All boys got promoted" for a particular school, this does not exclude the converse affirmation "All those who got promoted were boys" (case I), but neither does it exclude "Only some of those who got promoted were boys" (case II). Even so, in order to describe uniquely each of the five cases in predicative form, we can add, by means of the conjunction *and*, to the three predications over the pair *ba* those generated by the inverted predicative pair *ab* (see column c in Fig. 4).

Case A If every b is a, there are two possibilities: every a is also b (case I), or only some a is b (case II).

---

[4]He provided a more determined representation than the diagrams produced by men like Wilhelm Gottfried Leibniz (1646–1716), Johann Heinrich Lambert (1728–1777) or Leonard Euler (1707–1783).

[5]The first in modern history to design a QP system was Gottfried Ploucquet (1716–1790) in his study of 1763. Other QP logicians were Georg Johann von Holland (1742–1784), George Bentham (1800–1884), William Hamilton (1788–1856) and Charles Earl Stanhope (1753–1816). See [15, 16, 29].

Case Y If only some b is a, there are two possibilities: every a is b (case III) or only some a is b (case IV).
Case E If no b is a, then no a is b (case V).

A more synthetic, though also less grammatical, way to express such conjunctions is that of quantification of the predicate[6], as in column d of Fig. 4.

3 The Seven Relations

What is typical and relatively new in our approach is the simultaneous comparison not only of two sets but also of their two complementary sets. By definition, the latter collect the features that are present in the universe but were left out of consideration for the former. This way, the comparison of two terms will implicitly clarify the context of the universe of discourse. There exists, indeed, a more complete way of dealing with concepts, whereby the negation plays a role, as in *not-b*, shown in yellow in Fig. 5.

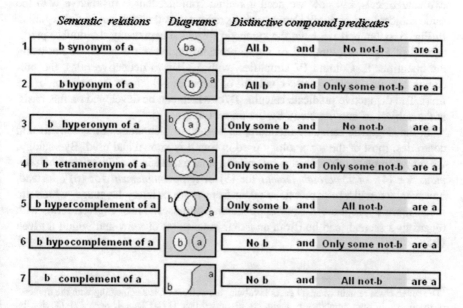

Semantic relations	Diagrams	Distinctive compound predicates			
1 b synonym of a	(ba)	All b	and	No not-b	are a
2 b hyponym of a	(b) a	All b	and	Only some not-b	are a
3 b hyperonym of a	b (a)	Only some b	and	No not-b	are a
4 b tetrameronym of a	b a	Only some b	and	Only some not-b	are a
5 b hypercomplement of a	b a	Only some b	and	All not-b	are a
6 b hypocomplement of a	(b) (a)	No b	and	Only some not-b	are a
7 b complement of a	a b	No b	and	All not-b	are a

Fig. 5 The seven relations

---

[6]To "quantify the predicate" we use the quantifier of the second categorical, which we have defined in terms of the inverted pair; E.g. case III can be read "only some b are all a". Various logicians have followed this path, committing various errors.

I 7 Cases	II Explicit forms	III Equivalent forms	IV D7c	V Tri-relational	VI Iconic
1	Aba ∧ Eb'a	Aba ∧ Ab'a'	AbaA,,	b Ọ a	b Ọ a
2	Aba ∧ Yb'a	Aba ∧ Yb'a'	AbaY,,	b))a	b))a
3	Yba ∧ Eb'a	Ab'a'∧Yba	Ab'a'Y,,	b')) a'	b((a
4	Yba ∧ Yb'a	Yba ∧ Yb'a'	YbaY,,	b)( )(a	b)( )(a
5	Yba ∧ Ab'a	Ab'a ∧Yba'	Ab'aY,,	b'))a	b( )a
6	Eba ∧ Yb'a	Aba' ∧ Yb'a	Aba'Y,,	b)) a'	b) (a
7	Eba ∧ Ab'a	Aba' ∧ Ab'a	Aba'A,,	b Ọ a'	bɕa

Fig. 6 Equivalent interpretations of the seven cases

The possible diagrammatic representations are then seven in number[7].

In the diagrams, the surrounding square represents the universe of discourse (UD). Only in case 5 is this missing, as the union of the two sets a and b form the UD.

We still have the possibility to characterize each case uniquely by conjoining two categoricals, but now we need a second conjunct that is distinctive with the same subject (and possibly also the same predicate) as the first, but in the negative. In Fig. 6, column II we have the synthetic formulae expressing the double predications of Fig. 5. Column III of Fig. 6 provides equivalent expressions not using the quantifier E. Column IV simplifies, with a view to deductive rules, the preceding column, implying the second pair of variables (which are the negation of the first). The distinctive predicate calculus **D7c**, which can be developed on this basis, is the subject of another study by the present author [2].

In the linguistic-semantic literature, as well as in dictionaries of synonyms and contraries, most of the terminology used in Fig. 4 is known and used. By analogy, we may complete this casuistry by coining new expressions such as *tetrameronymous* for (4), or *hypercomplement* for (5) or *hypocomplement* for (6)[8]. Instead, however, we will use expressions taken from more ordinary language, which are much more appropriate for our applicational purposes. Thus, a given term, with regard to a second, can be (from an extensional point of view and within a given UD):

---

[7]The seven cases remain when all cases involving the null set or sets coinciding with the universe of discourse, in pure combinatory logic, are discarded (see [17], Chap. 3, Sect. 27). Obviously, given the cases 2, 4 and 5, the set a must contain at least two members and likewise for b in the cases 3, 4 and 5. Augustus de Morgan [18] came to the seven cases on the basis of a more complex calculus. Further elements of our system converge on De Morgan syllogistic systems, from the concept of universe of discourse to the use of parentheses. Recently, Pieter Seuren [9] came to the seven cases (Chap. 8), on the basis of his theory of natural logic.

[8]In some particular contexts, some dictionaries of synonyms and antonyms call case 6 "cohyponymous", which would be an invitation to call, by analogy, case 5 "cohyperonymous". The latter, however, is found nowhere.

1. An **equivalent** of the second. Ex.: UD = triangles, equilateral–equiangular; or: UD = animals, donkeys–asses;
2. A **restriction** of the second. Ex.: UD = triangles, equilateral–isosceles; or: UD = animals, donkeys–equines;
3. An **expansion** of the second. Ex.: UD = quadrilaterals, rectangles–squares; or: UD = animals, equines–horses;
4. A **limited connection** of the second. Ex.: UD = quadrilaterals, rhombi–rectangles; or: UD = animals, oviparous–mammal (platypuses are both, protozoa are neither) or: UD = horses, females–foals; all four combinations are possible;
5. An **integrative connection** of the second. Ex.: UD = polygons, polygons with fewer than five edges–polygons with more than three edges; or: UD = hot-blooded animals, oviparous–mammals;
6. A **limited disconnection** of the second. Ex.: UD = polygons, triangles–squares; or: UD = African equines, donkeys–zebras;
7. An **integrative disconnection** of the second. Ex.: UD = triangles, isosceles–scalene; or: UD = autochthonous European equines, horses–donkeys.

One should consider the relevance of the specifications *integrative* and *limited*, meaning, respectively, that the two sets do or do not exhaust the universe of discourse, and thus permitting a distinction of the cases 5 and 6 from the cases 4 and 7, respectively.

A Boolean interpretation of the seven cases is illustrated below:

1. $$[(b \cap a) \neq \varnothing]*[(b \cap a') = \varnothing]*[(b' \cap a) = \varnothing]*[(b' \cap a') \neq \varnothing] \text{ that is } b = a$$

2. $$[(b \cap a) \neq \varnothing]*[(b \cap a') = \varnothing]*[(b' \cap a) \neq \varnothing]*[(b' \cap a') \neq \varnothing] \text{ that is } b \subset a$$

3. $$[(b \cap a) \neq \varnothing]*[(b \cap a') \neq \varnothing]*[(b' \cap a) = \varnothing]*[(b' \cap a') \neq \varnothing] \text{ that is } b \supset a$$

4. $$[(b \cap a) \neq \varnothing]*[(b \cap a') \neq \varnothing]*[(b' \cap a) \neq \varnothing]*[(b' \cap a') \neq \varnothing]$$

5. $$[(b \cap a) \neq \varnothing]*[(b \cap a') \neq \varnothing]*[(b' \cap a) \neq \varnothing]*[(b' \cap a') = \varnothing] \text{ that is } b \supset a'$$

6. $$[(b \cap a) = \varnothing]*[(b \cap a') \neq \varnothing]*[(b' \cap a) \neq \varnothing]*[(b' \cap a') \neq \varnothing] \text{ that is } b \subset a'$$

7. $$[(b \cap a) = \varnothing]*[(b \cap a') \neq \varnothing]*[(b' \cap a) \neq \varnothing]*[(b' \cap a') \neq \varnothing] \text{ that is } b = a'$$

For all cases, the condition [b ≠ ∅] * [b' ≠ ∅] * [a ≠ ∅] * [a' ≠ ∅] applies.

4   Semantic Prompter Engine

The Semantic Prompter Engine or SEM.PR.E (Italian *sempre* means "forever") that I propose [27] is a *program/database* that can be realized by an informatics programmer supported by a precise *archiving method* and a specific *question-answer algorithm*, designed by the present author[9].

When some search engine delivers unsatisfying or dubious results on a given word, SEMPRE can be queried on this term. It can be queried, for example, with regard to the restrictions, limited connections, etc., or with regard to possible other terms validating totally, partially or not at all the relation with the given term, within the terms of a given UD. From a logical point of view, such queries correspond to the multiple request for the truth value resulting from the application of all possible categorical predicates to the term in question compared with other terms in the database, in combination with their negative counterparts. The subject class is definite—that is, without a quantifier—but the predication and the resulting truth value, as shown above, are definable on the grounds of the mere comparison of the features (attributes) of the two terms in question, whereby it is specified, if necessary, in terms of what UD the query is processed, with or without its complement. This comparison is carried out by means of the Boolean operators over feature sets. Once the answer list has been obtained, one will proceed to the selection of the new input term for the same search engine. The answers provide the *alternative terms* for new searches ordered by type (and relative grade) of similarity (or difference) according to the seven typologies (weak similarity in cases 4 or 5 and no similarity for 6 or 7)[10]. So we generate the *intermediate values* with regard to pure but flexible synonyms typical of human cognition and human language. The seven relations are expressible linguistically or iconically (the bracket notation of R7), as one wishes.

The user will thus not have to memorize or call up all correlated terms, at the risk of forgetting important ones, but will make his or her choice among the most promising options suggested, perhaps, if one wishes, till all options are exhausted. The combined action of the prompter and the traditional search engine amounts to that of a *semantic search engine* with minimal human intervention. It represents both a limit and the opportunity to direct the search, as well as the possibility to discover unforeseen affinities or new associations based on objective common characteristics of concepts mostly thought to be far removed from each other.

To avoid the risk of lists becoming too long, answers will be organized hierarchically according to the level of generality desired or specified in the query. If so wished, the answers can be indexed according to level (0 horse, 1 equine, 2 mammal …). A further option, meant to simplify the selection of alternative terms, may consist in combining the output with a database that reorders the output terms on the basis of frequency coefficients in the language or texts concerned.

---

[9]There is already a rough prototype (a "demo"), resulting from the collaboration of the present author with Daniele Ingrassia, informatics engineer (University of Palermo, Italy).

[10]The five cases of Gergonne mean the loss of the distinction between contraries and complements.

5 Database Implementation

A preliminary condition of the prompter is well-structured archiving of the terms, notions and data in the database. It will not be easy to find the right person or persons to do is job, as a combination of interdisciplinary competences is required that is hardly ever realized in one single person or even in two or three. On the one hand, the implementer must be competent in the relevant aspects of logic, linguistics and philosophy; on the other, competence is required in programming and information technology; and finally, expertise is needed in the specific terminology of the discipline (or text sample) involved, so that a beginning can be made with the testing of the machinery. Take, for example, a logician, a computer specialist and a pharmacologist. The first two will establish a friendly interface that will allow video technicians to carry out hierarchical insertion, according to precise rules, of all the terms in the pharmacologist's discipline. The resulting semantic network is called an "ontology". Each term will be provided with its own features under the supervision of an expert of the discipline involved in his or her language.

To disambiguate polysemic (or homographic) words, namely those that allow more than one meaning (e.g. "set" in mathematics or in tennis; "function" in mathematics or in religion) one can make use of indexes, and matrices have been tested in projects such as the lexical ontology named "WordNet" (see Fig. 7).

Some ontologies (semantic networks) such as WordNet have a big limitation in terms of the relations, observable on two levels:

1. The logical relations between any pair of terms are just attributable to the simple or double inclusion/implication; there are no relations of intersection or complementation (if they appear, they do not interact with the other relations);
2. The totality of nodes or vertexes (= word—term—set) of the structure constitutes a hierarchical scheme, with each term being such a parent (or predecessor)–son (or successor): the topological structure is a tree graph; there are no circuits, no cyclical arcs or network structure. If there are more trees, they are not connected with each other (see [19]).

Such a tree structure, with logical relations of implication/equivalence, is useful in the early stages of cataloging, structuring data and acquisition of concepts. The transitivity of implication and equivalence allows one to deduce automatically any hierarchical knowledge which was not explicit during the insertion.

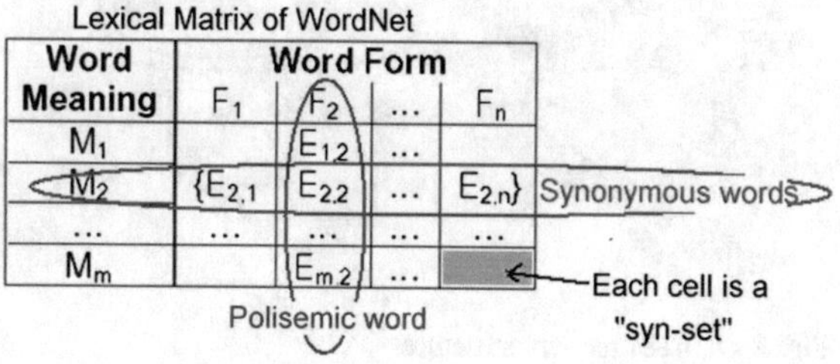

Fig. 7 A lexical matrix of WordNet

When we need to enrich the tree with new data, restructuring extension of the model, be neccessary. The tree becomes a network through the introduction of two new logical relations: intersection and complementation. In terms of natural language, they are similar to the concepts of "comparable/analogous/in-law" and "alternative/negative/outsider", respectively. These two relations together with simple/double inclusion constitute a powerful logical triad. It is formally solid in the classical bivalent logic, but also readable into fuzzy or trivalent logics.

The method for the construction of such a semantic (or ontological) network requires a category name for each node. There are two phases, both updatable:

(a) the tree (b) the creepers.

Phase a (inspired by Plato's method of *diaeresis* or analysis) allows one to *economize on time* and to avoid unnecessary repetitions in the data input, in that a feature (analytical key) inserted at any given node is inherited by all its dependent ramifications, all the way down to the terminal "leaves". In this phase, the model functions as a Porphyrian tree, as designed by Porphyry (233–305) or as contemporary cladistics (biological taxonomy) (see Fig. 8a). To enhance economy of use and avoid errors, the principle is to start from the most general or common categories among the concepts dealt with.

Phase b provides the opportunity to let other features "lean" on to the branches or to the final "leaves", thus fitting out each term with all its required essential features. As the creepers attach themselves to branches that may be quite distant from each other, we have the possibility to attach the same feature to leaves and branching nodes on different branches. Thus *Whale*, which, we assume, will be on the *Mammal* branch along with, say, *Kangaroo*, will share the feature *Aquatic* with *Shark*, which will be located on the *Fish* branch (see Fig. 8b).

There is obviously no pretence to render the metaphysical "essence" of things or thoughts. All we aim for is an appeal to pragmatic or conventional attitudes and beliefs: all that is meant with "ontology" is an underlying structuring in terms of a restricted "universe" for logical purposes. As a matter of principle, the interlaced network that comes into being via this method has no privileged "routes" and the subdivision into tree and creepers rests on an arbitrary choice made for the purpose of convenience, at both the insertion and the successive user stage.

To avoid ambiguities, the program must, in both phases, signal any possible inflow of data already present. In such a case, either the inflow is recognized as

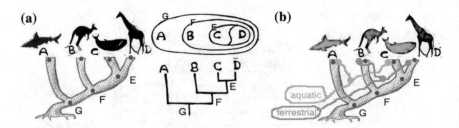

Fig. 8 A tree/creepers structure

being superfluous or the machine recognizes that it has made a "categorial" mistake, which can then be corrected at the highest hierarchical level so that corrections at lower levels are made automatically. The database will thus expand and update itself indefinitely, maintaining applicability in each phase. This allows it to integrate with other disciplines structured in analogous ways in other databases. The possibility of integrating a variety of disciplines will over time enhance the usage scope of the prompter, turning it into a continuously updatable and improvable encyclopedia.

6 Iconic Distinctive Calculus ID7

As a more iconic—that is, more visual and diagrammatic—alternative to the verbal-predicative system for the specific description of the seven cases, one may consider the iconic distinctive calculus ID7. Column VI in Fig. 6 shows a symbolic transposition of the seven topological situation classes. Symbols have been chosen (the brackets) that indicate, by their disposition and their direction, in an intuitively accessible way the extensional conditions of the terms to which they refer. Thus, the grapheme "))" evokes the situation in which the first set is an internal part of the second; ") (" stands for the situation in which two sets do not know each other but with something standing between them; in "( )" each set enters the other's territory, together occupying the entire UD; finally, ")( )(" creates not only a reciprocal territorial transgression but also a space not occupied by either.

Column III (Fig. 6) presents the translation from D7c expressions to a new iconic relational symbolism, through the equivalence based on this translation code (the dots stand for variables):

A..A becomes $\dot{\varphi}$ that means *equals*
A..Y becomes)) that means *enclosed in*
Y..Y becomes)( )(that means *tetraconnects*

The *immediate inferences* are obtained by the simple rule of *mirror rotation* of the parentheses or other grapheme together with the inversion of the quality of the adjacent term. Depending on the pair in question, we thus have four equivalent versions of diagrams and notation. For example, b))a = b) (a' = b' ((a' = b'( )a (see Fig. 9).

The $\dot{\varphi}$ (*equals*) relation consists of two hemicycles, the right and the left ones, each of which can refer to a term: if only one part rotates around its upper extremity, it results in a sort of $\}$ or mirror $\{$ (*integrates*) relation; if they both rotate, the *equals* relation $\dot{\varphi}$ is restored. Instead, the rotation of the (*tetraconnects*) relation) ( )(is not affected by free changes in the quality of the terms (Fig. 10).

This way our (distinctive) complex predicates, or even De Morgan's [18], as well as immediate inference rules, find easy translation into diagrams (Fig. 11).

In Fig. 12 we can see the mediate deductions of this system.

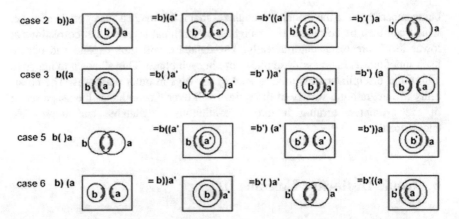

Fig. 9 Equivalent diagrams and notation of cases 2, 3, 5, 6

Fig. 10 Equivalent diagrams and notation of cases 1, 7, 4

7 cases		Distinctive Predicate Calculus		De Morgan (1847)	
(b' is yellow)		Iconic	compound predicates	compound predicates	synthetic
1	ba	bOa	Aba * Ab'a'	Aba * Ab'a'	D
2	(b) a	b))a	Aba * Yb'a'	Aba * Ob'a'	D,
3	b (a)	b((a	Yba * Ab'a'	Ab'a' * Oba	D'
4	b⚭a	b)()(a	Yba * Yb'a'	Oba * Ob'a' * Iba * Ib'a'	P
5	b a	b()a	Yba' * Ab'a	Eb'a' * Iba	C'
6	(b) (a)	b)(a	Aba' * Yb'a	Eba * Ib'a'	C,
7	b a	bꝺa	Aba' * Ab'a	Eba * Eb'a'	C

Fig. 11 Translation of the 7 cases in compound predicates (DPC/De Morgan)

		1	**2**	**3**	**4**	**5**	**6**	**7**
		b О a	b)) a	b((a	b)()(a	b( ) a	b) (a	b ◊ a
1	a О c	b О c	b)) c	b((c	b)()(c	b( ) c	b) (c	b ◊ c
2	a)) c	b)) c	b)) c	Ibc	Ycb	b ( ) c	Ib'c	b ( ) c
3	a((c	b((c	Ib'c'	b((c	Yc'b	Ibc'	b) (c	b) (c
4	a)()(c	b)()(c	Yb'c	Ybc		Ybc	Yb'c	b)()(c
5	a ( ) c	b ( ) c	Ib'c	b ( ) c	Ycb	Ibc	b)) c	b)) c
6	a) (c	b) (c	b) (c	Ibc'	Yc'b	b((c	Ib'c'	b((c
7	a ◊ c	b ◊ c	b) (c	b ( ) c	b)()(c	b((c	b)) c	b О c

Fig. 12 Table of mediate deductions of the iconic system

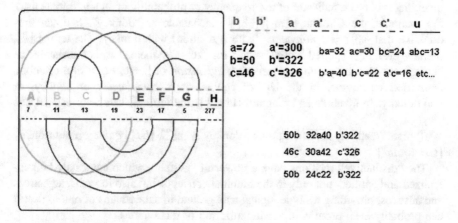

Fig. 13 A numerical diagram

A numerical development of the iconic system is also possible (the subject is work in progress by the author, see [8, 27]). We may assign a number to each of the sectors separated by the parentheses or other icons. For example:

b8 ọ a 5 => b and a are equivalent and 8 at all, the remaining are 5
b8)3)a 6 => b are 8, a are 11, not-b not-a are 6
6 b(3(8a => not-b not-a are 6, b are 11, a are 8
b5)(3)(7a 9 [or b(5(3)7)a 9] => b are 8, a are 10, not-a not-b are 9, ba are 3
b 5(3)7 a => b are 8, a are 10, not-a are 5, not-b are 7
b8) 2 (11a => b are 8, a are 11, 2 elements are neither a nor b
b4 ⟩ 3a => b are 4, a are 3

A numerical calculus can be inspired by Hacker and Parry [20, 21]: see [22, 23]. This can be considered the first step towards a fuzzy logic in the sense of [24, 25]: see [2, 27]. Numerical predicates and distinctive segment can be represented by diagrams, which can help numerical inferences (Fig. 13).

7 Distinctive Icons for Dictionaries of Synonyms, Contraries and Intermediates

ID7 can be applied to dictionaries of synonyms, contraries, with the possible addition of intermediates, as well as to machine translation programs, so as to counter cognitive anomalies in the translations. The adoption of the seven icons (or, if one prefers, only the five proposed by Gergonne) in an innovative bilingual dictionary will immediately bring to evidence which of the seven relations links the original term with each of the possible translations or descriptions (and a numerical quantifier will be an indicator of use frequency or probability of an appropriate use). For example, the Chinese term 屁股 is translatable as "donkey". We can thus establish that 屁股 is *synonymous* (with ọ̇ relation) with donkey. The term 斑馬, literally meaning "spotted horse", stands for "zebra", whereas 馬 is translatable as "horse" but also as "equine sui generis": *expansion* of horse, *expansion* of zebra, *restriction* of equine. In the UD of equines it is *complement* of donkey, or non-donkey. In symbols and diagram (Fig. 14) the term 馬 can be translated this way:

((horse; ((zebra ọ̇ 斑馬;)) equine; ⟨ donkey ọ̇ 屁股 [*UD equine*]; ọ̇ not-donkey [*UD equine*].

The "distinctive" relations have a powerful "gestalt", which can easily be generalized and applied, not only to the nominal concepts, but also to verbs, adjectives and adverbs, providing a single logical background to subdomains of ontologies. It can probably even promote the axiomatization of dictionaries.

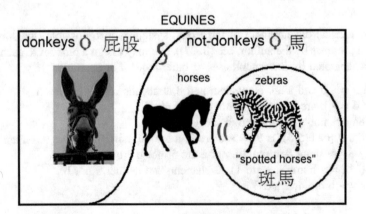

Fig. 14 A semantic diagram for zebras/equines in chinese terms

8 Cultural and Scientific Perspectives

In view of the recent multiplication of hyperspecialist languages and terminologies of subdisciplines, scientific institutes will feel the need to produce reliable translations for synthesis or interaction. The generalized adoption of the prompter will improve the quality of all translations in a wide sense, not only of texts in the general usage of a natural language but also for specialist jargons, interdisciplinary codes, glossaries of technical manuals, thesauri and library indexes of all kinds.

In cataloging of subjects in librarianship we find relations with the same logic structure as the five cases:

(I) HSF "**Head Subject For**" or UF "**Use For**" or " $=$ " all–all relation
(II) BT "**Broader Term**" all–some relation
(III) NT "**Narrower Term**" some–all relation [this and the former are called "hierarchical" relations]
(IV) "**Almost Generic**" or "[:]" some–some relation
(V) "**Related Term**" (but **if and only if** the related term is an *antonym*).

In many circumstances the **symbol** " $<>$ " encompasses the matter or field of interpretation of the term, playing a role similar to that of our **UD**, linguistically interpretable as "**conceptual background**" or "**lexicographical environment**".

In rhetoric/poetry [Liège's group, T. Todorov, see [26]], we can assimilate some "tropes" or analogies (logon) to the five cases: **Definitio** or periphrasis or tautology for I, **generalizing synecdoche** for II, **particularizing synecdoche** for III, **metaphor** for IV (product of II and III) and **litotes** or antithesis or irony for V.

The distinctive system will be helpful in the development of new approaches to numerous, often unexpected, problems, as it combines rigor with flexibility, and might even lead to a conceptual innovation. Precisely because of its closeness to natural language and the intuitiveness of its iconic-diagrammatical structure, it will be helpful not only in the more humanistic fields of philosophy, law, history, linguistics or psychology, but also, and perhaps even more so, in more scientific fields, such as logic or mathematics, computer science, engineering, the medical sciences, etc. We think that a semantic prompter can make a contribution to the organization of knowledge in all these areas. For example, it is possible to build a bridge between artificial (machine) and natural (human or animal) concepts, which are the alphabet of the representations of reality. In the case of animal representations, we should make a list of plausible elements that constitute an "animal" concept (reduced to set) analogous to that human one (the behavioral acts revealed by the animals will be the pragmatic test for the satisfactory choice of the elements). The concepts of an animal "mind" will be comparable with those of a human one, each highlighting terms of the distinctive or common features. On the same basis, we may create databases for these "mental" elements and programs to manage them.

References

1. Alesso, P.H., Smith, C.F.: Thinking on the web. In: Berners-Lee, T., Gödel, K., Turing, A. (eds.) Wiley & Sons (2006)
2. Cavaliere, F.: Fuzzy syllogisms, numerical square, triangle of contraries, interbivalence. In: Beziau, J.Y., Jacquette, D. (eds.) Around and Beyond the Square of Opposition, pp. 241–260. Springer, Basel (2012)
3. Suchon, W.: Vasil'iev: what did he exactly do? Log. Log. Phil. **7**, 131–141 (1999)
4. Blanché, R.: Structures Intellectuelles: Essai sur l'Organisation Systematique des Concepts. J. Vrin, Paris (1966)
5. Moretti, A.: The Geometry of Logical Opposition. PhD Thesis, Université de Neuchâtel, Switzerland (2009)
6. Smessaert, H.: On the 3D visualisation of logical relations. Log. Univ. **3**, 303–332 (2009)
7. Béziau, J.-Y.: The power of the hexagon. Log. Univers. **6**(1–2), 1–43 (2012)
8. Seuren, P.A.M.: The Logic of Language (Language from within, vol. 2). Oxford University Press (2010)
9. Seuren, P.A.M.: From Whorf to Montague: Explorations in the Theory of Language. Oxford: Oxford University Press (2013)
10. Aristotle: Metaphysics. Italian transl. by Reale, G., with the Greek text in front: Aristotele, Metafisica, Bompiani, Milano (2004)
11. Horn, L.R.: A Natural History of Negation, 2nd edn. UCP, Chicago (1989)
12. Moretti, A.: Why the hexagon? Log. Univers. **6**(1–2), 69–107 (2012)
13. Cavaliere, F.: A diagrammatic bridge between standard and non-standard logics: the numerical segment. In: Visual Reasoning with Diagrams, pp. 73–81. Springer, Basel (2013b)
14. Faris, J.A.: The Gergonne relations. J. Symb. Log. **20**, 207–231 (1955)
15. Gardner, M.: Logic machines and diagrams, 2nd edn. The University of Chicago Press (1982)
16. Kneale, W.M.: The Development of Logic. Claredon Press, Oxford (1962)
17. Bird, O.: Syllogistic and its Extensions. Prentice-Hall Inc, Englewood Cliffs, New Jersey (1964)
18. De Morgan, A.: Formal Logic, or the Calculus of Inference, Necessary and Probable. London (1847)
19. Kashyap, V., Bussler, C., Moran, M.: The semantic web: semantics for data and services on the web (Data-Centric Systems and Applications). Springer (2008)
20. Hacker, E.A., Parry, W.T.: Pure numerical boolean syllogisms. NDJFL **VIII**(4) (1967)
21. Murphree, W.A.: The numerical syllogism and existential presupposition. NDJFL = Notre Dame Journal of Formal Logic **38**(1) (1997)
22. Pfeifer, N.: Contemporary syllogistics: comparative and quantitative syllogisms. In: Kreuzbauer, G., Dorn, G. (eds.) Argumentation in Theorie und Praxis: Philosophie und Didaktik des Argumentierens, pp. 57–71. Vienna (2006)
23. Pratt-Hartmann, I.: On the complexity of the numerically definite syllogistic and related fragments. Bull. Symb. Log. **14**(1), 1–28 (2008)
24. Zadeh, L.A.: A computational approach to fuzzy quantifiers in natural languages. Comput. Math. Appl. **9**(1), 149–184 (1983)
25. Kosko, B.: Fuzzy Thinking: the New Science of Fuzzy Logic. Hyperion (1993)
26. Cerisola, P.L.: Trattato di retorica e semiotica letteraria. La Scuola, Brescia (1983)
27. Cavaliere, F.: Suggeritore Semantico per Motori di Ricerca e Traduttori, La Feltrinelli (2013a)
28. Hamilton, W.: Lectures on Logic. Edinburg (1860)
29. Vasil'ev, N.A.: Logica Immaginaria, Carocci (2012)

Being Aware of Rational Animals

Jean-Yves Beziau

> *Joyful is the person who finds wisdom,*
> *the one who gains understanding.*
> King Solomon, Proverbs 3:13

Abstract Modern science has qualified human beings as *homo sapiens*. Is there a serious scientific theory backing this nomenclature? And can we proclaim ourselves as wise (*sapiens*)? The classical *rational animals* characterization has apparently the same syntactic form (a qualificative applied to a substantive) but it is not working exactly in the same way. Moreover the semantics behind is more appropriate, encompassing a pivotal ambiguity. In the second part of the paper, we further delve into this ambiguity, relating rationality with three fundamental features of these creatures: ability to laugh, sexuality and transformation.

1 The Primitive Soup

Once upon a time human beings were conceived as *rational animals*. What does this mean exactly? Can we still use this definition? Are there alternative ways to understand what a human being is, to qualify, to describe this entity?

Nowadays "rational animal" looks old fashioned, but what is the new fashion, if any? The common idea is that human beings are animals, but it is not clear what exactly distinguished them from other animals (see[10]). There is in the air a mixture of different things: physiological (size of the brain, erect posture, hands), cultural (religion, art), sociological and political (cities, states, money).

Visiting scholar at the Department of Philosophy of the University of California, San Diego, invited by Gila Sher and supported by a CAPES grant (BEX 2408/14-07).

J.-Y. Beziau (✉)
University of Brazil, Rio de Janeiro, Brazil
e-mail: jyb.logician@gmail.com

J.-Y. Beziau
University of San Diego, San Diego, CA, USA

© Springer International Publishing AG 2017
G. Dodig-Crnkovic and R. Giovagnoli (eds.), *Representation and Reality in Humans,*
Other Living Organisms and Intelligent Machines, Studies in Applied Philosophy,
Epistemology and Rational Ethics 28, DOI 10.1007/978-3-319-43784-2_16

People are trying to make a connection between these different features but they are not gathered into a single idea characterizing human beings. *La mayonnaise ne prend pas*, as they say in Marseilles, and we are left in a rather chaotic situation, a kind of primitive soup. It is not clear that something will ever emerge from this soup, maybe it will keep boiling and after evaporation we shall return to dust.

Although what prevails to study human beings is a scientific approach—a mix of biology, sociology, anthropology, etc.—paradoxically science is not highlighted at the meta-level, i.e. science is not considered as a critical feature characterizing these creatures. People are not talking of human beings as "scientific animals". This would be an expression close to "rational animals", since "rational" is the Latin translation of "logos", and one of the meanings of "logos" is science. We find this meaning of "logos" in many neologisms like "anthropology", literally the science of humans, or "biology", literally the science of life.

Maybe on the one hand for post-modern philosophers science is considered as a by-product of society and culture, while on the other hand for scientists, "Rational animals" seems a mythological way to conceive human beings, they would perhaps prefer to talk about "brainy animals".

2 The Homo Sapiens Baptism

Since 1758 *homo sapiens*[1] has been used to talk about human beings, it has become a kind of official name. "Homo sapiens" is a formal expression built according to the binomial nomenclature promoted by Carl Linnaeus. Using this tool Linnaeus and his followers have baptized thousands of creatures.

"Binomial nomenclature" is a rather technical expression, more simply people will say that it is a scientific name. Nomenclature is a system to build names, related to taxonomy. Here is how taxonomy is presented in Wikipedia[2]:

> Taxonomy (from Ancient Greek: τάξις taxis, "arrangement", and νομία nomia, "method") is the science of defining groups of biological organisms on the basis of shared characteristics and giving names to those groups. Organisms are grouped together into taxa (singular: taxon) and given a taxonomic rank; groups of a given rank can be aggregated to form a super group of higher rank and thus create a taxonomic hierarchy. The Swedish botanist Carolus Linnaeus is regarded as the father of taxonomy, as he developed a system known as Linnaean classification for categorization of organisms and binomial nomenclature for naming organisms.

---

[1]Shall we write "Homo sapiens" or "homo sapiens"? Is it a proper name? This is an open question for philosophers of language.

[2]Certainly we must be cautious citing Wikipedia but it is a rather snobbish academic attitude to systematically reject it. The Wikipedia phenomenon is an important advance in the development of human rationality.

In this presentation, like in the shorter definition of the Cambridge Dictionary (Taxonomy: *a system for naming and organizing things, especially plants and animals, into groups that share similar qualities*), there is a mixture of two things: language and classification. This mixture makes sense in the perspective of establishing a parallel or isomorphism between language and classification. But this can be confused, promoting the idea that taxonomy is crucially dependent on nomenclature. This perspective was promoted by Linnaeus himself giving too much emphasize to nomenclature.

The classification used by Linnaeus is based on trees, one of the most popular forms of classification[3]; one could also say one of the most *natural* forms of classification, since it is inspired, as the very name indicated, by trees. The formal definition of tree as a mathematical object is very close to the image that we have of a tree, leaving aside leaves, fruits and other decorations—an abstract tree is completely naked. Human beings as *homo sapiens* are within a tree like other apes but also like cows, dogs, ... all the menagerie. *Cada macaco no seu galho* as they say in Piracicaba. But what kind of understanding of human beings are we getting by placing them in a tree?[4]

Taxonomy has been mainly developed in biology but it has also been exported to other areas. However taxonomy does not play a fundamental role in sciences like physics and mathematics and it is also not a key feature of contemporary biology. Taxonomy is not nowadays the basis of science. It rather appears as a primitive aspect of science going back to Aristotle. Linnaean nomenclature is based on species and genus, the two ultimate ramifications of the tree. If Linnaeus can be considered as the father of taxonomy, Aristotle can be considered as its grandfather. Aristotle had a strong interest in biology and his conception of science is much related with this science for which he promoted the idea of classification. Although the notion of tree does not appear explicitly in the work of Aristotle, he established the distinction between species and genus and the corresponding branching via difference (*diaphora*)—about this topic, see e.g. Granger [6].

Can we consider classification as the basis of science? In many senses classification is limited. It looks superficial and childish. It can be considered as a first step towards understanding but also it can be misleading, taking us in the wrong direction, obscuring things, especially when it is strongly attached with the naming game of nomenclature, like in the case of the Linnaean methodology. If we want to understand what a giraffe is, and someone tells us that it is a *Giraffa Camelopardalis* what have we learnt? Latin names are good for "épater le bourgeois" but they do not necessarily promote deep understanding.

---

[3]About the theory of classification see for example the recent book of Parrochia and Neuville [16].

[4]Wittgenstein wrote the *Tractatus Logico-Philosophicus* using the structure of a tree. Such a process is much more than simple classification; it is an articulation of the thought. A bright idea since the standard linearity is not necessarily the best way to think.

In "*Giraffa Camelopardalis*",[5] *Giraffa* is the genus and *Cameloperdalis* is the species. This means that what we usually simply called "giraffe" is part of a generic group of similar animals: giraffes have sisters or/and cousins quite similar to them (none of them are still alive). What are the common features of the giraffe genus? It is not given by the name, which is a latinization of the French word "giraffe", itself coming from the Arabic word "zaraffa". It is not by latinizing a word like "hazard", also supposedly from Arabic origin, transforming it into "hazardus", that we will have a better understanding of the idea beyond the word.[6] "Homo sapiens" works in the same way as "*Giraffa Cameloperdalis*". This suggests that human beings are part of a tribe, they are not isolated beings lost in the universe, but the characteristic of this tribe is not given by the word "homo".

Maybe we can get understanding with the difference giving birth to the species. In the case of "*Giraffa Cameloperdalis*", the species is given through "*Camelo- pardalis*", quite a beautiful name. It is a portmanteau name, a combination of "camel" and "leopard". This name is based on appearances, a phenomenon of approximation and association. In Africa some people call planes "fire birds". We can transform this primitive conceptualization into a scientific name: *ignis volu- cress*. This primitive approach also reminds us of "featherless bipeds", a definition of human beings attributed to Plato based on external features which was mocked by Diogene bringing a plucked chicken to Plato.

This is a funny story but probably not true. Plato was part of a school of people rejecting appearances, himself promoting the appraisal of the true nature of things: the "ideas" which are beyond the illusory appearances and can only be grasped by our reason. The limitation of the primitive approach can be stressed by putting parrots and human beings together in the same species, saying they are speaking animals. Nature itself, with parrots, shows us how appearances can be misleading. If we try to have a conversation with a parrot, instead of answering the question, he will repeat the question.[7]

To what kind of differences is "sapiens" pointing at, the word defining the species of human beings within the genus "homo"? Strangely enough it is nothing like feather, brain or bipedism. "Sapiens" is a Latin word meaning wisdom, translation of the Greek word "Sophia". Why Linnaeus chose this qualificative? Is it a good description of what human beings are?[8]

The technical biological contemporary characterization of human beings is not at all connected with wisdom: It is difficult to find wisdom among cells, nerves, genomes. At best we can say that biological beings are complex, that there is a kind

---

[5]The original name given by Linnaeus was *Cervus Cameloperdalis*.

[6]About the origin of the word "hazard", see the interesting philosophical analysis of Clément Rosset in his book *La logique du pire* (1971).

[7]We are exaggerating a bit. Parrots, like Alex the grey, can somewhat answer questions, but also machines can do.

[8]Human beings are in fact nowadays considered as a subspecies of *homo sapiens* called *homo sapiens sapiens*, they are the wisest among the wise ...

of intelligence in action (see [14] about this topic). But wisdom is not part of the biological machinery. What is wisdom?

Today a standard meaning of wisdom is "The ability to use knowledge and experience to make good decisions". It is not clear that this characterizes human beings. Can we say that the destruction of millions of living beings, including human beings, is a manifestation of wisdom? It is rather pretentious to claim oneself as wise. This pretension is more typical of sophists by opposition to philosophers. "Sophists" means *the Wise*, "Philosophers" means those who are *fond of wisdom*.

Blaise Pascal said: "Je ne discute pas du mot pourvu qu'on m'explique le sens qu'on lui donne" (I never quarrel about a name, provided I am apprised of the sense in which it is understood—*Provincial Letters* 1657). This claim is nice but controversial. On the one hand we can agree with Pascal that words are not so important, we don't want to be lost in wordy discussions. On the other hand to give the right name to the right thing is a wonderful art.

The arbitrariness of naming can be supported by the axiomatic method. In his famous essay on *the geometrical spirit and the art of persuasion* (1657) Pascal gave a new shape to the axiomatic method, articulating clearly its three aspects: axioms, definitions and deductions. But Pascal in this essay claims that it does not make sense to define everything and he gives as a typical example human beings. It is true that it is easier to define a circle or infinity than a pig or a human being.

From the point of view of the modern axiomatic approach, in principle everything can be defined. Pascal's essay was a source of inspiration for Tarski (see in particular his 1937 essay), one of the main promoters of the modern axiomatic method and the developer of model theory. Tarski, as other people of the movement for the unity of science in the 1930s, had the idea to apply the axiomatic method to any kind of science. This was done mainly in physics despite the fact that Tarski's first love was biology. He did not work himself in that direction but encouraged other people to do so, in particular Woodger (see his 1937 book). However this line of research did not grow and flourish to give birth to an axiomatic theory of life, which could be the basis of a definition of human beings, and we are left with the chaos of an empirical science.

3 The Syntax of Rational Animal

"Rational animal" has the same syntactic appearance as "homo sapiens", but if we have a closer look we can see that it works quite differently. Rational animal is a mixture of two notions, one expressed through a substantive, the other one through a qualificative.[9]

---

[9]We are not writing here "rational animals" because our discussion is not about words. And we say that rational, not "rational", is a qualificative because we are talking about the notion expressed by the word.

In the same way we can speak about

- round squares
- black cows
- sports cars
- prime numbers
- needled trees
- strange ideas
- big cats

Sometimes it is possible to commute the qualificative with the substantive. For example instead of talking about a round square, we can talk about a squared circle. Both have quite the same meaning, and the same denotation. In the other above examples it is not so easy, perhaps impossible, to commute. Most of the qualificatives have a substantive correspondent. For example to strange corresponds strangeness, but the problem is rather with substantives that cannot so easily be transposed into qualificatives. Idea can drive us to ideal, but to which qualificative can we go with a car?

To rational corresponds rationality, and to animal corresponds, as qualificative, animal or bestial. But rationality is like humanity, it is rather singular than plural, not to speak about gold or other mass terms. There are no rationals (except in mathematics) like there are animals. And if there were, would bestial rationals be synonymous with rational animals? Bestial rationals could be considered as human beings by opposition to computers or other nice machines having no emotions or brutal behaviors.

"Rational animals" looks like "acronymic words": we have a class of things given by the substantive and the qualificative is delimiting a subclass of those things to which it applies, not applying to something outside of the ground class. There are no acronymic beings besides words. There are no rational beings besides animals. At least this was the idea before computers came into existence and if we consider that God does not exist or is an animal or is not rational.[10]

Can we say for this reason that the qualificative is dealing with an essential feature? A qualificative which applies to many different substances is necessarily superficial: there are black cats, black holes, black skirts. Acronyms are a specific species of words, but they can be considered as a kind of abbreviation. It means that the idea beyond the notion is not so original; it is a particular aspect of a more general phenomenon from which it is not radically different. In the case of rational animal we have something more essential. Rational is not part of a broader quality.

And considering other groups of animals, is it possible to describe them in a similar way: dogs, snakes, birds? Let us have a look at some standard definitions:

Bird: an animal that has wings and feathers and is usually able to fly
Snake: a long, thin creature with no legs that slides along the ground
Dog: an animal with hair, four legs and a tail, often kept as a pet.

---

[10]Let us remember here that according to the *Bible*, God = Logos (cf. John 1:1).

None is really working like rational animals, maybe the closest we get is with the definition of birds as flying animals. Flying is a remarkable feature, not as superficial as color or size. However:

(1) some birds are flightless, like penguins
(2) some beings like mosquitos or butterflies are considered animals, but not birds, despite the fact they fly
(3) some things which are not animals, like planes, fly.

Making a parallel between rational animals and flying animals (3) is not necessarily fatal: one may also argue that nowadays there are rational things which are not human beings. In both cases, planes and computers, these are machines built by human beings. But maybe it is more "natural" to say that planes fly than computers reason, even if we are supporting the idea of "artificial" intelligence based on penguins, the mascot of non-monotonicians see e.g. [7].

Regarding (2), one may argue that the flight of mosquitos and butterflies is quite different from the flight of swallows, eagles or flamingos, in the same way that monkeys or dolphins are not really reasoning. About (1), one may say that penguins have wings, a potential capacity to fly but that for some reasons they are not using them. We can say that some human beings have some capacity to reason but are not reasoning. It is not clear how we can characterize this subspecies. In any case this is not a breeding class, i.e. a group of animals that can breed among each other and with no others.

Something which is maybe more similar to "rational animals" than "flying animals" is "prime numbers" if we consider "prime" in its technical sense, but these are abstract entities.

4 The Semantic Network of Logos

"Rational animals" is the translation/adaptation of the Greek expression "logical animals". If "zoion" and "animals" can be considered as semantically equivalent both extensionally and intensionally, this is not exactly the case with "rational" and "logical". "Logical" is connected with "logos", a central word of classical Greece. This word has four main meanings: language, science, relation and reason. The linguistical dimension of "logos" appears in particular in "neologism" which means *new word* and "syllogism", which means *sentences put together*. As we have already pointed out, "logos" also means science, as shown in neologisms like "biology" or "psychology". The relational aspect clearly appears in mathematics with the notion of irrational numbers ("a-logical" numbers), numbers which are not relations between natural numbers. The fact that there are such numbers was put forward in the proof of the irrationality of the square root of two. This is considered by some as the starting point of mathematics [5], because it is the first non-trivial proof. A mathematical proof is a prototype of reasoning. And reasoning is the fourth aspect of the "logos".

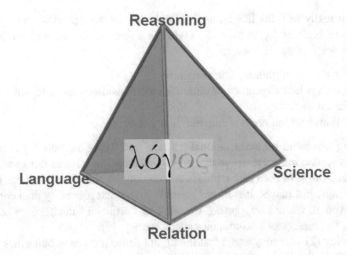

Let us emphasize that the above diagram which is centered on a tetrahedron (a 3-dimensional simplex), can be seen as metaphorical or/and "pittoresque", but nevertheless has an intrinsic value.[11] This is not because many people use diagrams and images in a rather superficial way in business and new age philosophy that they cannot be used in a positive and intelligent way. Mathematics which is considered as the climax of reason is much connected with visual thinking. This has been made clear in recent decades in many ways, through fractals,[12] diagrams as logical tools [11] and proofs without words [15].

Can we define human beings as "reasoning animals"? It is not obvious that human beings have always reasoned in the strong sense of the word. The reasoning leading to the irrationality of the square root of two is a reasoning by absurdity. Although it seems that the idea of reasoning by absurdity emerged from Eleatic philosophy [20], it is not clear that this kind of reasoning appeared in other parts of the world before, and that in the Western world it really developed and firmly took shape outside of mathematics, although Plato in some sense tried to use it in a broader context.

The understanding of what is reasoning by absurdity took many centuries and was really clarified only at the beginning of the twentieth century at the time of modern logic. Modern logic gave birth to computation and a computer can perform reasoning by absurdity.

Logical in this sense does not reduce to a biological feature, because on the one hand the birth of the reduction to the absurd is not related to a biological mutation

---

[11]It is better to use a tetrahedron because here the four corners have the same distance between each other, which is not the case in a square. Such a feature was taken into consideration by Alessio Moretti to generalize the square of opposition using simplexes see e.g. [12].

[12]Let us remember here that fractals were developed by the nephew of the Bourbachic mathematician Szolem Mandelbrojt, Benoît Mandelbrojt.

or the size of the brain, and on the other hand a purely physical device, like a computer, can perform it.

We will explore next in which sense rationality in a strong sense is related to other crucial aspects of human beings. First we start with a rather light feature before we go to some more tragic issues.

5 Laughing Animals

Laughter is a distinctive feature of human beings. Who has seen a cat, a dog or even a monkey laughing? A computer can laugh. But this is an artificial laugh. Note that sometimes people also laugh rather mechanically.

Can we characterize animals as laughing animals? Or is laughing just a superficial feature of these creatures? We can have a better understanding of this question by examining the relation between laugh and rationality.

There are different connections between laugh and rationality, which have been described by different authors in different ways, in particular Bergson [2]. One of the basic ideas is that laughing is the result of an unusual and surprising mix of concepts. We can therefore qualify laughing as conceptual. And conceptualization is an important feature of rationality.

Morris [13] has pointed out that the child first starts to laugh when he is able to recognize his mother. The relation between this recognition and laughing is as follows: the mother is considered by the child as the basis of his security, when she surprises him by jokingly creating fear this appears to him as a contradiction and this contradiction is the cause of his laugh. This can be called "rational laugh". It has many aspects and is related more generally to absurdity of which contradiction is a part.

But there is also a kind of irrational laugh, also proper to human beings. It manifests as fits of laughter. In Paris, people say *fou rire*, literally "crazy laugh". What is crazy in this laugh is not necessarily the object/cause of the laugh but its expression. Someone starts to laugh and cannot stop. It is called "une crise (crisis) de fou rire". This can be contagious and can lead to an epidemic like the one in Tanzania in 1962 (see [9]). Something lighter is "rire aux éclats", "éclater de rire" (burst of laughter). These are rather irrational features of laughing contrasting with conceptual laughing. But when we are talking about irrationality we are still connected with rationality.

Beyond rational and irrational laughing, there is a third different phenomenon, which is smiling. Smiling is related with a kind of intelligence, expressed by "cleverness" in English, "Metis" in Greek,[13] "ruse" (cunning) or "malin" in French. The fox is the symbol of the *ruse*, a famous character of Aesop's tale *The fox and the crow*, and of the medieval *Roman de renart*. This kind of intelligence is quite

---

[13]See the excellent book of Detienne and Vernant, *Les ruses de l'intelligence*, 1974.

ambiguous and can sometimes appear as dangerous not to say fiendish, as it appears in the French word "malin" and in English, "devilish".

On the other hand the smile of Mona Lisa seems to escape this perspective, having only a positive aspect, but which is difficult to explain. This smile is indeed qualified as mysterious or enigmatic. "Foxy lady" is a legendary mix of the two.

6 Sexual Animals

In 2014 I organized the *5th World Congress on Paraconsistency* with Mihir Chakraborty at the Indian Statistical Institute in Kolkata, India and had the idea to launch the contest "Picturing contradiction". The contest was open to anyone, even to those not able to come to Kolkata. The contenders had to send us a picture of a contradiction with a few words of explanation. Among the proposals, we received a picture by Sharon Kaye, John Carroll University, called "The Orgasmic Calculator" presenting a woman having a sexual relation and at the same time performing a complicated computation. S. Kay makes the following comments: "This is a picture of contradiction because the woman is both rational and irrational at the same time. Rationality and irrationality are mutually exclusive and jointly exhaustive opposites. So her state is logically impossible. She is rational because she is performing a complex mathematical operation. Complex mathematical operations are instances of pure rationality. She is irrational because she is in the middle of a sexual orgasm. A sexual orgasm is an instance of pure irrationality".

This proposal did not win the contest for two reasons: on the one hand the quality of the picture was rather weak, on the other hand her explanation although interesting is not necessarily convincing. A first point is that it is possible to operate highly rational activities under emotional circumstances—whether connected to pleasure or pain. Blaise Pascal while having intense physical tooth pain was occupying his mind with abstract reasoning to escape the pain. A second very different point is that the mental state of sexual orgasm may be connected with some intellectual activities. Here is a quotation by the famous Bourbachic mathematician André Weil: "Every mathematician worthy of the name has experienced, if only rarely, the state of exaltation in which one thought succeeds another as if miraculously, and in which the unconscious (however one interprets the word) seems to play a role … Unlike sexual pleasure, this feeling may last for hours at a time, even for days" ([23], p. 91).

This should not be confused with "mental masturbation", an intellectual selfish pleasure. Like orgasm, what Weil is talking about is the contrary of selfishness, because on the one hand one loses his self, and on the other hand this unification with the whole is related to creation, creation of a new human being or new ideas and understanding.

Human sexuality is much stronger than the sexuality of other animals. This manifests in two ways, one being the intensity and duration of the orgasm, the other one being love. Desmond Morris, in his famous book *The naked ape*, has argued that the sensibility of the skin is strongly sexual and that sexuality takes the dimension of love among human beings because a strong attachment has to be developed between the father and the mother so that they will stay together long enough to take care of the child whose development needs more time than other animals, mainly due to the complexity of his brain, the source of rationality.

Before Morris, Schopenhauer had emphasized the specificity of the sexuality of human beings in a quite similar way, i.e. making an explicit connection between sexuality and love, but also in a much more philosophical way. For Schopenhauer human sexuality is the climax of the Will, the Will being the basis of the manifestation of reality. Human sexuality is therefore the symbol of being. But in a strange dialectical reversal, Schopenhauer then promotes an asceticism strongly attached to the rejection of sexuality, since according to him the negation of the Will is the negation of the illusory reality, leading to nirvana.

Sexuality, for Schopenhauer, like for Freud later on,[14] is a profound unconscious power. One can argue that rejection of the will is intelligence. To tame this power, the human being has to be highly intelligent. But there is a less dramatic way to see the situation. Sexuality leads to orgasm which is a fruitful and creative state of mind related to the outmost form of rationality, wisdom. Without wisdom human beings are at best transforming animals.

7 Transforming Animals

Human beings can be considered as transforming animals. Transformation is a striking feature of these animals. These creatures have the capacity to transform nearly everything: themselves and their surrounding reality. Transforming matter into a space rocket, a computer, a movie … transforming ideas into reality. The earth has been seriously transformed by human beings. This is completely different from what happens with other animals. Although in nature everything is changing all the time, standard animals are evolving in rather stable ecological systems.

Human beings are transforming not only one ecological system but the whole ecological system in particular through global warming. Even before global warning, they had destroyed thousands of species, hunting without survival necessity. "Reasons" ranging from pleasure, business or simplification. Systematically killing wolves or other animals big or small which could have been dangerous for them. The word "reason" is used here in a general sense. When someone says "the reason of X is Y", she is trying to explain the relation between X and Y, sometimes saying that X is the *cause* of Y. We find back here one of the four aspects of the logos, *relation,* but mixed in a quite ambiguous way with another aspect, the *rational* aspect. This mixing can lead to a justification of everything: a war has a cause, so it makes sense …

---

[14]About the relations between Schopenhauer and Freud, see e.g. [8].

Extinction of species is a natural phenomenon, but the quality and quantity of species extinction caused by human beings has no equivalent in the animal world. It has an equivalent maybe only at a pure physical level, considering for example the supposed extinction of dinosaurs as a result of the shaking of the earth by a meteorite. But an absurd blind force projecting a meteorite on the earth, or an earthquake like that in Lisbon in 1755, destroying one of the most important cities of that time, is not the same as this destructive transformation by human beings against nature or/and themselves.

Transformation is certainly connected with destruction, as expressed by the figure of Shiva considered as the transformer-destroyer, in the classical Hinduist *trimurti*, by opposition to Brahma, the creator, and Vishnu, the preserver.[15] Human beings can transform reality for better or for worse through creation and destruction. This capacity for transformation is amplified by science based on rationality, and in modern times with the creation of the atomic bomb, humanity has reached a crucial level, the level of auto-destruction.

If humanity is annihilated by the atomic bomb, this could be interpreted as a kind of suicide (about suicide, see[3]). Suicide is also a striking feature of human beings, other animals don't commit suicide. "Managing" death is a fundamental aspect of human beings, and rationality can be a pivotal tool for suicide, for individuals: just like taking pills or using a revolver makes suicide much easier, at the general level, science with its atomic bombs allows humanity to commit suicide.

Now that science has made life much easier, eliminating dangerous wolves, cold winter, the need to fight for food and shelter, people find themselves in a much more comfortable situation and then boredom appears and the main cause of death is suicide.

A striking characteristic of rationality as it appears in science is that it is disconnected with moral issues. On the one hand someone may be a bad guy and be very "intelligent", for example a very good mathematician—the proof of the theorem by X will be true forever even if X's behavior is not politically correct. On the other hand products of science and technology can be used in good or bad ways. The father of aviation, Santos Dumont committed suicide when he saw that planes were used for war see [22].

Can we rationally use rationality? That is a difficult question. This is directly related with philosophy (cf. Plato's cave) and religion. The ultimate science would be the science of distinguishing the bad from the good, but it is not clear that it is a human science, cf. *Genesis* Basset [1].

Acknowledgements Thanks for comments by Catherine Chantilly, Jeremy Narby and six anonymous referees.

---

[15]About this triangle of opposition and other figures of opposition see our paper "The Power of the hexagon" (2012).

References

1. Basset, L.: Langage symbolique de Genèse 2–3. In: Beziau, J.-Y. (ed.) La pointure du symbole. Petra, Paris (2014)
2. Bergson, H.: Le Rire. Essai sur la signification du comique - Laughter, an essay on the meaning of the comic. Paris (1900)
3. Beziau, J.-Y.: O suicídio segundo Arthur Schopenhauer. Discurso **28**, 127–143 (1997)
4. Detienne, M., Vernant, J.-P.: Les ruses de l'intelligence. La mètis des grecs. Flammarion, Paris (1974)
5. Dieudonné, J.: Pour l'honneur de l'esprit humain. Hachette, Paris (1987). (English trans.: Mathematics—The Music of Reason. Springer, Berlin (1992))
6. Granger, H.: Aristotle and the genus-species relation. South. J. Philos. **18**, 37–50 (1980)
7. Ginsberg, M.L. (ed.): Readings in Nonmonotonic Reasoning. Morgan Kaufmann, Los Altos (1987)
8. Gupta, R.K.: Freud and Schopenhauer. J. Hist. Ideas **36**, 721–728 (1975)
9. Hess, C.W., Dvorák, C.: Neurologie du rire. Rev. Méd. Suisse **179**, 2473–2478 (2008)
10. Hurley, S., Nudds, M. (eds.): Rational Animals?. Oxford University Press, Oxford (2006)
11. Moktefi, A., Shin, S.-J.: Visual Reasoning with Diagrams. Birkhäuser, Basel (2013)
12. Moretti, A.: Why the logical hexagon? Logica Universalis **6**, 69–107 (2012)
13. Morris, D.: The Naked Ape: A Zoologist's Study of the Human Animal. Jonathan Cape, London (1967)
14. Narby, J.: Intelligence in Nature—An Inquiry into Knowledge. J.P. Tarcher/Penguin, New York (2005)
15. Nelsen, R.B.: Proofs without Words: Exercises in Visual Thinking I and II. Mathematical Association of America, Washington (1983, 2000)
16. Parrochia, D., Neuville, P.: Towards a General Theory of Classification. Birkhäuser, Basel (2013)
17. Pascal, B.: De l'esprit géométrique et de l'art de persuader (1657). (trans.: On the Geometrical Spirit and the Art of Persuasion)
18. Rosset, C.: La logique du pire. PUF, Paris (1971)
19. Schopenhauer, A.: Die Welt als Wille und Vorstellung. (1818, 1844). (trans.: The World as Will and Representation)
20. Szabó, A.: Anfänge der griechischen Mathematik. Akademiai Kiádo, Budapest (1969). (English trans.: The Beginnings of Greek Mathematics. Kluwer, Dordrecht (1978))
21. Tarski, A.: Sur la méthode déductive. In: Travaux du IXe Congrès International de Philosophie, VI, pp. 95–103. Hermann, Paris (1937)
22. Wykeha, P.: Santos Dumont: A Study in Obsession. Harcourt, Brace & World, New York (1962)
23. Weil, A.: Souvenirs d'apprentissage. Birkhäuser, Basel (1991). (trans.: The apprenticeship of a mathematician, Birkhäuser, Basel (1992))
24. Woodger, J.H.: The Axiomatic Method in Biology. Cambridge University Press, Cambridge (1937)

Part VI
Machine Perspectives

Simple or Complex Bodies? Trade-offs in Exploiting Body Morphology for Control

Matej Hoffmann and Vincent C. Müller

Abstract Engineers fine-tune the design of robot bodies for control purposes; however, a methodology or set of tools is largely absent, and optimization of morphology (shape, material properties of robot bodies, etc.) is lagging behind the development of controllers. This has become even more prominent with the advent of compliant, deformable or 'soft' bodies. These carry substantial potential regarding their exploitation for control—sometimes referred to as 'morphological computation'. In this article, we briefly review different notions of computation by physical systems and propose the dynamical systems framework as the most useful in the context of describing and eventually designing the interactions of controllers and bodies. Then, we look at the pros and cons of simple versus complex bodies, critically reviewing the attractive notion of 'soft' bodies automatically taking over control tasks. We address another key dimension of the design space—whether model-based control should be used and to what extent it is feasible to develop faithful models for different morphologies.

This article is a substantially revised version of [20].

M. Hoffmann (✉)
Faculty of Electrical Engineering, Department of Cybernetics, Czech Technical
University in Prague, Karlovo Namesti 13, 121 35 Prague 2, Prague, Czech Republic
e-mail: matej.hoffmann@fel.cvut.cz

M. Hoffmann
iCub Facility, Istituto Italiano di Tecnologia, Via Morego 30, 16163 Genoa, Italy

V.C. Müller
Anatolia College/ACT, PO Box 21021, 55510 Pylaia, Greece
e-mail: vmueller@act.edu

V.C. Müller
Department of Philosophy, IDEA Centre, University of Leeds, Leeds LS2 9JT, UK

© Springer International Publishing AG 2017
G. Dodig-Crnkovic and R. Giovagnoli (eds.), *Representation and Reality in Humans,
Other Living Organisms and Intelligent Machines*, Studies in Applied Philosophy,
Epistemology and Rational Ethics 28, DOI 10.1007/978-3-319-43784-2_17

1   Introduction

It has become increasingly common to explain the intelligent abilities of natural agents through reference to their bodily structure, their morphology and to make extended use of this morphology for the engineering of intelligent abilities in artificial agents, e.g. robots—thus 'offloading' computational processing from a central controller to the morphology. These uses of morphology for explanation and engineering are sometimes referred to as 'morphological computation' (e.g. [15, 16, 39]). However, in our view, only some of the characteristic cases that are embraced by the community as instances of morphological computation have a truly computational flavor. Instead, many of them are concerned with exploiting morphological properties to simplify a control task. This has been labeled 'morphological control' in [14]; 'mechanical control' could be an alternative label. Developing controllers that exploit a given morphology is only a first step. The space of possible solutions to a task increases dramatically once the mechanical design is included in the design space: imagine having a hand with 10 instead of 5 fingers: there will be completely new ways of grasping things. At the same time, the search space of controllers and mechanical design combinations also becomes enormous.

In this work, we want to take a close look at these issues. First, we will borrow the 'trading spaces' landscape from [41] that introduces a number of characteristic examples and distributes them along a metaphorical axis from 'informational computation' to 'morphological computation'. Second, we will analyze under what circumstances physical bodies can be said to compute and then propose the dynamical systems description as the most versatile framework to deal with brain–body–environment interactions. Third, we will critically look at the pros and cons of simple versus complex (highly dimensional, dynamic, nonlinear, compliant, deformable, 'soft') bodies. Fourth, we will address another key dimension of the design space— whether model-based control should be used and to what extent it is feasible to develop faithful models for different morphologies. We will close with an outlook into the future of 'soft' robotics.

2   Design 'Trading Spaces'

Pfeifer et al. [41] offer one possible perspective on the problem in Fig. 1. In traditional robots—as represented by industrial robots and Asimo in the figure—control is essentially confined to the software domain, where a model of the robot exists and the current state of the robot and the environment is continuously being updated in order to generate appropriate control actions sent to the actuators. In biological organisms, on the other hand, this does not seem to be the case: the separation between 'controllers' and 'controlled' is much less clear, and behavior is orchestrated through a distributed network of interactions of informational (neural) and physical processes. Furthermore, there is no centralized neural control, but a multitude of recurrent loops

informational
computation ⟷ morphological
computation

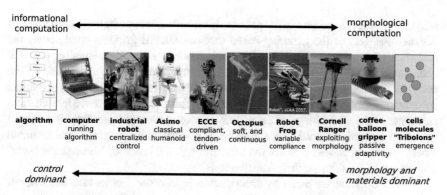

algorithm	computer	industrial	Asimo	ECCE	Octopus	Robot	Cornell	coffee-	cells
	running algorithm	robot centralized control	classical humanoid	compliant, tendon-driven	soft, and continuous	Frog variable compliance	Ranger exploiting morphology	balloon gripper passive adaptivity	molecules "Tribolons" emergence

control
dominant ⟷ morphology and
materials dominant

Fig. 1 The design trading space. This figure illustrates the degree to which each system relies on explicit control or self-organization of mechanical dynamics. On the *left-hand side* of the spectrum, computer algorithms and commercial computers rely on physical self-organization at the minimum level, while towards the *right-hand side*, more embodied, more soft and smaller scale systems require physical interactions as driving forces of behaviors. The design goal then is to find a proper compromise between efficiency and flexibility, taking into account that a certain level of flexibility can also be achieved by changing morphological and material characteristics. (Figure and caption from [41])

from the lowest level (e.g. reflexes and pattern generators in the spinal cord) to different subcortical and cortical areas in the brain. At the same time, the bodies themselves tend to be much more complex in terms of geometrical as well as dynamical properties. This has motivated the design of compliant, tendon-driven robots such as ECCE [51] or Kenshiro [35] and soft, deformable robots such as Octopus (e.g. [29]) (we are moving from left to right in Fig. 1). However, compared with humans or biological octopus, a comparable level of versatility and robustness in the orchestration of behavior has not yet been achieved in the robotic counterparts. In more restricted settings, the design and subsequent exploitation of morphology is easier, as the jumping and landing robot frog [36], the passive dynamic based walker (Cornell Ranger [4]), or the coffee-balloon gripper demonstrate. The predecessors of the Cornell Ranger, the original passive dynamic walkers [33], are a powerful demonstration that appropriate design of morphology can generate behavior in complete absence of software control. Yet, there is only a single behavior, and the environmental niche is very narrow. The coffee-balloon gripper [8] employs a similar strategy, but achieves surprising versatility on the types of objects that can be grasped. Body designs that follow this guideline were also labeled 'cheap designs' [42].

3 Is the Body Really Computing?

The systems toward the right-hand side of Fig. 1 rely on physical interactions rather than computer programs to orchestrate their behaviors. This end of the spectrum is labeled 'morphological computation'. However, in which sense can these systems be said to compute? In the case of the passive dynamic walker and its active descendants (Cornell Ranger) or the jamming-based (coffee-balloon) grippers cited above, the body is ingeniously contributing to its, perhaps primary, function: enabling physical behavior in the real world. This is often interpreted in the 'offloading sense': the body design takes over computation from the brain (e.g. [39])—the hypothetical computation that is needed for walking has been fully off-loaded from a hypothetical controller to the morphology of the walker. However, we argue that this view is hard to defend beyond the level of a metaphor. It is difficult to imagine a real example where one could choose to solve the task 'through the brain' or 'through the body' and smoothly interchange their contributions.

A word on what we mean by computation is in order. Let us take the Cornell Ranger example—a robot based on the passive dynamic walker with a simple controller on top. The robot is certainly not performing abstract digital computation (as represented by the Turing model, for example). Borrowing the terminology from [13, p. 5–6], the part of the controller can be said to perform online and embedded computation—such computation is interactive rather than batch, as it relies on a continuous stream of inputs (from sensors in this case) for its execution and produces a continuous stream of outputs (control actions). However, it is the physical interaction, not the controller, that plays the key part in accomplishing the task here. Some authors would subsume this type of interaction under a computational framework as well—e.g. "embodied computation should be understood as a physical process in an ongoing interaction with its environment" [13, p. 6]. Other authors pose much stricter requirements on physical computation: according to Horseman et al. [22], a physical system can be said to compute only if it was designed as such. That is, there needs to be a user that has an abstract computational problem that he wants to solve by a physical machine. This machine (the computer) needs to be designed and its model derived that allows for encoding of abstract inputs into the machine and decoding them again after physical evolution of the machine's state. In this view, computation cannot be assigned *ex post*, and physical systems with interesting computational properties, 'intrinsically computing' [11], do not fulfill these requirements.

It is not central to practitioners whether the controlled system is 'computing'. However, a unified theory or level of description is desirable. The dynamical systems framework seems to be the most versatile in this context, as it (i) fits the informational and physical processes equally well, (ii) copes with continuous (in time) streams of continuous input and output signals, and (iii) is already used by control theory.

4 Dynamical Systems Perspective

Let us look at the concept of self-stabilization, which is often cited in the "morphological computation community". While maximally exploiting the interaction of the body with the environment can lead to 'pure physics walking' like in the passive dynamic walker case [33], what if the agent is perturbed out of this preferred regime? It seems that corrective action needs to be taken. However, it can be the very same mechanical system that can generate this corrective response. This phenomenon is known as self-stabilization and is a result of a mechanical feedback loop. To use the dynamical systems description, certain trajectories (such as walking with a particular gait) have attracting properties and small perturbations are automatically corrected, without control, or one could say that 'control' is inherent in the mechanical system. Examples of this phenomenon are a self-stable bike, driving alone after being pushed and compensating for major disturbances [27], or the contribution of biological muscles to human walking as reviewed by Blickhan et al. [5] in a paper entitled 'Intelligence by mechanics' (more examples in this line can be found in [19] or [18] with videos of the bicycle and other material).

A general formulation of a control problem in control theory is making a dynamical system follow a desired trajectory. For our purposes, we will consider the cases where the dynamical system is physical—the body of the agent; in control theory, this is the so-called 'plant'. There are numerous control schemes and branches of control theory, and the reader is referred to abundant literature on the topic (e.g. [3, 12, 25]). The performance of the controller can be evaluated on various grounds: precision of a trajectory with respect to a reference trajectory, or energy expenditure, for example. In addition, performance, stability, and robustness guarantees are required by industry. Control theory typically deals with the design of controllers that optimize these criteria. Some control schemes with appropriate cost functions will automatically result in minimal control actions and thus "optimize the contribution of the morphology". For example, Moore et al. [34] used discrete mechanics and optimal control to steer a satellite while exploiting its dynamics to the maximum. Carbajal [10] developed related methods for reaching, plus offered a formalization of the concept of 'natural dynamics'. Nevertheless, the plant is treated as fixed in these approaches. Yet, the properties of the physical body obviously have a key influence on the final performance of the whole system (plant + controller), which calls for including them into the design space.

5   Simple or Complex Bodies?

The spirit of the morphological computation literature that follows the 'offloading' or 'trade-off' perspective, is that complex (highly dimensional, dynamic, nonlinear, compliant, deformable, 'soft') bodies are advantageous for control because they can take over the 'computation' that a controller would otherwise have to perform (e.g. [15, 16, 39] or [9] explicitly in Fig. 1). Complex nonlinear bodies certainly give rise to more complex dynamical landscapes where the location of attractors *could*—if properly exploited— facilitate the performance on a given task.

This view is in stark contrast to the views prevalent in control theory. There, linear time-invariant systems are the ideal plants to control. Solutions for nonlinear systems are much more difficult to obtain, and they often involve a linearization of the system of some sort. In fact, human-like bodies are a nightmare for control engineers ([43] is an interesting case study), and highly complex models and controllers would be required.

What would be an ideal body then? And, does a complex body imply simple or complex control? Recent attempts at quantifying the amount of morphological computation shed more light on this issue. Zahedi and Ay [52] propose two concepts for measuring the amount of morphological computation by calculating the conditional dependence of future world states W' (encompassing the body state) on previous world states W and action A taken by the agent. According to concept 1, the amount of morphological computation is inversely proportional to the contribution of the agent's actions to the overall behavior. That is, if action of the agent's motors (A) has little influence on the future physical state of the agent in the environment (W'), morphological computation is high. Concept 2 calculates the amount of morphological computation by isolating the positive contribution of the world to the overall behavior (effect of W on W'), obtained from the 'difference' between conditional probability distributions with and without the action variable, $p(w' \mid w, a)$ versus $p(w' \mid w)$ (see [52] for details). Here, systems with high morphological computation would be those with strong 'body dynamics' or 'natural dynamics' (see, e.g. [23] or [10] for a formal definition). However, optimizing for morphological computation in the above sense, one would arrive at systems with strong internal dynamics (concept 2), resisting control actions (concept 1), which seems very impractical for engineers. In fact, Klyubin et al. [26] proposed a different measure relying on information theory, empowerment, which is equivalent to the opposite of morphological computation under concept 1; maximizing empowerment amounts to maximizing the effect of the agent's actions.

Rückert and Neumann [45] study learning of optimal control policies for a simulated four-link pendulum which needs to maintain balance in the presence of disturbances. The morphology (link lengths and joint friction and stiffness) is manipulated, and controllers are learned for every new morphology. They show that: (1) for a single controller, the complexity of the control (as measured by the 'variability' of the controller) varies with the properties of the morphology: certain morphologies can be controlled with simple controllers; (2) optimal morphology depends on the controller used; (3) more complex (time-varying) controllers achieve much higher performance than simple control across morphologies.

In summary, the performance on a task will always depend on a complex interplay of the controller, body and environment: taking out the controller is just as big a mistake as taking out the body was. The tasks that can be completely solved by appropriate tuning of the body, such as passive dynamic walking, are the exception rather than the rule. A controller will thus be needed too. A complex body may have the potential to partially solve certain tasks on its own; yet, it may present itself as difficult to control, model (if the controller is relying on models), design and manufacture. An optimal balance thus needs to be found. For that, however, new design methodologies that would encompass complex cost functions (performance on a task, versatility, robustness, costs associated with hardware whose parameters can be manipulated, etc.) are needed. Hermans et al. [17] very recently proposed such a method that uses machine learning to optimize physical systems; an approximate parametric model of the system's dynamics and sufficient examples of the desired dynamical behavior need to be available though—which leads us to the next section.

6 With or Without a Model?

Including the parameters of the body into the design considerations may give rise to better performance of the whole system; these may be solutions involving a simpler controller, but also solutions that were previously unattainable when the body was fixed. Following the dynamical systems perspective, [14] provides an illustration of the possible goals of the design process: (1) To design the physical dynamical system such that desired regions of the state space have attracting properties. Then it is sufficient to use a simple control signal that will bring the system to the basins of attraction of individual stable points that correspond to target behaviors. (2) More complicated behavior can be achieved if the attractor landscape can be manipulated by the control signal.

If a mathematical formulation of the controller and the plant is available, this design methodology can be directly applied. The first part is demonstrated by McGeer [33] on the passive dynamic walker: The influence of scale, foot radius, leg inertia, height of center of mass, hip mass and damping, mass offset and leg mismatch is evaluated. In addition, the stability of the walker is calculated. Recently, Jerrold Marsden and his coworkers presented a method that allows for co-optimization of the controller and plant by combining an inner loop (with discrete mechanics and optimal control) and an outer loop (multiscale trend optimization). They applied it to a model of a walker and obtained the best position of the knee joints ([38], Chap. 5).

However, typical real-world agents are more complex than simple walkers. Holmes et al. [21] provide an excellent dynamical systems analysis of the locomotion of rapidly running insects and derive implications for the design of the RHex robot. Yet, they conclude that "a gulf remains between the performance we can elicit empirically and what mathematical analyses or numerical simulations can explain. Modeling is still too crude to offer detailed design insights for dynamically stable autonomous machines in physically interesting settings". Hermans et al. [17] similarly note that applying their method to robotics, which is known to suffer from lack of accurate models, is a challenge. The modeling and optimization of more complicated morphologies—such as compliant structures—is nevertheless an active research topic (e.g. Wang [50] and other work by the author). The second point of Füchslin et al. [14]—achieving 'morphological programmability' by constructing a dynamical system with a parametrized attractor landscape—remains even more challenging though.

One of the merits of exploiting the contributions of body morphology should be that the physical processes do not need to be modeled, but can be used directly. However, without a model of the body at hand, several body designs need to be produced and—together with the controller—tested in the respective task setting. The design space of the joint controller–body system blows up, and we may be facing a curse of dimensionality. This is presumably the strategy adopted by the evolution of biological organisms that could cope with the enormous dimensionality of the space. In robotics, this has been taken up by evolutionary robotics [37]. The simulated agents of Karl Sims [47] demonstrate that co-evolving brains and bodies together can give rise to unexpected solutions to problems. More recently, Bongard [6] showed that morphological change indeed accelerates the evolution of robust behavior in such a brain–body co-evolution setting. With the advent of rapid prototyping technologies, physics-based simulation could be complemented by testing in real hardware [31], but this reintroduces the modeling through the back door: the phenotypes in the simulator now become models, and they need to sufficiently match their real counterparts. Yet, a 'reality gap' [24, 28] always remains between simulated and real physics. The only alternative is to optimize in hardware directly, which is in general slow and costly. Brodbeck et al. [7] provide an interesting illustration showing how locomoting cube-like creatures can be evolved in a model-free fashion through automated manufacturing and testing. However, in summary, the design decisions—which parameters to optimize—are based on heuristics, and a clear methodology is still missing. Furthermore, with the absence of an analytical model of the controller and plant, no guarantees on the system's performance can be given.

7   Conclusion and Outlook

'Morphological computation' and 'morphological control' are very attractive concepts, receiving significant attention and carrying great potential. The rich properties of 'soft' bodies (highly dimensional, dynamic, nonlinear, compliant, and deformable) have been largely overlooked or deliberately suppressed by classical mechatronic designs, as they are largely incompatible with traditional control frameworks, where linear plants are preferred. This is definitely a missed opportunity. On the other hand, while complex bodies carry a lot of 'auto-control' potential, this property does not come for free. In this article, we provided a critical review of the design 'trading spaces', an imaginary landscape from 'control–dominant' systems whose natural dynamics is suppressed to designs that capitalize on self-organization of the physical system interacting with the environment. We conclude that the contributions of the body to the task are not computational in any substantial sense and proposed the dynamical systems descriptions as the most versatile in order to facilitate description, understanding, control, and design of brain–body–environment systems. The pros and cons of simple versus complex bodies were illustrated on examples. It has to be said that the exploitation of truly complex bodies to accomplish tasks is still mostly at a 'proof-of-concept' stage. A closely connected issue is that of modeling of these systems—soft bodies are notoriously difficult to model. The model may not be necessary for the system to perform the task; however, without a model, the understanding and design are more complicated and performance guarantees are limited. The field, which has been dominated by heuristics so far, needs to embrace more systematic approaches that allow one to navigate in this complex landscape.

In terms of applications, the most relevant area where exploitation of morphology is and will be the key is probably robotics, and in particular soft robotics (see [2, 40, 41, 49] and the first issue of the journal *Soft Robotics* [48]). 'Soft' robots, with the robot Octopus (e.g. [29]) serving as a good representative, break the traditional separation of control and mechanics and exploit the morphology of the body and properties of materials to assist control as well as perceptual tasks. Pfeifer et al. [40] even discuss a new industrial revolution. Appropriate, 'cheap', designs lead to simpler control structures, and eventually can lead to technology that is cheap in a monetary sense and thus more likely to impact on practical applications. Yet, a lot of research in design, simulation, and fabrication is needed (see [30, 32, 46] for reviews).

The area of soft robotics and morphological computation seems to be rife with different trading spaces [41]. As we move from the traditional engineering framework with a central controller that commands a 'dumb' body toward delegating more functionality to the physical morphology, some convenient properties will be lost. In particular, the solutions may not be portable to other platforms anymore, as they will become dependent on the particular morphology and environment (the passive dynamic walker is the extreme case). The versatility of the solutions is likely to drop as well. To some extent, the morphology itself can be used to alleviate these issues— if it becomes adaptive. Online changes of morphology (such as changes of stiffness or shape) thus constitute another tough technological challenge (see also the project LOCOMORPH [1]). Completely new, distributed control algorithms that rely on self-organizing properties of complex bodies and local distributed control units will need to be developed [32, 44].

Acknowledgements M.H. was supported by the Czech Science Foundation under Project GA17-15697Y and by the Marie Curie Intra European Fellowship iCub Body Schema (625727) within the 7th European Community Framework Programme. M.H. also thanks Juan Pablo Carbajal for fruitful discussions and pointers to literature. Both authors thank the EUCogIII project (FP7-ICT 269981) for making us talk to each other.

References

1. Project LOCOMORPH. FP7-ICT-231688
2. Albu-Schaffer, A., Eiberger, O., Grebenstein, M., Haddadin, S., Ott, C., Wimbock, T., Wolf, S., Hirzinger, G.: Soft robotics. IEEE Robot. Autom. Mag. **15**(3), 20 –30 (2008)
3. Aström, K.J., Murray, R.M.: Feedback Systems: An Introduction for Scientists and Engineers. Princeton University Press (2008)
4. Bhounsule, P.A., Cortell, J., Grewal, A., Hendriksen, B., Daniël Karssen, J.G., Paul, C., Ruina, A.: Low-bandwidth reflex-based control for lower power walking: 65 km on a single battery charge. Int. J. Robot. Res. **33**(10), 1305–1321 (2014)
5. Blickhan, R., Seyfarth, A., Geyer, H., Grimmer, S., Wagner, H., Guenther, M.: Intelligence by mechanics. Phil. Trans. R. Soc. Lond. A **365**, 199–220 (2007)
6. Bongard, J.: Morphological change in machines accelerates the evolution of robust behavior. Proc. Nat. Acad. Sci. **108**(4), 1234–1239 (2011)
7. Brodbeck, L., Hauser, S., Iida, F.: Morphological evolution of physical robots through model-free phenotype development. PloS one **10**(6), e0128444 (2015)
8. Brown, E., Rodenberg, N., Amend, J., Mozeika, A., Steltz, E., Zakin, M.R., Lipson, H., Jaeger, H.M.: From the cover: Universal robotic gripper based on the jamming of granular material. Proc. Natl. Acad. Sci. U.S.A. **107**(44), 18809–18814 (2010)

9. Caluwaerts, K., D'Haene, M., Verstraeten, D., Schrauwen, B.: Locomotion without a brain: physical reservoir computing in tensegrity structures. Artificial Life **19**(1), 35–66 (2013)
10. Carbajal, J.P. : Harnessing nonlinearities: behavior generation from natural dynamics. PhD thesis, University of Zurich (2012)
11. Crutchfield, J.P., Ditto, W.L., Sinha, S.: Introduction to focus issue: intrinsic and designed computation: information processing in dynamical systems—beyond the digital hegemony. Chaos **20**(3), 037101_1–037101_6 (2010)
12. Emami-Naeini, A., Franklin, G.F., Powell, J.D.: Feedback Control of Dynamic Systems. Prentice Hall (2002)
13. Fresco, N.: Physical Computation and Cognitive Science. Springer (2014)
14. Füchslin, R.M., Dzyakanchuk, A., Flumini, D., Hauser, H., Hunt, K.J., Luchsinger, R.H., Reller, B., Scheidegger, S., Walker, R.: Morphological computation and morphological control: steps towards a formal theory and applications. Artificial Life **19**(1), 9–34 (2013)
15. Hauser, H., Ijspeert, A.J., Füchslin, R.M., Pfeifer, R., Maass, W.: Towards a theoretical foundation for morphological computation with compliant bodies. Biol. Cybern. **105**, 355–370 (2011)
16. Hauser, H., Ijspeert, A.J., Füchslin, R.M., Pfeifer, R., Maass, W.: The role of feedback in morphological computation with compliant bodies. Biol. Cybern. **106**, 595–613 (2012)
17. Hermans, M., Schrauwen, B., Bienstman, P., Dambre, J.: Automated design of complex dynamic systems. PLOS ONE **9**(1), e86696 (2014)
18. Hoffmann, M., Assaf, D., Pfeifer, R.: A tutorial on embodiment (2011). http://www.eucognition.org/index.php?page=tutorial-on-embodiment
19. Hoffmann, M., Pfeifer, R.: The implications of embodiment for behavior and cognition: animal and robotic case studies. In: The Implications of Embodiment: Cognition and Communication, pp. 31–58. Exeter: Imprint Academic (2011)
20. Hoffmann, M., Müller, V.C.: Trade-offs in exploiting body morphology for control: from simple bodies and model-based control to complex bodies with model-free distributed control schemes. In: Hauser, H., Füchslin, R.M., Pfeifer, R. (eds.) E-book on Opinions and Outlooks on Morphological Computation, chap. 17, pp. 185–194 (2014)
21. Holmes, P., Full, R.J., Koditschek, D., Guckenheimer, J.: The dynamics of legged locomotion: models, analyses and challenges. SIAM Rev. **48**(2), 207–304 (2006)
22. Horsman, C., Stepney, S., Wagner, R.C., Kendon, V.: When does a physical system compute? Proc. R. Soc. A **470**(2169), 20140182 (2014)
23. Iida, F., Gómez, G., Pfeifer, R.: Exploiting body dynamics for controlling a running quadruped robot. In: Proceedings of the 12th International Conferences on Advanced Robotics (ICAR05), pp. 229–235, Seattle, U.S.A. (2005)
24. Jakobi, N., Husbands, P., Harvey, I.: Noise and the reality gap: the use of simulation in evolutionary robotics. In: Advances in Artificial Life, pp. 704–720. Springer (1995)
25. Kirk, D.: Optimal Control Theory: An Introduction. Dover Publications (2004)
26. Klyubin, A.S., Polani, D., Nehaniv, C.L.: All else being equal be empowered. In: Advances in Artificial Life, pp. 744–753. Springer (2005)
27. Kooijman, J.D.G., Meijaard, J.P., Papadopoulos, J.M., Ruina, A., Schwab, A.: A bicycle can be self-stable without gyroscopic or caster effects. Science **332**(6027), 339–342 (2011)
28. Koos, S., Mouret, J.-B., Doncieux, S.: The transferability approach: crossing the reality gap in evolutionary robotics. IEEE Trans. Evol. Comput. **17**(1), 122–145 (2013)
29. Laschi, C., Cianchetti, M., Mazzolai, B., Margheri, L., Follador, M., Dario, P.: Soft robot arm inspired by the octopus. Adv. Robot. **26**(7), 709–727 (2012)
30. Lipson, H.: Challenges and opportunities for design, simulation, and fabrication of soft robots. Soft Robot. **1**, 21–27 (2013)
31. Lipson, H., Pollack, J.B.: Automatic design and manufacture of robotic lifeforms. Nature **406**(6799), 974–978 (2000)
32. McEvoy, M.A., Correll, N.: Materials that couple sensing, actuation, computation, and communication. Science **347**(6228), 1261689 (2015)
33. McGeer, T.: Passive dynamic walking. Int. J. Robot. Res. **9**(2), 62–82 (1990)

34. Moore, A., Ober-Blöbaum, S., Marsden, J.E.: Trajectory design combining invariant manifolds with discrete mechanics and optimal control. J. Guid. Control Dyn. **35**(5), 1507–1525 (2012)
35. Nakanishi, Y., Asano, Y., Kozuki, T., Mizoguchi, H., Motegi, Y., Osada, M., Shirai, T., Urata, J., Okada, K., Inaba, M.: Design concept of detail musculoskeletal humanoid Kenshiro—toward a real human body musculoskeletal simulator. In: 12th IEEE-RAS International Conference on Humanoid Robots (Humanoids) 2012, pp. 1–6. IEEE (2012)
36. Niiyama, R., Nagakubo, A., Kuniyoshi, Y.: Mowgli: a bipedal jumping and landing robot with an artificial musculoskeletal system. In: IEEE International Conference Robotics and Automation (ICRA), pp. 2546–2551. IEEE (2007)
37. Nolfi, S., Floreano, D.: Evolutionary Robotics: The Biology, Intelligence, and Technology of Self-organizing Machines. MIT Press Cambridge (2000)
38. Pekarek, D.N.: Variational methods for control and design of bipedal robot models. PhD thesis, California Institute of Technology (2010)
39. Pfeifer, R., Bongard, J.C.: How the Body Shapes the Way We Think: A New View of Intelligence. MIT Press, Cambridge, MA (2007)
40. Pfeifer, R., Lungarella, M., Iida, F.: The challenges ahead for bio-inspired 'soft' robotics. Commun. ACM **55**(11), 76–87 (2012)
41. Pfeifer, R., Marques, H.G., Iida, F.: Soft robotics: the next generation of intelligent machines. In: Proceedings 23rd International Joint Conference on Artificial Intelligence, pp. 5–11. AAAI Press (2013)
42. Pfeifer, R., Scheier, C.: Understanding Intelligence. MIT Press (1999)
43. Potkonjak, V., Svetozarevic, B., Jovanovic, K., Holland, O.: The puller-follower control of compliant and noncompliant antagonistic tendon drives in robotic systems. Int. J. Adv. Robot. Syst. **8**(5), 143–155 (2011)
44. Rieffel, J.A., Valero-Cuevas, F.J., Lipson, H.: Morphological communication: exploiting coupled dynamics in a complex mechanical structure to achieve locomotion. J. R. Soc. Interface **7**(45), 613–621 (2010)
45. Rückert, E., Neumann, G.: Stochastic optimal control methods for investigating the power of morphological computation. Artificial Life **19**, 115–131 (2013)
46. Rus, D., Tolley, M.T.: Design, fabrication and control of soft robots. Nature **521**(7553), 467–475 (2015)
47. Sims, K.: Evolving 3D morphology and behavior by competition. Artificial Life **1**(4), 353–372 (1994)
48. Trimmer, B.: A journal of soft robotics: why now? Soft Robot. **1**, 1–4 (2013)
49. Trivedi, D., Rahn, C.D., Kier, W.M., Walker, I.D.: Soft robotics: biological inspiration, state of the art, and future research. Appl. Bionics Biomech. **5**(3), 99–117 (2008)
50. Wang M.Y.: A kinetoelastic formulation of compliant mechanism optimization. J. Mech. Robot. **1**(2), 021011 (2009)
51. Wittmeier, S., Alessandro, C., Bascarevic, N., Dalamagkidis, K., Devereux, D., Diamond, A., Jäntsch, M., Jovanovic, K., Knight, R., Marques, H.G., et al.: Toward anthropomimetic robotics: development, simulation, and control of a musculoskeletal torso. Artificial life **19**(1), 171–193 (2013)
52. Zahedi, K., Ay, N.: Quantifying morphological computation. Entropy **15**(5), 1887–1915 (2013)

On the Realism of Human and Machine Representational Constraints: A Functionalist Account on Cognitive Ontologies

David Zarebski

Abstract This paper is concerned with the primitive constraints on Information Systems' and Humans' representations of the reality. We intend to support the idea that a proper understanding of what is at stake in Information Systems Ontologies (ISO) and its relation to the representational constraints in human cognition may solve a recent issue about the status of philosophical intuitions in metaphysics.

1 Introduction

Information comes with degrees of generality. While some information, like the location of my bike at a certain time, is factual, the fact that, as a physical object, my bike *has to* occupy a certain location *at any time* is structural. One historical account on such structural information is to be found in Aristotle's theory of *categories*. Initially understood as the real, mind-independent primitive components of the world, collections of categories—a.k.a. *ontologies*—have also been a useful tool in cognitive sciences and data engineering for the last three decades.

Despite many interactions, the very meaning of *ontologies* varies across these fields, for they do not share the same goal—theoretical versus practical—the same domain generality nor the same attitude toward the correlate of categories—physical world [45], human cognitive structure or electronic information structure. Most of the interactions between these fields consist either in explaining how the human cognitive structure may affect metaphysical practices (*meta-metaphysics*) or why Information Systems Ontologies (ISO) should be based on the results of formal philosophy.[1]

However, the possibility that some of the reflections around ISO might solve deep metaphysical issues has not been taken for granted. In this paper, I plan to show how

---

[1]*"The ontological problem for computer and information science is thus identical to many of the problems of philosophical ontology, and it is becoming more and more clear that success in the former will be achievable, if at all, only by appeal to the methods, insights and theories of the latter"* [69]:7.

D. Zarebski (✉)
IHPST, University Paris 1 Pantheon Sorbonne, Paris, France
e-mail: zarebskidavid@gmail.com

© Springer International Publishing AG 2017

G. Dodig-Crnkovic and R. Giovagnoli (eds.), *Representation and Reality in Humans, Other Living Organisms and Intelligent Machines*, Studies in Applied Philosophy, Epistemology and Rational Ethics 28, DOI 10.1007/978-3-319-43784-2_18

reflexions around ISO may disentangle a purely philosophical and speculative issue, namely the putative incompatibility of *ontological realism* and *naturalized meta-physics* as lately advocated in [2, 12]. More precisely I will focus on Alvin Gold-man's variant of *naturalized metaphysics* [23, 25]: the *Cognitive-Sciences-based Meta-Metaphysics* (CSMM).

After a clarification on the different meanings of *ontologies* (Sect. 2) and what is meant by CSMM (Sect. 3), I will present Allen's and Chalmers' arguments for this incompatibility (Sect. 4). Finally, Sect. 5 provides arguments for a machine-based approach on this issue and explores its consequences for the different varieties of realism.

2 Ontology, Ontologies: The Structure of Reality and Its Representations

Though this partition might seem sharp given the great number of multidisciplinary approaches, there are three ways of conceiving ontologies. They can be either conceived as the mind independent structure of the reality to be investigated by philosophy and formal ontology [73, 74]—Sect. 2.1—as the structure of the human representation of the world explored by cognitive sciences—Sect. 2.2—or as the structure of knowledge representation inquired by data engineering—Sect. 2.3.

2.1 The Philosophical Realist Account on Categories

For centuries, ontologies have merely been philosophical and theoretical objects of investigation. Initially understood as real primitive components of the world, the list of categories that constitute an ontology typically includes

individuals or time-proof entities (say Fritz, my cat)
tropes properties borne by individuals (say the unique color of Fritz)
universals or natural kinds (cats, tables)
relations

Though this nomenclature might not be an exhaustive one, it is however complete enough to present some of the key features of philosophical ontologies (PO). First of all, PO are domain general. This means that the same scheme holds for both "*Fritz the cat is on the table*" and "*electron 1 is attracted by electron 2*": two individuals which instantiate natural kinds hold a certain relation. This has a crucial influence on the kind of inferences to be made from the premises, for this level of abstraction does not inform on factual details about feline life or sub-atomic particles, but rather synthesizes common characteristics of individuals as the fact that individuals are countable entities—e.g. one cat and two electrons.

Finally, it should be acknowledged that the tacit realism together with the putative domain generality of PO imply that *there must be only one* genuine Ontology. Let us call this principle the *uniqueness assumption* which states that the existence of certain types of entities—e.g. Universals as in XIIIth century's problem of Universals—is not a matter of whether we could *conceive, make sense of* or *explain* a world without it but whether there is a *fact of the matter* about it. These metaphysical issues should be carefully distinguished from epistemological ones, for what is at stake is not whether there are such things as *cats* or whether *Fritz* or *Gandalf* exist but whether reality is fragmented into different *kinds of beings* with their respective properties.

2.2 Conceptualism and the Human Cognitive Structure

A challenging '*conceptualist tradition*' has stressed the necessity to understand ontologies as the structure of our cognitive scheme rather than the structure of the world. This designation is a broad one, for Conceptualism this way conceived includes account as different as the Kantian transcendental idealism or Strawson's *Descriptive Metaphysics* which "[...] *is content to describe the actual structure of our thought about the world* [...]" [70].

Nonetheless, despite the great variability of philosophical and psychological accounts, the core idea might be stated as follows. Our mental—i.e. perceptual or conceptual—representations of the world follow some primitive rules which segment the reality in different *types of things*—a.k.a. Categories. As a perceptual example, to acknowledge that "*Fritz the cat is on the table*" is to be able to split this complex object in two independent individuals rather than their mereological sum—i.e. some table-cat. To express it in another way, being able to perform such a task implies to possess a category of time-proof, spatially unified, individual object in our *Cognitive Ontology* (CO). Before I turn to cognitive sciences-based account on conceptual structures, I would like to avoid a common confusion about the meaning to be given to '*Categories*' in cognitive sciences.

There is a "*rampant ambiguity about categories*" [31], for this concept has sometimes been used in developmental psychology to denote the psychological counterparts of natural kinds as in [35, 43] or in Medin's folk-biology [47, 48]. Though relevant as descriptions of the kind of constraints on our domain specific representations of natural kinds, '*Categories*' in a more high level context of abstract representational constraints is closer to Barry Smith's perspective on the so called *Common Sense Ontology* [62] or the notion of *meta-level categories* [30]. While some psychological enquiries focus on the *content* of human representation, the primitive components of our CO would be our criteria for the individuation of objects (individuals) or the general structure of the representation of natural kinds rather that the actual content of concepts such as '*cats*' or '*tables*'. According to Smith, the categories that constitute our *Common Sense Ontology* are also the ones responsible for a great number of pre-theoretical and intuitive systems of belief such as:

Folk physics   the putatively innate system of knowledge about solid-based physical phenomena [33]—see also [64]²

Common sense   the fact that, as early expressed by Köhler, our perception "[...]
*consists first of all of objects, their properties and changes, which appear to exist
and to happen quite independently of us* [...]" [39] based on

Gestalt principles   driven by visual features such as good continuation, similarity
or symmetry.

To put it in other words, psychological investigations on CO are concerned with
the general stratification of our representations of the world we live in.³ To turn
to CO's specificities, it should first be pointed out that CO are neutral toward realism. Whether the existence of a particular cognitive mechanism implies some mind-
independent property of the world [58] or not [19, 47] does not impact on the descriptive approach endorsed by Conceptualism nor our cognitive scheme. Secondly, CO
and PO differ in their respective domains, for CO apply to the ecological context of
middle sized entities—i.e. Fritz rather than the particles he is composed of.

2.3   Ontologies as Information Structuring

In the last 15 years, ontologies became also relevant in the context of knowledge representation in information systems (IS). To illustrate the way categories intervene in
information systems, let's say that I need to add in a certain data base the fact that
"Fritz the cat is on the table". Sure, I could describe this fact in a merely relational,
XML-styled way without any considerations for the common nature of Fritz and the
table—i.e. two three-dimensional individuals—nor the specificities of the relation
'is on'. However, a more *efficient way* to do it seems to be to type every component of this fact to indicate what they have in common with the other entities of my
database. For instance I could follow the categories of the *Descriptive Ontology for
Linguistic and Cognitive Engineering* (DOLCE [20, 27]) and use different types for
Agentive Physical Object (APO, e.g. Fritz) and non-agentive ones (the table). Likewise, I could type the relation *'being on'* as an endurant-endurant relation—i.e. a
relation which holds between individuals. In this artificial intelligence-based context, the fitness or realism of an ISO should be understood as the way it impacts on
the computational cost of the operations performed by an inference engine.

From this point of view, ISO share some common features with CO. First of
all, ISO are also domain specific though in a plural fashion, for different ontologies
have been proposed for many different domains from geographical representations
[8, 46] to biomedical data [66] or even philosophical positions [29]. However,

---

²It should be acknowledged that Folk Physics also drew the attention of researchers in Artificial
Intelligence. This interest led to PROLOG or LISP-based implementations of various automatic systems of physical deductions known as the *Naive Physics*—see [1, 60] for details.

³"*And this recognition leads straightaway to one of the fundamental theses of ontology: to be is to
be an item of a certain type or kind*" [55]:26.

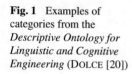
Fig. 1 Examples of
categories from the
*Descriptive Ontology for
Linguistic and Cognitive
Engineering* (DOLCE [20])

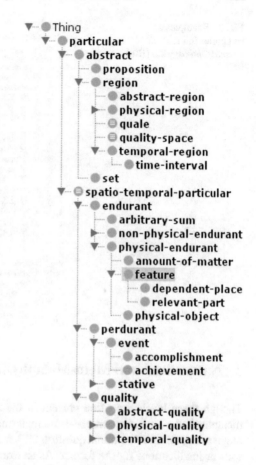

research around formal ontology has also developed *upper level ontologies*—see
Figs. 1 and 2—supposed to be both *domain general* and efficient on a pragmatical
point of view. Because of their similarity with CO categories [62], I will focus on
this kind of ISO in the rest of the article.

Finally, as CO, ISO are also neutral toward realism and common sense. While
authors have advocated a purely computational argument for ontological pluralism,
it has also been suggested that the realism of information structure impacts the relia-
bility of computer-based IS tasks [28] despite its possible contradiction of common
sense. [4]

---

[4]*"the primary concern of knowledge engineering is modeling systems in the world, not replicating
how people think"* [14]:34.

Fig. 2 Examples of
categories in the *Basic
Formal Ontology* 2.0 (**Bfo**)
[65]

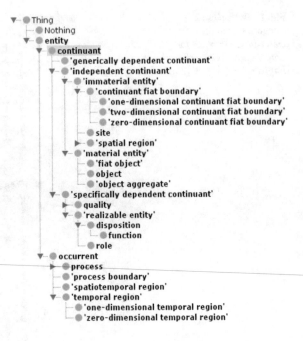

3   Naturalized and Meta-Metaphysics

Though these levels are often present in the same thesis, one should distinguish metaphysical—a.k.a. first order—from meta-metaphysical claims. While the former ought to answer the traditional question *"What is there?"*, the latter deal with the reasons or justifications for the former. As an example of this dichotomy, while Quine advocated for an individual-based metaphysics [53], his meta-metaphysics is the well known thesis that we are ontologically committed to the kind of entity our best scientific theories quantify over [54].

3.1   The Issue of Intuitions' Epistemological Status

Among the meta-metaphysical choices to be made, one must judge the legitimacy of intuitive methods in metaphysics. Another related question to be raised is whether metaphysics should use the methodology or the results of empirical sciences—i.e. whether metaphysics should be *naturalized*. Even if *naturalized metaphysics* is often defective toward the use of intuitions in metaphysics [42], the Cognitive Science Meta-Metaphysics (CSMM) supported by Goldman provides a positive role for intuitions [26]. Broadly, Goldman's CSMM can be summarized in three points.

α Intuitions do play a role in first order metaphysics [26]

β The origins of these intuitions are to be found in the most primitive features of our common sense and investigated by cognitive sciences [25]

γ These intuitions remain reliable enough to support some varieties of realism [24, 26]

The point α is a parsimonious one, acknowledged for a long time by both detractors [18, 77] and partisans of intuitive arguments and methodology for first order metaphysics [3, 4]. As a kind of meta-metaphysics, point β deserves an illustration. Metaphysical approaches to event individuation do not agree about the identity of the following events:

1. Boris pulling the trigger
2. Boris firing the gun
3. Boris killing Pierre

While the spatiotemporal based *unifier approach* [16, 17] conceives 1, 2 and 3 as one and the same event, the *property exemplification view* [21, 36, 37] distinguishes these events based on the fact that firing a gun differs from killing somebody— one could well fire a gun without killing and *vice versa*. Goldman suggests that the two different intuitions that support these approaches might correspond to different *formats of mental representations* of events—see [25]. Alongside the individuation of objects, empirical investigations on the individuation of event suggest that, while children tend to use a spatiotemporal criterion in both tracking-based [78] and counting-based experiments [59], they start to develop a membership strategy for goal oriented actions after the age of 5 [75]—e.g. they would count *"draw a flower"* as one event instead of counting every sub-actions of the drawing.

Though [26] does not explicitly mention it, the kind of primitive representational components involved in metaphysical intuition are interestingly similar to those responsible for CO as understood earlier [23–25]. The dependencies of metaphysical intuitions toward these formats of representation is explicitly endorsed by Goldman who concludes for a case of ontological pluralism which seems *prima facie* incompatible with the tacit ontological realism of PO. [5]

3.2 Goldman's Defense of Intuitions

Yet, to come to point γ, Goldman is nonetheless committed to some kind of realism. The previous case was one where grounded strong intuitions contradicted each other. But had these intuitions converged, would they have allowed stronger assumptions?

---

[5] *"The best solution is to countenance two metaphysical categories of events, EVENTS 1 and EVENTS 2. This is how cognitive science can play a role in the conduct of metaphysicalizing."* [25]:475.

First of all, it should be acknowledged that Goldman agrees that the ultimate aim of metaphysics is to investigate mind-independent entities.[6] The question is whether intuitions are a good tool to do it. As pointed out above, intuitions may be contradictory. However, it is precisely the job of cognitive sciences to disentangle the most primitive components of our conceptual schemes so as to distinguish genuine from apparent or language-based contradictions. As stated in Goldman's answer to Kornblith's critique about universals [40, 41], this methodology does not rule out the possibility of moving from individual to shared, embodied concepts.[7]

However, despite the facts that Goldman is known for his reliabilist account of epistemic justification of belief [22] and advocates that the justification of belief lies in its causal relation with the psychological process it is produced by,[8] it is not clear how these shared mechanisms provide evidence for a certain mind-independent stratification of reality—e.g. the existence of different kinds of being such as individuals, properties, relations among universals, *etc*.

I plan to show that the main reason for this limitation is because Goldman's view on ontological realism remains dependent on his adaptationist conception of truth *modulo* human resources and is thus exposed to the same limitations of any *naturalized epistemology*—see [44, 51]. Even if Goldman advocates that intuitions are both *a priori* justified and compatible with a naturalist framework,[9] the only kind of argument moved forward for this "*evidence-conferring power of intuitions*" relies on an introspective capability to *simulate* what would be the intuitions of anybody else—a.k.a. *third person conceptual investigation*.

4 CSMM and Ontological Realism Putative Incompatibility

The idea that the existence of robust categories of our CO entails or provides evidence for their real counterparts (in PO) has recently been challenged by two skeptic accounts.

---

[6]"*Metaphysics seeks to understand the nature of the world as it is independently of how we think of it*" [25]:457.

[7]"*I am not saying that the analysis of personal concepts is the be-all and end-all of, philosophy, even the analytical part of philosophy. But perhaps we can move from concepts 2 to concepts 3, i.e. shared (psychological) concepts*". [26]:16-7.

[8]"*My favored kind of epistemological naturalism holds that warrant, or justification, arises from, or supervenes on, psychological processes that are causally responsible for belief (Goldman 1986, 1994).*" [26]:19.

[9]"*A first reply is that, in my view, there is no incompatibility between naturalism and a priori warrant*" [26]:19.

4.1 The Ontological Limitations of Naturalized Metaphysics

First of all, it has been lately suggested in [2] that, because of its very methodology, *naturalized metaphysics* cannot meet the requirements for ontological realism. In a nutshell, because *naturalized metaphysics* shares its methodology with science— i.e. a justified explanations based inquiry—and it is known that different theoretical commitments could be equally justified, *naturalized metaphysics* could not fulfill the *uniqueness assumption* induced by PO realism.[10]

As an example of the ontological plurality induced by the naturalistic framework, it has been argued that an ontology can avoid the category of individuals given some adjustments on tropes—see [61] for details. From a purely empirical point of view, despite the different ontological commitment, these bundles of tropes behave the same way as individuals. As a kind of *naturalized metaphysics*, Goldman's CSMM is supposed to suffer from the same incapability to choose one ontology over another inasmuch as the existence of competing mechanisms—a.k.a. format of mental representation—rules out the possibility to choose between two competing ontologies:

> There seems to be no point at which a theory chosen in this manner puts us in better epistemic contact with the ontology of the natural world than any of the rejected candidate theories do; there seems to be no reason why intuition would happen to begin with knowledge about the objective nature of the world, or else why this systematized reflection should result in such knowledge. [2]:227

4.2 The Issue of the Conceptual Scheme Indetermination

Another kind of argument against classical ontological realism, initially proposed in [52] in support of his *internal realism* and lately generalized in [12], suggests that even if there were strong shared concepts as understood in Goldman's CSMM, the very possibility that an intelligent organism could possess a different conceptual structure challenges the idea that our own fits with the structure of reality.

As an example, while anyone would count two objects on the table (mug1 and mug2), we could perfectly imagine that a Martian with a different conceptual scheme would consider mereological sums of distinct objects as genuine entities and thus count three objects (mug1, mug2 and mug1+mug2). In a similar fashion, this Martian would consider a *table-cat* as a genuine entity he observes when Fritz is on the table.

---

[10]"*Having attempted to establish premises (1)–(4), it is now time to draw conclusions from the incompatibility of naturalized metaphysics with a commitment to robust realism about the entities in metaphysical theories*" [2]:292.

5   Is Our Cognitive Scheme Arbitrary?

Both arguments rely on the same components:

uniqueness assumption   if there are mind-independent categories, there must be
only one correct ontology
ontological pluralism   There is or can be different cognitive ontologies (CO)
CSMM   PO depend on, fit, or are supported by CO, or at least by some of their
by-products (i.e. intuitions)
ontological antirealism   thus ontological realism is false

As a *sine qua non* condition for the realism of PO, we have no choice but to maintain the *uniqueness assumption*. Likewise, it seems parsimonious to defend that our
actual cognitive scheme produces intuitions and supports first order metaphysical
claims about the most basic components of the world—a.k.a. CSMM. As a consequence, the best way to oust *ontological antirealism* seems to question the *ontological pluralism* premise of this argument. As detailed above, the *ontological-pluralism*
component comes with two variants: an actual one—based on the contradiction of
intuitions—and a potential one—the conceivability of an *alien conceptual structure*.

5.1   Robust versus Weak Primitive Cognitive Components of
Mental Representations

To start with the actual contradiction of some of our intuitions, Allen's conflation of
naturalized metaphysics with Goldman's CSMM misses the crucial point that meta-
metaphysics differs from first order metaphysics. CSMM is not a first order science-
based metaphysics but a second order scientific investigation on the intuitions that
provide first order metaphysical claims and their cognitive bases. Thus, CSMM does
not contradict the *a priori* metaphysics but explains it.

This said, the issue whether intuitions and their cognitive origins are robust
enough remains. However both classical and contemporary research have underlined
a great number of primitive representational invariants from colours perception [5]
to causal induction [15, 68] or individuation criterion [75]. Though the question
remains an empirical one to be decided by science rather than philosophy, I would
like to dispel a fallacy. It has been argued that categories change from one culture to
another. However, such an account confuses the *content* level categories—i.e. categories as the cognitive correlate of natural kinds such as *cats* or '*tables*' see [38]—
and what Guarino called *meta-level categories* [30]—i.e. different types of being. In
other words, this account conflates epistemological and ontological issues. Different people may well *categorize* (in the first sense) differently, but what is at stake in
the realism of cognitive ontology is whether the underling cognitive components are
consistent.

As an example, Sheppard has supported the idea that the generalisation laws that underline categorisation are universal [57, 58], thus robust enough to maintain a realist view: our primitive knowledge about solids or colours reflects the physical properties of the world. More recently, the Theory-based Bayesian models [71, 72] have suggested that concept learning relies on inferences, success of which depends on a correct inference about the hidden structure of the data set—i.e. inferences about the relationship that second order entities (putative natural kinds) entertain. In other words, to be able to learn the meaning of concepts requires some prior knowledge about the existence of entities types which differ from those of stimuli—i.e. natural kinds instead of individuals. Without enumerating further the robust primitive components acknowledged in the psychological literature, it seems parsimonious to claim that our actual cognitive scheme may be more stable than initially thought by Goldman and Allen.

5.2 Cognitive Ontologies Beyond Our Actual Conceptual Scheme

However, the challenge raised in [12] remains. One way to overcome this issue consists in wondering whether some intelligent agent, an alien, could make sense of the world without distinguishing, say, perdurant (events) from endurant (individuals) beings, or genuine objects from any arbitrary mereological sum.

Here is where Information System Ontologies (ISO) occurs. In this final section, I would like to support the idea that (1) reflexions around ISO might rule out the very possibility of such an intelligent agent and thus (2) provide support for ontological realism.

Critical realism in IS It should first be acknowledged that the issue of the realism of knowledge representation in IS is not a new one. As an example, it has been argued that to adopt *critical realism* as the underlying philosophy of IS should overcome practical inconsistencies [49, 67]. Here is how the argument is conceived.

As general views on knowledge and scientific practices, *empiricism*—i.e. a regularity-based view of science—and *conventionalism*—i.e. the cultural dependency of theoretical objects—impact the way IS are conceived and realized. While mere statistical inferences supported by *empiricism* may fail to explain regularities, arbitrary reifications supported by the relativist part of *conventionalism* may also fail to explain the success of some reifications rather than others. Mingers proposes then a form of *no-miracle argument*,[11] based on the idea that the stratification of IS ontological domains should be understood as the way the *structure of reality* dictates the conditions of possibility of knowledge.

---

[11]*"The argument is that neither empiricism nor idealism can successfully explain these occurrences and that they necessitate some form of realist ontology"* [49]:92.

In contrast to this [Kant's transcendental idealism] critical realism asserts that the conditions for knowledge do not arise in our minds but in the structure of reality, and that such knowledge will not be universal and ahistorical. [49]:92

Information and belief: a functionalist account Though philosophically relevant, this attempt at *real-izing* information systems [49] remains limited, for (1) it takes PO implicit realism as a premise rather than a conclusion, and (2) does not rule out the possibility that another stratification of IS ontological domain might solve representational issues as well. Another strategy is required.

The present account endorses an idealization formulated a decade ago by cognitive informatics [7, 76]: every single cognitive issue possesses a computational counterpart. From a purely functional point of view, every human intentional act—perception, communication or recollection—can be conceived as information retrieval tasks which make use of structural properties of information. From a great variety of retinal patterns, I will use criteria such as continuity to perceive Fritz crossing my desktop rather than a succession of *"here and now catness"* as in the classical Quinean argument for ontological plurality [54]. Likewise, with *"Fritz the cat is crossing my desktop"*, the grammatical structure of this utterance indicates the kind of being involved in this state of affairs.

The same holds for an inference engine. However, an expert system—i.e. an inference engine with a data base [28]—lives in a world far different from our own: *a sea of information* [69] some of which differs crucially from the ecological entities homo sapiens and other animals are confronted with. Some of the most striking examples of non-ecological domains for humans are the domains of biomedical information systems such as SNOMED CT [11] or the Open Biological Ontologies [63] populated by genes, organs, illnesses and their respective properties and families. Yet, the same fitness-based arguments hold, though formulated in terms of:

computational cost   a well organized database speeds up information processing [28] the same way early pattern recognition enhances neural encoding

inferential success   inferences based on a well founded ontology are more reliable than merely first order, descriptive systems [28]

re-usability   enhance knowledge sharing across different systems so as to infer new facts about the given domain—a.k.a. inter-operability [9]

For at least two reasons, it should be stressed that this last argument should not merely be conceived as the adoption of a common vocabulary. [12] Firstly, adopting a certain ontology is not a referential issue. As pointed out earlier, to adopt a certain ontology is not to fix the meaning of particular entities—say *Fritz* or *cats*—but to consider that a certain hierarchical structure of types of beings—together with their respective properties such as *is-part-of*, *contains-process* or *has-specific-dependent-at-some-time*—is a good way to represent knowledge. [13] This is even more striking

---

[12] *"The ontological problem of information repository construction and management is not, however, simply the problem of agreeing on the use of a common vocabulary"* [69]:7.

[13] *"Rather, it is the problem of adopting a (sometimes very general) set of basic categories of objects, of determining what kinds of entities fall within each of these categories of objects, and of determining what relationships hold within and amongst the different categories in the ontology"* [69]:7.

when we consider *upper level ontologies*, categories of which —see Figs. 1 and 2—
are supposed to represent adequately any fact, no matter what the domain is.

Secondly, the probability that any arbitrary *'common vocabulary'* might solve
computational issues the way BFO increased the inferential power of biomedical
IS—e.g. the Foundational Model of Anatomy (FMA [56]) or the Gene Ontology (GO
[6, 32])—seems very remote [10]. Computers—understood as inference engines—
are *'dumb beasts'* [69]. What makes them smart is the way data is encoded, the way
data—i.e. *"raw pieces of abstract items or things"*—becomes information or *"data
that has been assigned attributes along with limited logical relationship between
data"* [13].

To what extent does it solve the issue of ontological realism? But what about a
less parsimonious conclusion? The account about ISO realism sketched above may
remind one of the classical *naturalization of truth* as useful belief—see Sect. 3.2.
However, it differs in three ways from this problematic framework.

First of all, as an ontological account on representations rather than an epistemological one, the naturalization of ontology proposed above does not directly concern
particular beliefs but their categorical constraints. This ontological turn has practical consequences for the historical anti-realist arguments. As an example, Putnam's
argument for the *inscrutability of the reference* by permutation—i.e. the fact that
"The cat is on the doormat" might be true in a world in which *cat* means cherry and
doormat means cherry tree [50, Chap. 2]—becomes harmless in such a context, for
what is at stake concerns the existence of a common type of beings able to entertain
a certain kind of relations instantiated here by *'being-on'*.

Secondly, while a biological entity cannot change its cognitive structure, the same
data base can be organized by means of many different ISO. The crucial practical
aftermath of this is that the respective fitness of different ontologies could thus be
compared in the same way it has been done for Biomedical IS [10]. In other words,
the adaptation of the most primitive cognitive constraints on the representations of a
generic intelligent agent may become an empirical investigation rather than an arm-
chair or intuitive *third person conceptual investigation* as advocated by Goldman.
While some ontological distinctions like *endurant–perdurant* seem indispensable
because they are shared by the most efficient ISO, some other categories like *Event*
may *salva veritate* be differently conceived from one ISO to another. [14]

The reader may retort that the very notion of *fitness* remains dependent on the
human cognitive scheme, given that human agents enter data in the database and
thus transmit or, at least, expect ISO to be consistent with their own CO. However,
the idea that IS structures capture some of the very grounded primitive components
of the human representational system—e.g. *endurant, continuant, process, disposi-
tional properties*, etc.—does not mean that ISO are expected to fit entirely with CO.
While categories from some upper ontologies such as DOLCE [20] were built so as to
explicitly fit with common sense, the *zero-dimensional temporal region* of the BFO
[65]—i.e. temporal region without extent—does not seem intuitive *prima facie*.

---

[14]As an example, *'Event'* is a category explicitly spatio-temporally interpreted in DOLCE while it
does not exist *per se* in BFO 1.1 and 2.0.

Finally, the ability to produce ontological-based predictions, inferences of which would not have been anticipated by any human agent, rules out an important argument for CO dispensability. As an example, the classical debate around *naturalized epistemology* suggests that categories of CO may be conceived as a way to make the physical world predictable by by-passing the computational limitations of the human brain—see [44]. However, given the computational power currently available by expert systems together with ISO impacts on their capacities, it seems highly improbable that an intelligent agent could make sense of the world without discriminating *kinds of beings* as primitive as *individuals*—i.e. *Sortals* or *Endurants* in BFO. To put it in other words, the ISO impact on IS efficiency suggests some kind of necessity behind our own cognitive scheme (CO).

6   Conclusion

Among the different available options in the debate about the compatibility of CSMM [25, 26] with ontological realism [2, 12], I choose to question the putative indetermination of our *Cognitive Ontology*. From what has been suggested above, it remains possible to support both the idea that first order metaphysical assumptions reflect some primitive, embodied, representational constraint with a certain variety of ontological realism. By means of a *no-miracle argument* at the level of categories rather than belief toward particulars, *upper level ontologies* provide evidence about the existence of generic and putatively universal cognitive constraints for both humans and machines representations, and hence a reason to be realistic about these upper level categories.

Though the attempt to reconcile realism and the representational dimension of knowledge in cognitive sciences [57, 58] and information systems [62] is not a new one, the present account differs in two ways. While the cognitive science-based realist position rarely distinguish ontological issues from epistemological ones, I advocated that it is possible to be a realist about the formes without facing the difficulties raised by the naturalization of the latter. Secondly, while supporters of information systems realism take the philosophically implicit realism as a foundational premise [34, 62, 67], the present account advocates that considering machine-based information systems as generic cognitive agents might support this strongly debated premise.

References

1. Allen, J.: Towards a general theory of action and time. Artif. Intell. **23**(2), 123–154 (1984)
2. Allen, S.R.: What Matters in (Naturalized) Metaphysics? Essays Philos. **13**(1), 211–241 (2012)
3. Bealer, G.: Modal epistemology and the rationalist renaissance. In: Szabo Gendler, T., Hawthorne, J. (eds.) Conceivability and Possibility, pp. 71–126. Oxford University Press (2004)

4. Bealer, G., Strawson, P.F.: The incoherence of empiricism. In: Proceedings of the Aristotelian Society, Supplementary Volumes, vol. 66, pp. 99–143 (1992)

5. Berlin, B., Kay, P.: Basic Color Terms: Their Universality and Evolution. University of California Press, Berkeley, CA (1969)

6. Botstein, D., et al.: Gene Ontology: tool for the unification of biology. Nat. Genet. **25**(1), 25–29 (2000)

7. Bryant, A.: Cognitive informatics, distributed representation and embodiment. Brain Mind **4**(2), 215–228 (2003)

8. Casati, R., Smith, B., Varzi, A.: Ontological tools for geographic representation. In: Formal Ontology in Information Systems, pp. 77–85 (1998)

9. Cerovsek, T.: A review and outlook for a 'Building Information Model' (BIM): a multistandpoint framework for technological development. Adv. Eng. Inf. **25**(2), 224–244 (2011)

10. Ceusters, W., Smith, B.: A realism-based approach to the evolution of biomedical ontologies. In: AMIA Annual Symposium Proceedings, vol. 2006, p. 121 (2006)

11. Ceusters, W., Smith, B., et al.: Ontology-based error detection in SNOMED-CT. In: Proceedings of MEDINFO, vol. 2004, pp. 6–482 (2004)

12. Chalmers, D.: Ontological Anti-Realism. In: Chalmers, D., Manley, D., Wasserman, R. (eds.) Metametaphysics: New Essays on the Foundations of Ontology, pp. 77–129. Oxford University Press, New York (2009)

13. Chiew, V.: A software engineering cognitive knowledge discovery framework. In: Proceedings of the First IEEE International Conference on Cognitive Informatics, vol. 2002, pp. 163–172 (2002)

14. Clancey, W.J.: The knowledge level reinterpreted: modeling sociotechnical systems. Int. J. Intell. Syst. **8**(1), 33–49 (1993)

15. Danks, D.: Theory unification and graphical models in human categorization. In: Gopnik, A., Schulz, L. (eds.) Causal learning, pp. 174–189. Oxford University Press, New York (2007)

16. Davidson, D.: Actions, reasons, and causes. J. Philos. **60**(23), 685–700 (1963)

17. Davidson, D.: The individuation of events. In: G. Hempel (ed.) Essays in Honor of Carl, pp. 216–234. Reidel, Dordrecht (1969)

18. Devitt, M.: The methodology of naturalistic semantics. J. Philos. **91**(10), 545–572 (1994)

19. Dölling, J.: Commonsense ontology and semantics of natural language. Sprachtypologie und Universalienforschung **46**(2), 133–141 (1993)

20. Gangemi, A., et al.: Sweetening ontologies with DOLCE. In: Knowledge engineering and knowledge management: Ontologies and the semantic Web, pp. 166-181. Springer (2002)

21. Goldman, A.: The individuation of action. J. Philos. **68**(21), 761–774 (1971)

22. Goldman, A.: Epistemology and Cognition. Harvard University Press (1986)

23. Goldman, A.: Cognitive science and metaphysics. J. Philos. **84**(10), 537–544 (1987)

24. Goldman, A.: Liaisons: Philosophy Meets the Cognitive and Social Sciences. The MIT Press (1992)

25. Goldman, A.: A Program for naturalizing metaphysics, with application to the ontology of events. In: The Monist, vol. 90(3), pp. 457–479 (2007)

26. Goldman, A.: Philosophical intuitions: their target, their source, and their epistemic status. In: Beyer, C., Burri, A. (eds.) Philosophical Knowledge: Its Possibility and Scope, pp. 1–26. Rodopi, Amsterdam (2007b)

27. Gómez, J., et al.: Naturalized epistemology for autonomous systems. In: Kazimierz Naturalised Epistemology Workshop (2007)

28. Grenon, P.: A primer on knowledge representation and ontological engineering. In: Smith, B., Munn, K. (eds.) Applied Ontology: An Introduction, Vol. 9, pp. 57–84. Ontos Verlag. Metaphysical Research 8. Heusenstamm (2008)

29. Grenon, P., Smith, B.: Foundations of an ontology of philosophy. In: Synthese, vol. 182(2), pp. 185–204 (2011)

30. Guarino, N., Carrara, M., Giaretta, P.: An Ontology of meta-level categories. In: KR 94, pp. 270–280 (1994)

31. Hacking, I.: Aristotelian categories and cognitive domains. In: Synthese, vol. 126(3), pp. 473–515 (2001)
32. Harris, M.A., et al.: The Gene Ontology (GO) database and informatics resource. In: Nucleic Acids Research 32. Database issue, pp. D258–61 (2004)
33. Hayes, P.: The second naive physics manifesto". In: Hobbs, J., Moore, R. (eds.) Formal Theories of the Commonsense World, pp. 1–20. Ablex Publishing Corporation, Norwood, NJ (1985)
34. Hennig, B.: What Is Formal Ontology? In: Smith, B., Munn, K. (eds.) Applied Ontology: An Introduction. Ontos Verlag. Vol. 9, pp. 39–56. Metaphysical Research 8. Heusenstamm (2008)
35. Keil, F.C.: The growth of causal understanding of natural kinds. In: Sperber, D., Premack, D., Premack, J. (eds.) Causal Cognition, pp. 234–262. Oxford University Press, New York (1995)
36. Kim, J.: Events and their descriptions: some considerations. In: Rescher N. (ed.) Essays in Honor of Carl G. Hempel, pp. 198–215. Reidel, Dordrecht (1969)
37. Kim, J.: Events as Property Exemplifications. In: Brand, M., Walton, D. (eds.) Action Theory, pp. 159–177. Reidel, Dordrecht (1976)
38. Kistler, M.: On the content of natural kind concepts. In: Acta Analytica, vol. 16, pp. 55–79 (1996)
39. Köhler, W.: Intelligenzprüfungen an Anthropoiden. Royal Prussian Society of Sciences, Berlin (1917)
40. Kornblith, H.: Précis of knowledge and its place in nature. Philos. Phenomenol. Res. 71(2), 399–402 (2005a)
41. Kornblith, H.: Replies. Philos. Phenomenol. Res. 71(2), 427–441 (2005b)
42. Ladyman, J., Ross, D.: Every thing must go metaphysics naturalized. In: Spurrett, D., Collier, J. (eds.) Oxford University Press, New York (2007)
43. Lakoff, G.: Women, Fire, and Dangerous Things: What Categories Reveal About the Mind. University of Chicago Press (1987)
44. Levine, M.: Alvin I. Goldman's epistemology and cognition: an introduction. Philosophia 19(2–3), 209–225 (1989)
45. Lowe, E.J.: The Four-Category Ontology; A Metaphysical Foundation for Natural Science. Oxford University Press, New York (2006)
46. Mark, D., Smith, B., Tversky, B.: Ontology and geographic objects: an empirical study of cognitive categorization. In: Spatial Information Theory. Cognitive and Computational Foundations of Geographic Information Science, pp. 283–298. Springer (1999)
47. Medin, D.L.: Concepts and conceptual structure. Am. Psychol. 44, 1469–1481 (1989)
48. Medin, D.L.: Folkbiology. Atran, S. (ed.). MIT Press, Cambridge, MA (1999)
49. Mingers, J.: Real-izing information systems: critical realism as an underpinning philosophy for information systems. Inf. Org. 14(2), 87–103 (2004)
50. Putnam, H.: Reason, Truth, and History. Cambridge University Press, New York (1981)
51. Putnam, H.: Why reason can't be naturalized. Synthese 52(1), 3–23 (1982)
52. Putnam, H.: The many faces of realism: the Paul Carus lectures. English. Open Court, La Salle, Ill (1987)
53. Quine, W.V.O.: Word and Object. MIT Press, Cambridge, MA (1960)
54. Quine, W.V.O.: Ontological Relativity and Other Essays. Columbia University Press, New York (1969)
55. Rescher, N.: Ontology in cognitive perspective. Axiomathes 18(1), 25–36 (2008)
56. Rosse, C., Mejino, J.: The foundational model of anatomy ontology. In: Burger, A., Davidson, D., Baldock, R. (eds.) Anatomy Ontologies for Bioinformatics, pp. 59–117. Springer, New York (2008)
57. Shepard, R.N.: Toward a universal law of generalization for psychological science. Science 237(4820), 1317–1323 (1987)
58. Shepard, R.N.: Perceptual-cognitive universals as reflections of the world. Psychon. Bull. Rev. 1(1), 2–28 (1994)
59. Shipley, E., Shepperson, B.: Countable entities: developmental changes. Cognition 34(2), 109–136 (1990)

60. Shoham, Y.: Naive kinematics: one aspect of shape. In: Proceedings of the International Joint Conference on Artificial Intelligence, pp. 436–442. Los Angeles (1985)
61. Simons, P.: Particulars in particular clothing: three trope theories of substance. In: Laurence, S., McDonald, C. (eds.) Contemporary Readings in the Foundations of Metaphysics, pp. 364–385. Blackwell (1998)
62. Smith, B.: Formal ontology, common sense, and cognitive science. Int. J. Hum. Comput. Stud. **43**, 641–667 (1995)
63. Smith, B., Ashburner, M., et al.: The OBO Foundry: coordinated evolution of ontologies to support biomedical data integration. Nat. Biotechnol. **25**(11), 1251–1255 (2007)
64. Smith, B., Casati, R.: Naive physics: an essay in Ontology. Philos. Psychol. **7**(2), 227–247 (1994)
65. Smith, B., Grenon, P., Goldberg, L.: Biodynamic Ontology: Applying BFO in the Biomedical Domain. In: Smith, B., Munn, K. (eds.) Applied Ontology: An Introduction. Ontos Verlag, vol. 9, pp. 21–38, Metaphysical Research 8. Heusenstamm (2004)
66. Smith, B., Klagges B.: Philosophy and Biomedical Information Systems. In: Smith, B., Munn, K. (eds.) Applied Ontology: An Introduction. Ontos Verlag, vol. 9, pp. 21–38, Metaphysical Research 8. Heusenstamm (2008)
67. Smith, M.L.: Overcoming theory-practice inconsistencies: Critical realism and information systems research. Inf. org. **16**(3), 191–211 (2006)
68. Sobel, D.M., Tenenbaum, J.B., Gopnik, A.: Children's causal inferences from indirect evidence: backwards blocking and Bayesian reasoning in preschoolers. Cogn Sci **28**(3), 303–333 (2004)
69. Spear, A.D.: Ontology for the Twenty First Century: an introduction with recommendations. Institute for Formal Ontology and Medical Information Science, Saarbrücken (2006)
70. Strawson, P.F.: Individuals: An Essay in Descriptive Metaphysics, 2005th edn. Routledge, New York (1959)
71. Tenenbaum, J.B., Griffiths, T.L.: Generalization, similarity, and Bayesian inference. Behav. Brain Sci. **24**(4), 629–640 (2001)
72. Tenenbaum, J.B., Griffiths, T.L., Kemp, C.: Theory-based Bayesian models of inductive learning and reasoning. Trends Cogn. Sci. **10**(7), 309–318 (2006)
73. Varzi, A.: Words and objects. In: Individuals, Essence and Identity, pp. 49–75. Springer (2002)
74. Varzi, A.: Spatial reasoning and ontology: parts, wholes, and locations. In: Handbook of Spatial Logics, pp. 945–1038. Springer (2007)
75. Wagner, L., Carey, S.: Individuation of objects and events: a developmental study. Cognition **90**(2), 163–191 (2003)
76. Wang, Y.: On cognitive informatics. Brain Mind **4**(2), 151–167 (2003)
77. Weinberg, J., Nichols, S., Stich, S.: Normativity and epistemic intuitions. Philos. Top. **29**, 429–460 (2001)
78. Xu, F., Carey, S.: Infants' metaphysics: the case of numerical identity. Cogn. Psychol. **30**(2), 111–153 (1996)

Would Super-Human Machine Intelligence Really Be Super-Human?

Philip Larrey

Abstract Given recent advances in the field of artificial intelligence, the notion of creating a digital machine capable of not only logical operations, but also of complex inferences and other processes usually associated with human thought, must now be considered. This chapter attempts to provide a speculative basis for such a consideration. The chapter attempts to defend the position that although machine "intelligence" will always differ from human intelligence in nature, it will exceed human intelligence in significant ways that will require a serious and profound reflection on the meaning of thought itself.

1 Introduction

Recently, several renowned personalities have weighed in on the theme of "super-human artificial intelligence" in the popular press. The famous astrophysicist from Britain, Stephen Hawking, thinks that artificial intelligence could end mankind[1]; the founder of Tesla, CEO of SpaceX and Silicon Valley guru, Elon Musk, warned us at MIT that artificial intelligence is our biggest existential threat[2]; a meeting of the minds took place at the beginning of 2016 in Puerto Rico called: "The Future of AI: Opportunities and Challenges",[3] and James Barrat worries us with the very title of his recent work, *Our Final Invention*.[4]

---

[1]Cf. http://www.bbc.com/news/technology-30290540.

[2]Cf. http://webcast.amps.ms.mit.edu/fall2014/AeroAstro/index-Fri-PM.html. Centennial Symposium, at the 1:07:26 mark.

[3]Cf. their website: http://futureoflife.org/misc/ai_conference. Organized by the Future of Life Institute (Boston), many of the key people developing artificial intelligence attended.

[4]Cf. James Barrat, *Our Final Invention. Artificial Intelligence and the End of the Human Era*, St. Martin's Press, New York 2013.

P. Larrey (✉)
Pontifical Lateran University, Vatican City, Italy
e-mail: plarrey@uni.net

© Springer International Publishing AG 2017
G. Dodig-Crnkovic and R. Giovagnoli (eds.), *Representation and Reality in Humans,
Other Living Organisms and Intelligent Machines*, Studies in Applied Philosophy,
Epistemology and Rational Ethics 28, DOI 10.1007/978-3-319-43784-2_19

   Aside from the apocalyptic scenarios which make for very good sci-fi films, the issue of super-human artificial intelligence is now on contemporary man's table. Specialists in the field have known for some time that in a very real way, we are sharing existence with other types of non-human intelligence. Aristotle understood that animals are intelligent (some exhibit more intelligence than others)[5] and that they also had 'souls', in the sense of a life principle. Thus, we have the term 'animate objects' connoting those objects which are alive, i.e. contain a soul (*anima* in Latin). The difference between animal intelligence and human intelligence is due to the degree of *being* (*esse* in Latin) of the vital principle in humans: the life principle ('form') in humans is capable of operations that exceed the potentiality of the body ('matter'), and therefore is capable of existing without the body, whereas the forms of animals cannot exist without the body. Such was Aristotle's philosophical argument in favor of the immortality of the human soul.

   Aristotle also postulated the existence of purely spiritual beings, as forms which have no material substratum. These are the pure forms that inhabit the celestial domain and are responsible for the motion of the heavenly bodies. Thomas Aquinas will call these pure forms 'angels' and will conclude that they are not only intelligent, but also much more intelligent than human beings because they do not need to turn to the senses in order to possess knowledge: they 'know' by virtue of their essence.

   With the birth of computers operating on binary logical systems, the term 'artificial intelligence' was coined, and was meant to connote logical operations achieved by software programs running on silicon chips. For a while, it was fashionable in philosophical circles (especially in the cognitive sciences) to conceive the relationship between the mind and the brain as similar to the relationship between software and hardware: the brain acts like a hard drive for the mind's software (read: program). Hilary Putnam called this *functionalism*: a view that he once held and later abandoned (as have almost all philosophers). It was the American philosopher from UC Berkeley, John Searle, who devised the very useful distinction between *strong AI* and *soft AI*, in order to draw clearer boundaries between what the human intellect does and what computers do.[6]

   Perhaps it was unfortunate that computer engineers and philosophers called what a binary computer achieves (applying concepts created by the great British mathematician, Alan Turing) 'intelligence', albeit 'artificial'. Yet, the name has stuck. Such was the rationale behind the term "super-human machine intelligence" to connote the future evolution of AI which, it is assumed, will surpass or exceed human intelligence. How we get to this level of super-human intelligence is usually explained through a series of extrapolations, starting with what AI is capable of doing now, and assuming that as computers get faster, more powerful and cheaper

---

[5]The "smartest" animals in the animal kingdom may in fact not turn out to be the closest biologically to humans, but rather birds. Cf. Noah Strycker "In almost any realm of bird behavior—reproduction, populations, movements, daily rhythms, communication, navigation, intelligence, and so on—there are deep and meaningful parallels with our own", xii.

[6]John R. Searle, *Mind. A Brief Introduction*, Oxford University Press, 2004 65 ff.

(thanks to Moore's Law of accelerating returns), they will eventually achieve "super-human level intelligence".

Everyone knows that it is easy to extrapolate. The more difficult question is simply this: what is intelligence, and what is *human* intelligence? This more profound question has perplexed philosophers and non-philosophers for as long as we can remember.

A very opportune place to begin to address such a question in this context is with Nick Bostrom, from Oxford University.[7] In 2014, he published his very insightful work, *Superintelligence. Paths, Dangers, Strategies.*[8] In chapter three of that book, Bostrom identifies "Forms of superintelligence" and opens the chapter with this startling affirmation: "We also show that the potential for intelligence in a machine substrate is vastly greater than in a biological substrate. Machines have a number of fundamental advantages which will give them overwhelming superiority. Biological humans, even if enhanced, will be outclassed" (52). So, just what are the 'forms' that such a superintelligence could take?

1. Speed superintelligence. "The simplest example of speed superintelligence would be a whole brain emulation running on fast hardware. An emulation operating at a speed of ten thousand times that of a biological brain would be able to read a book in a few seconds and write a PhD thesis in an afternoon. With the speedup factor of a million, an emulation could accomplish an entire millennium of intellectual work in one working day" (53). This type of super-human capacity is easy to extrapolate by simply doing the math on speed. Yet it assumes that a super-human level of intelligence can be achieved by speeding everything up. If intelligence is measured by speed, then of course it is obvious that if an artificial intelligence can do it faster than biology-based intelligence, it is by definition 'super-human'. Yet there are underlying presuppositions which are controversial.

Simply emulating a human brain will not necessarily produce intelligence, and many AI experts agree because we do not understand how the brain produces intelligence (and much less how the brain would 'cause' consciousness—if, in fact, it does). Ben Goertzel, the founder and leading intellectual at the Open Cog Project (perhaps the best known group specifically dedicated to the development of artificial *general* intelligence), states the following: "My current feeling is that brain emulation won't be the fastest or best approach to creating human-level AGI. One 'minor problem' with this approach is that we don't really understand how the brain works yet, because our tools for measuring the brain are still pretty crude. Even our theoretical models of what we should be measuring in the first place are still hotly debated".[9] Although we have made much progress in the speed of computers (and more will come), it is clear that for general intelligence, speed is not a panacea.

---

[7] At Oxford, Nick Bostrom is Director of the Future of Humanity Institute and has recently received $10 million from Elon Musk, who expressed deep interest in such research.

[8] Nick Bostrom, *Superintelligence. Paths, Dangers, Strategies*, Oxford University Press, 2014.

[9] Ben Goertzel, *Ten Years To the Singularity If We Really, Really Try ... and Other Essays on AGI and Its Implications*, CreateSpace Independent Publishing Platform, 2014, 112.

For very limited and narrow AI uses such as playing chess, data mining and hugely powerful search engines, fast computers are extremely important and they already are better at accomplishing their tasks than human beings. This comes as no surprise. But from these applications to extrapolate to a superintelligence is unwarranted.

The problem is really philosophical in nature.

Let us look at the two specific examples which Bostrom offers: reading a book and writing a PhD thesis. If 'reading' a book means digitizing the content and having it reside in some sort of memory (like RAM or on a hard drive), then computers can already do this and very quickly. But Bostrom knows that the real issue is deeper. The assumption is that a computer simulation of a human brain would in fact do everything the brain does, i.e. read and write a PhD thesis. Yet there is a decisive difference between *understanding* something and *simulating an understanding* of something. With the advent of the 'semantic web', sufficiently fast computers with proper software are going to achieve this simulation of *understanding meaning*, yet they will not really understand meaning.

John Searle's Chinese Room thought experiment is very illustrating in this sense.[10] The hypothetical scenario is a man in a room who does not speak Chinese. Chinese speakers outside the room slide slips of paper with Chinese characters on them, asking questions to the man inside. The man inside then consults a series of rule-books which indicate to him which Chinese characters he must write down on slips of paper to properly answer the question … in Chinese, which of course he does not understand. The Chinese speakers on the outside receive the slips of paper and are convinced that they were written by a Chinese speaker.

The man in the room does not understand Chinese at all, he has no idea what the characters *mean*, but he uses the rule-books to answer the questions. This is exactly what a computer is achieving. Searle concludes: "[T]he implemented syntactical or formal program of a computer is not constitutive of nor otherwise sufficient to guarantee the presence of semantic content; and secondly, simulation is not duplication".[11] To drive home the point, he also recalls his famous example of digestion: a commercial computer can certainly simulate the digestive process that happens in the body, but it is not really digesting anything. There is a big difference.

Although the Chinese Room experiment is quite dated today, Searle believes it is still valid as describing the essential difference between artificial intelligence and human intelligence. "My reason for having so much confidence that the basic argument is sound is that in the past 21 years I have not seen anything to shake its fundamental thesis. The fundamental claim is that the purely formal or abstract or syntactical processes of the implemented computer program could not by themselves be sufficient to *guarantee* the presence of mental content or semantic content

---

[10]Cf. the insightful work edited by John Preston and Mark Bishop, *Views into the Chinese Room. New Essays on Searle and Artificial Intelligence*, Oxford 2002.

[11]John Searle, "Twenty-One Years in the Chinese Room", in *Views into the Chinese Room, cit.*, 52.

of the sort that is essential to human cognition".[12] Could the brain emulation trick a human observer into believing that it is, in fact, *understanding* the meaning in the text? The short answer, I believe, is yes, at least for a sufficiently fast computer with the proper software. At this point, we would have a machine that successfully passes the Turing Test (which to date–no computer has yet achieved, even though there have been news-worthy attempts[13]).

However, as Searle concludes, "The 'system', whether me in the Chinese Room, the whole room, or a commercial computer, passes the Turing Test for understanding Chinese but it does not understand Chinese, because it has no way of attaching any meaning to the Chinese symbols. The appearance of understanding is an illusion".[14] Returning to Bostrom's examples, we would be assuming that "reading a book" or "completing a PhD dissertation" implies *understanding*, that very subtle, complex cognitive activity unique to human beings. On just about any comprehension of a theory of meaning, the human cognitive process captures meanings, evaluates them, compares and contrasts them and interprets. This is why a good text to read is not simply the product of rote memory or the repetition of things already stated. It goes much further: it implies a cognitive activity that *understands* and advances understanding in some significant way. The human intellect is capable of this kind of activity because the human being is *conscious*, it is aware and even further it is *self-aware*.

David Chalmers, when he was teaching in Arizona in 1994, called this the "hard problem of consciousness", and still today leaders in cognitive sciences do not seem to have progressed very much.[15] Perhaps one reason why progress does not seem to occur in this field is due to a philosophical option: that of reductionism. Reductionism, simply stated, proposes that consciousness is *reducible* to brain states (patterns of neurons firing and synapses exchanging information), and that the brain *causes* consciousness. If reductionism turns out to be the correct assumption, then we should be able to solve the 'hard problem' with more sophisticated technology and software.[16] As Searle states, "[t]he point, however, is that any such artificial machine would have to be able to duplicate, and not merely simulate, the causal powers of the original biological machine … An artificial brain would have to do something more than simulate consciousness, it would have to be able to *produce* consciousness. It would have to cause consciousness".[17] "In order to create

---

[12]*Ibid.*, 51.

[13]Cf. for example, a computer program called Eugene Goostman: http://www.bbc.com/news/technology-27762088. Most specialists in the AI field contested the published results.

[14]John Searle, *cit.*, 61.

[15]Cf. Oliver Burkeman's insightful article in the *Guardian* quite recently: "Why can't the world's greatest minds solve the mystery of consciousness?", http://www.theguardian.com/science/2015/jan/21/-sp-why-cant-worlds-greatest-minds-solve-mystery-consciousness.

[16]Such would seem to be the goal of Ray Kurzweil who now works for Google as head of engineering. Cf. his recent work, *How to Create a Mind. The Secrets of Human Thoughts Revealed*, Penguin Books, 2013.

[17]John Searle, *cit.*, 56.

consciousness you have to create mechanisms which can duplicate and not merely simulate the capacity of the brain to create consciousness".[18] Here, of course, Searle is assuming that the brain *causes* consciousness, which may or may not be true. The philosophical jury is still out on the issue. According to one long-standing tradition in philosophy, the brain 'houses' consciousness, but consciousness itself would be caused by the soul, or by the principle of being which gives existence to the subject.[19] As stated above, such a principle resides in all animate objects (which are composed of form and matter), and therefore on such an assumption, we can attribute consciousness also to non-human life forms (such as birds, dogs and cats). Ask any owner of a dog if their pet is conscious, and the answer will be obviously yes. The level of consciousness would be less than for humans, yet it would be present nonetheless.

One of the most impressive demonstrations of computer generated 'intelligence' from an historical view point was IBM's *Deep Blue*, which beat the world's number one Grand Master of chess, Gary Kasparov on May 11, 1997: a feat that many at the time considered impossible. Although the event was historical in many senses, *Deep Blue* still had not given evidence of "super-human intelligence": it was simply better at playing chess. "In the case of Deep Blue, the machine did not know that it was playing chess, evaluating possible moves, or even winning and losing. It did not know any of these things, because it does not know anything".[20] Ordinary computer chess programs on your laptop now reach *Deep Blue* levels at playing chess. Yet we would generally not say that the program "knows how to play chess".

Another IBM experiment recently challenged our conception of artificial intelligence, namely the super computer called *Watson*, which in 2011 defeated the two most successful players of *Jeopardy!* and was awarded a million dollars. The intriguing element here is that in the quiz show, one must come up with the questions to the answers which are given. It would seem that being able to achieve this would require the machine to *understand* human language. When faced with

---

[18]John Searle, *cit.*, 68.

[19]Cf. Thomas Aquinas, *Summa Theologica*, Q. 75, art. 2: "Therefore, the intellectual principle, which we call the mind or the intellect, has an operation in which the body does not share. Now only that which subsists in itself can have an operation in itself … We must conclude, therefore, that the human soul, which is called intellect or mind, is something incorporeal and subsistent." Also, "Now it is clear that the first thing by which the body lives is the soul. And as life appears through various operations in different degrees of living things, that whereby we primarily perform each of all these vital actions is the soul. For the soul is the primary principle of our nourishment, sensation, and local movement; and likewise of our understanding. Therefore this principle by which we primarily understand, whether it be called the intellect or the intellectual soul, is the form of the body. This is the demonstration used by Aristotle (*De Anima* ii, 2). But if anyone says that the intellectual soul is not the form of the body he must first explain how it is that this action of understanding is the action of this particular man; *for each one is conscious that it is himself who understands.* But one cannot sense without a body: therefore the body must be some part of man. It follows therefore that the intellect by which Socrates understands is a part of Socrates, so that in some way it is united to the body of Socrates." Id., *ST*, Q. 76, art. 1. My emphasis.

[20]John Searle, *cit.*, 65.

this clue given during competition: "A long tiresome speech delivered by a frothy pie topping", *Watson* came up with the correct question: "What is meringue harangue?" This was quite impressive. The builders of *Watson* even admitted that they are not sure *how* the machine arrived at the proper answers, given the various subroutines operating within the software program. Yet again, using Searle's distinction, one can still suggest that *Watson* is simulating having understood, yet it really does not understand anything. In no way does this diminish the amazing capability of the machine, which is now being used at the Memorial Sloan Kettering Cancer Center in Manhattan to help diagnose cancer in patients. It has been reported that 90% of the nurses concur with its analysis. It simply tells us that there is a difference between what *Watson* does and what the human intellect does.

It also serves as a warning, already mentioned by Bostrom: machines are better than human beings in many cognitive functions, and they are getting even better, even without solving the "hard problem of consciousness". He concludes: "Although these systems [such as *Watson*] do not understand what they read in the same sense or to the same extent as a human does, they can nevertheless extract significant amounts of information from natural language and use that information to make simple inferences and answer questions. They can also learn from experience".[21]

The second type of possible superintelligence analyzed by Bostrom is called 'Collective superintelligence', and consists of "[a] system composed of a large number of smaller intellects such that the system's overall performance across many very general domains vastly outstrips that of any current cognitive system".[22] From a theoretical point of view, this second type of superintelligence does not differ radically from the first type: instead of only one, there are many, and they collaborate with each other in order to solve problems. Within the field of AI research, this is a plausible outcome of the many actors who are developing general super-human artificial intelligence.[23] However, because of the different platforms currently being developed, it is not clear how 'separate' intelligent machines would communicate harmoniously with each other. As an example, much development is being carried out in the field of quantum computing, using qubits instead of binary bits, capable of housing information in superimposed states (and not simply as ones and zeros). A working prototype is already being used at the NASA Ames Research center, sponsored in part by Google, and is called *D-Wave Two*. The Chief Scientist there is Eric Ladizinsky, who is very articulate, and states that this machine is a thousand times more capable of computing than a traditional machine. It utilizes

---

[21]Bostrom, *cit.*, 71.

[22]*Ibid.*, 54.

[23]The current leader in the field of general (as opposed to specific) AI is probably *DeepMind*, owned by Google. There may be covert programs in different parts of the world, about which we know little or nothing. However, with the enormous resources at their disposal, Google is positioned as the likely leader in the race to produce a general super-human level of artificial intelligence. In his book presentation at UC Berkeley several months ago, Bostrom concurred that Google would likely be the first to create true superintelligence.

quantum states on a macroscopic level. This development will probably ensure that Moore's Law will continue to function into the foreseeable future.

As an interesting note, Bostrom admits: "nothing in our definition of collective superintelligence implies that a society with greater collective intelligence is necessarily better off. The definition does not even imply that the more collectively intelligent society is *wiser*".[24] This is a non-obvious truth that must be highlighted. Human history is replete with episodes of very intelligent people making very poor decisions and creating very harsh conditions for millions of people. Yet, the point refers to intelligence in general, and specifically to the ability to solve cognitive problems: not to construct a better society for human beings, laudable as that goal is.

The third and final form of superintelligence is called 'Quality superintelligence' and consists in "[a] system that is at least as fast as a human mind and vastly qualitatively smarter".[25] Here the difficulty is to define the term 'qualitatively' when referring to cognitive activity. Bostrom offers some examples to make the point. The first consists of non-human animal intelligence, which we know exists and which is 'qualitatively' inferior to human intelligence (and which has already been mentioned above, for example the case of bird intelligence). Interestingly, "[i]n terms of raw computational power, human brains are probably inferior to those of some large animals, including elephants and whales. And although humanity's complex technological civilization would be impossible without our massive advantage in collective intelligence, not all distinctly human cognitive capabilities depend on collective intelligence".[26] Thus, human intelligence is qualitatively superior to animal intelligence, and not because of pure computational power. In fact, a very important factor is often neglected when referring to computational power and that is computational architecture. The human brain has a more complex computational architecture than other animal brains, that might be actually much larger. In this sense, a superintelligence would need to exhibit intelligence qualitatively superior to humans.

Another example given deals with human intelligence's capability for complex linguistic representations, which gives humans an enormous evolutionary advantage over nonhumans. Most probably, linguistic capabilities were developed for communicative purposes. Thus, linguistic skills are part of human collective intelligence. Humans were constituted with the cognitive modules that enable linguistic representations and therefore became superior to the brutes. Furthermore, "were we to *gain* some new set of modules giving an advantage comparable to that of being able to form complex linguistic representations, we would become superintelligent".[27] The implication here is that were a machine to gain a similar set of modules, it would be considered 'super-human'.

---

[24]*Ibid.*, 55.

[25]*Ibid.*, 56.

[26]*Ibid.*, 57.

[27]*Ibid.*, 57.

It is not clear in what a similar set of modules would consist, in terms of cognitive activity. In many respects, certain machines are already 'super-human' in terms of brute strength (we can think of the large industrial machines or those used in agriculture, which have replaced millions of people in recent decades), speed, precision, sensing ability (like military-grade satellites which can 'see' hundreds of miles away or in the dark), etc. It would seem logical to eventually add 'cognition' to the list of things which machines do better than we do. And, as already mentioned, in some respects they already out perform us in many, isolated tasks, from the perspective of cognitive activity. Yet, to achieve *general* super-human intelligence, something more will be needed. And a sufficiently complex and powerful computer just may be able to come up with those 'extra modules' necessary to give the machine a cognitive advantage over humans. Ben Goertzel states: "As every software engineer knows, the design and implementation of complex software is a process that constantly pushes against the limitations of the human brain—such as our limited short term memory capacity, which doesn't allow us to simultaneously manage the states of more than a few dozen variables or software objects. There seems little doubt that a human-level AGI, once trained in computer science, would be able to analyze and refine its own underlying algorithms with a far greater effectiveness than any human being".[28]

2 Common Sense

One area in which artificial intelligence systems have yet to make much progress is something so natural and so spontaneous that all humans exercise effortlessly, i.e., common sense. Although at times it is said that 'common sense' is not very common (because of the insane behavior of which many human beings are capable every day, which Erasmus writes of in his *The Praise of Folly*), this characteristic of human cognition is extremely difficult to replicate in a digital computer. With his groundbreaking paper, "Programs with Common Sense",[29] John McCarthy ushered in the era of attempting to formalize common sense knowledge so as to be used by a digital computer. "We shall therefore say that *a program has common sense if it automatically deduces for itself a sufficiently wide class of immediate consequences of anything it is told and what it already knows*".[30] In that paper, McCarthy described the theoretical basis for a program to 'learn' from experience and from

---

[28]Ben Goertzel, *Ten Years To the Singularity If We Really, Really Try ... and other Essays on AGI and its Implications*, CreateSpace Independent Publishing Platform, 2014, 19. Cf. the recent article in the *New Scientist* which analyzes exactly this type of technology: http://www.newscientist.com/article/mg22429932.200-computer-with-humanlike-learning-will-program-itself.html#. VMpnW9KG9vA.

[29]Cf. McCarthy, J., "Programs with Common Sense". *Proc. of Conference on the Mechanization of Thought Processes*, 1959, 75–91.

[30]*Ibid.*, 78. Italics in original.

the world as such. And he further stipulated that the mechanism through which the program 'learns' would also have to be *improvable*, i.e., as more and more information becomes available and assimilated, the way in which the program utilizes it has to become more complex and applicable. This was a profound insight, and has led to what is commonly known as 'machine learning' which is found in many software applications in use by Amazon, Netflix, Google and Facebook. Of course, the program is not 'learning' anything, for it is simply calculating (and recalculating) relationships based on acquired patterns in order to produce suggestions about what books or films you may like, or what kind of advertising would be best suited for you. Such software has been described as 'creepy', precisely because it gives you the impression that it is 'learning' on the basis of information given by you.

Producing authentic common sense in a digital format has proven to be quite elusive. There have been efforts in this direction, yet perhaps the most serious difficulty arises from the fact that common sense reasoning is not reducible to sheer logic (for programs are much better than humans from a strictly logical point of view). That is the whole point about the uniqueness of common sense knowledge: it is not based simply on laws or logical inferences. Another difficulty arises from the ambiguity of just what is common sense knowledge. What do we mean by common sense knowledge?

In his work, Antonio Livi has attempted to provide some clarification.[31] However, his primary interest lies with the relationship between common sense knowledge and belief in a divine being, which of course does not seem too applicable to a software program (although one could make the case for its importance). Thomas Reid was the 'grandfather' of the Scottish school of common sense, attempting to oppose David Hume's skepticism, founded on assumptions carried over from John Locke's notion of 'ideal theory'. Between the two world wars, Cambridge saw a flourishing of the philosophy of common sense,[32] and, more recently, Noah Lemos has returned to the question in his *Common Sense. A Contemporary Defense*.[33]

Seeing how important common sense knowledge is for human beings, a crucial task for AI engineers working on superintelligence would necessarily imply giving such an ability to machines. Can machines *simulate* common sense knowledge? To answer this question, a group of intellectuals have come up with the Winograd Schema Challenge, named after Terry Winograd, which would replace the Turing Test in order to assess the ability of a digital computer to pass as a human.[34] Every 2 years, a major convention is held to allow participants to try and win the

---

[31]Cf. Antonio Livi, *A Philosophy of Common Sense. The Modern Discovery of the Epistemic Foundations of Science and Belief*, trans. Peter Waymel, Davies Group Publishers, Aurora, 2013.

[32]Cf. John Coates, *The Claims of Common Sense. Moore, Wittgenstein, Keynes and the Social Sciences*, Cambridge University Press, 1996/2001.

[33]Noah Lemos, *Common Sense. A Contemporary Defense*, Cambridge University Press, 2004/2010.

[34]Cf. http://commonsensereasoning.org/winograd.html.

challenge. So far, no one has been able to. Let us look at one of the examples that has been set forth in order to test common sense knowledge in a computer. The machine would be asked the following question and would have to answer either 0 or 1, with the option of using either 'big' or 'small': "The trophy would not fit in the brown suitcase because it was too big (*small*). What was too big (*small*)? Answer 0: the trophy. Answer 1: the suitcase". For a human intellect, it is clear that if 'big' is used as the main adjective, then it refers to the trophy (answer 0); whereas if the preferred adjective is 'small', then it refers to the suitcase (answer 1). According to the promoters of the challenge, this is an example of common sense knowledge that humans routinely and effortlessly achieve, and AI has yet to be able to tackle. A human who answers these questions correctly typically uses his abilities in spatial reasoning, his knowledge about the typical sizes of objects, as well as other types of common sense reasoning, to determine the correct answer.[35] Such abilities are common to all conscious humans who do not suffer from some sort of mental disability. Computers (at present) lack such abilities. Common sense knowledge is vital for humans to be able to interact in the world, and without it, people would surely die (and quickly). "Standing in front of a speeding train" goes against common sense (unless of course one wanted to commit suicide), and one does not need to experience the effects of doing so before concluding that it would be unwise. This shows one of the special characteristics of common sense: it is not the result of experience (or trial and error), but rather it is connatural with human intelligence.

Some software engineers are attempting to 'teach' a computer common sense knowledge by entering lists of common sense statements. Perhaps the most successful thus far has been the MIT Media Lab's project called Open Mind Common Sense (OMCS) which uses their ConceptNet as an engine to make connections among the millions of common sense phrases that have been introduced by more than 15,000 participants. The project is directed by Catherine Havasi, one of the original founders who worked with Marvin Minsky.

The efforts of the people at Media Lab may prove to be successful at having an AI *simulate* common sense knowledge, yet the AI will still not possess common sense. One might respond by saying that common sense is not necessary for AI to achieve excellent results in interacting in the real world, and this would be an important distinction to bear in mind. But it also shows the limitations of AI in the real world which must utilize formal models of the world in order to interact with our reality. Up until now, human programmers have furnished these formalized models to the AI and have been able to achieve remarkable results.[36] It does seem clear that the human intellect does not use formalized models in the same way as AI does. The human intellect has direct contact with reality and objects of medium

---

[35]http://commonsensereasoning.org/winograd.html.

[36]It will be interesting to see if a sufficiently advanced AI will become capable of coming up with its own formalized models of reality and use those instead of the ones provided by human programmers. This is perhaps one of the goals of the DeepMind project, directed by Demis Hassabis and located in London.

range (i.e. those not too large—like planets—and those not too small—like atoms, the knowledge of which requires special tools and analysis) and through a process often known as 'abstraction', it is able to *understand the nature of things*. From a philosophical perspective, our knowledge of reality begins with common sense, for it is the inescapable initial relationship between our minds and reality, and it is one that is not learned through experience but rather is 'hardwired' in us.

The renowned historian of science, Steven Shapin, goes even further and claims that even the best scientific knowledge begins with common sense. He states: "In the 1850s, T.H. Huxley wrote that 'Science is, I believe, nothing but trained and organized common sense'. The whole of science, according to Albert Einstein, 'is nothing more than a refinement of every day thinking'. Max Planck agreed: 'Scientific reasoning does not differ from ordinary reasoning in kind, but merely in degree of refinement and accuracy'. And so did J. Robert Oppenheimer: 'Science is based on common sense; it cannot contradict it'."[37] Therefore, the basis of our most sophisticated scientific reasoning is actually something ordinary and practiced by all human beings, common sense. Proverbs such as "A stitch in time saves nine", or "Great oaks from little acorns grow", or "Stolen apples are sweetest" are phrases which contain great knowledge, expressed in simple form and applicable to the world at large. Such propositions would be difficult for an AI to harness, in part because they are not *always* true (albeit almost always) and in part because they are phrases which are applicable to reality through the intermediacy of the human thinker who captures the proverbial meaning of the sentence and applies it to a concrete situation (i.e. when debating whether to sew a sock or wait a couple of more days to do so).

Strangely enough, perhaps it will be precisely that type of knowledge which we as human beings take so much for granted that will ensure our uniqueness and importance in cognitive activity when seriously challenged by AI. "Many people suppose that computing machines are replacements for intelligence and have cut down the need for original thought", Norbert Wiener once wrote. "This is not the case".[38] "The more powerful the computer, the greater the premium that will be placed on connecting it with imaginative, creative, high-level human thinking".[39] Instead of the popular scenario of 'us versus them', the more probable outcome of advanced AI systems will be one of collaboration, where each type of 'intelligence' is able to maximize its own qualities in order to achieve its goals. This also makes sense from a business point of view. Stephen F. DeAngeli, President and CEO of the cognitive computing firm *Enterra Solutions* writes: "Although concerns remain that intelligent computers will continue to put workers out on the street, we believe

---

[37]Steven Shapin, *Never Pure. Historical Studies of Science as If It Was Produced by People with Bodies, Situated in Time, Space, Culture, and Society, and Struggling for Credibility and Authority*, The Johns Hopkins University Press, Baltimore 2014, 349–350.

[38]Norbert Wiener, "A Scientist's Dilemma in a Materialistic World" (1957), in *Collected Works*, vol. 4 (MIT Press, 1984), 709.

[39]Walter Isaacson, *The Innovators. How a Group of Hackers, Geniuses, and Geeks Created the Digital Revolution*, Simon & Schuster, New York 2014, 222.

that computers working with (not in place of) humans creates the most effective, efficient, and profitable working environment".[40]In his book on the protagonists of the digital era, Walter Isaacson shows complete agreement: "These ideas formed the basis for one of the most influential papers in the history of postwar technology, titled 'Man–Computer Symbiosis', which Licklider published in 1960. 'The hope is that, in not too many years, human brains and computing machines will be coupled together very tightly', he wrote, 'and that the resulting partnership will think as no human brain has ever thought and process data in a way not approached by the information-handling machines we know today'. This sentence bears rereading, because it became one of the seminal concepts of the digital age".[41]

If at such a point we call this type of intelligence 'super-human', then the answer to the question that this paper asked at the beginning, "Would Super-human-machine intelligence really be super-human?" would be yes. However, if the intelligence doing the thinking is human, the answer would be no, simply because it is still human. Perhaps we will choose to refer to such intelligence as 'augmented', and yet perhaps not. For centuries we have used telescopes and microscopes to augment our knowledge, yet we usually do not refer to such knowledge as 'super-human'. In any event, as AI continues to develop, it will be fascinating to see what happens. Ours is truly an 'unknown future'.

References

1. Aquinas, T: Summa Theologica, pp. 75–76
2. Barrat, J.: Our Final Invention: Artificial Intelligence and the End of the Human Era. St. Martin's Press, New York (2013)
3. Bostrom, N.: Superintelligence: Paths, Dangers, Strategies. Oxford University Press, Oxford (2014)
4. Burkeman, O.: Why can't the world's greatest minds solve the mystery of consciousness?. http://www.theguardian.com/science/2015/jan/21/-sp-why-cant-worlds-greatest-minds-solve-mystery-consciousness
5. Coates, J.: The Claims of Common Sense: Moore, Wittgenstein, Keynes and the Social Sciences. Cambridge University Press (1996/2001)
6. Goertzel, B.: Ten Years To the Singularity If We Really, Really Try … and Other Essays on AGI and Its Implications. CreateSpace Independent Publishing Platform (2014)
7. Isaacson, W.: The Innovators: How a Group of Hackers, Geniuses, and Geeks Created the Digital Revolution. Simon & Schuster, New York (2014)
8. Kurzweil, R.: How to Create a Mind: The Secrets of Human Thoughts Revealed. Penguin Books (2013)
9. Lemos, N.: Common Sense: A Contemporary Defense. Cambridge University Press (2004/2010)

---

[40]http://innovationinsights.wired.com/insights/2014/08/ai-systems-will-prove-useful-long-become-self-aware/.

[41]Walter Isaacson, *cit.*, 226.

10. Livi, A.: A Philosophy of Common Sense: The Modern Discovery of the Epistemic Foundations of Science and Belief, translated by Peter Waymel. Davies Group Publishers, Aurora (2013)
11. McCarthy, J.: Programs with Common Sense. In: Proceedings of Conference on the Mechanization of Thought Processes, pp. 75–91 (1959)
12. Preston, J., Bishop, M.: Views into the Chinese Room: New Essays on Searle and Artificial Intelligence, Oxford (2002)
13. Searle, J.R.: Mind: A Brief Introduction. Oxford University Press (2004)
14. Shapin, S.: Never Pure: Historical Studies of Science as if It Was Produced by People with Bodies, Situated in Time, Space, Culture, and Society, and Struggling for Credibility and Authority. The Johns Hopkins University Press, Baltimore (2010)
15. Strycker, N.: The Thing with Feathers: The Surprising Lives of Birds and What They Reveal About Being Human. Riverhead Books (2014)
16. Wiener, N.: A Scientist's Dilemma in a Materialistic World (1957). in Collected Works, vol. 4. MIT Press (1984)

Printed in the United States
By Bookmasters